UNGULATE MANAGEMENT IN EUROPE: PROBLEMS AND PRACTICES

This book considers a number of problems posed by ungulates and their management in Europe. Through a synthesis of the underlying biology and a comparison of the management techniques adopted in different countries the book explores which management approaches seem effective – and in which circumstances.

Experts in a number of different areas of applied wildlife biology review various management problems and alternative solutions, including

- impact of large ungulates on agriculture, forestry and conservation habitats in Europe
- road traffic accidents involving ungulates and measures for mitigation
- large herbivores as agents of deliberate habitat change
- the impact of predation on wild ungulate populations
- the role of disease in regulating large ungulate populations
- wild ungulates as vectors of disease.

This book is directed at practising wildlife managers; those involved in research to improve methods of wildlife management; and policy-makers in local, regional and national administrations.

RORY PUTMAN worked for many years within the Biology Department of the University of Southampton, where he established and led the University's highly regarded Deer Management Research group; latterly he moved to become Research Professor of Behavioural and Environmental Biology at the Manchester Metropolitan University. He now works as a freelance environmental consultant and wildlife adviser based in Scotland. He has worked widely in the UK and overseas, with research efforts focused on the population ecology of ungulates and their interaction with their vegetational environment – always with the explicit focus of helping to develop more sensitive and more effective methods of managing those same ungulate populations and their impacts on agriculture, forestry or conservation interests.

MARCO APOLLONIO is Full Professor at the University of Sassari, Italy, where he is presently Director of the Department of Zoology and of the PhD school in Natural Science.

His main interests are in ungulate behaviour, ecology and genetics, with a specific focus on mating and social behaviour and on predator–prey relationships involving wolves and ungulates. He is Past-President of the Italian Mammalogical Society and was involved in conservation activities as CITES Scientific Commission Member for Italy and as a member of the board of directors of two national parks in the last 15 years.

REIDAR ANDERSEN worked for many years in the Norwegian Institute for Nature Research. In 2005 he became Professor in Conservation Biology at the Museum of Natural History and Archaeology at the Norwegian University of Science and Technology. He has been leading several cross-disciplinary research projects focusing on ungulates and large carnivores, always aiming to produce applied knowledge, securing sustainable management of the species involved. Since 2009 he has become part of the Directorate team at the Norwegian Directorate for Nature Conservation.

UNGULATE MANAGEMENT IN EUROPE: PROBLEMS AND PRACTICES

Edited by

RORY PUTMAN

MARCO APOLLONIO

REIDAR ANDERSEN

CAMBRIDGE
UNIVERSITY PRESS

CAMBRIDGE UNIVERSITY PRESS
Cambridge, New York, Melbourne, Madrid, Cape Town, Singapore,
São Paulo, Delhi, Dubai, Tokyo, Mexico City

Cambridge University Press
The Edinburgh Building, Cambridge CB2 8RU, UK

Published in the United States of America by Cambridge University Press, New York

www.cambridge.org
Information on this title: www.cambridge.org/9780521760591

First published 2011

Printed in the United Kingdom at the University Press, Cambridge

A catalogue record for this publication is available from the British Library

Library of Congress Cataloging-in-Publication Data

Ungulate management in Europe : problems and practices / edited by Rory Putman,
Marco Apollonio, Reidar Andersen.
p. cm.
ISBN 978-0-521-76059-1 (Hardback)
1. Ungulates–Ecology–Europe. 2. Wildlife management–Europe. I. Putman, Rory.
II. Apollonio, Marco. III. Andersen, Reidar.
QL737.U4U54 2011
639.97′96094–dc22

2010038359

ISBN 978-0-521-76059-1 Hardback

Contents

Contributors

Professor Reidar Andersen
Directorate for Nature Management, N-7485 Trondheim, Norway

Professor Marco Apollonio
Department of Zoology and Evolutionary Genetics, University of Sassari, Via Muroni 25, I-07100 Sassari, Italy

Professor Luděk Bartoš
Ethology Group, Research Institute for Animal Production, Pratelstvi 8, CZ-104 00 PRAHA 10, Uhrineves, Czech Republic

Dr Bruno Bassano
Alpine Wildlife Research Center, Gran Paradiso National Park, Fraz. Degioz – Valsavarenche, Italy

Dr Marion L. East
Evolutionary Ecology Research Group, Leibniz Institute for Zoo and Wildlife Research, Alfred-Kowalke-Str. 17, D-10315 Berlin, Germany

Professor Ezio Ferroglio
University of Turin, Faculty of Veterinary Medicine, Department of Animal Production, Epidemiology and Ecology Via L. Da Vinci, 44 10095 Grugliasco (TO), Italy

Professor Christian Gortázar
Instituto de Investigación en Recursos Cinegéticos (IREC), Ronda de Toledo 13071, Ciudad Real, Spain

Dr Stefano Grignolio
Department of Zoology and Evolutionary Genetics, University of Sassari, Via Muroni 25, I-07100 Sassari, Italy

Professor Bogumiła Jędrzejewska
Mammal Research Institute, Polish Academy of Sciences, 17–230 Białowieża, Poland

Professor Włodzimierz Jędrzejewski
Mammal Research Institute, Polish Academy of Sciences, 17–230 Białowieża, Poland

Professor Robert Kenward
NERC Centre for Ecology and Hydrology, Benson Lane, Crowmarsh Gifford, Wallingford OX10 8BB, UK

Dr François Klein
Office National de la Chasse et de la Faune Sauvage, Direction des Etudes et de la Recherche, 85bis Avenue de Wagram, 75017 Paris, France

Dr Ilpo Kojola
Finnish Game and Fisheries Research Institute, Oulu Game and Fisheries Research, Tutkijantie 2A, FIN-90570 Oulu, Finland

Dr Jochen Langbein
Langbein Wildlife Associates, Chapel Cleeve, Minehead, Somerset, TA24 6HY, UK

Dr John D. C. Linnell
Norwegian Institute for Nature Research, Tungasletta 2, NO-7485, Trondheim, Norway

Dr Nicolas Morellet
Comportement et Ecologie de la Faune Sauvage, Institut National de la Recherche Agronomique, BP 52627, Castanet-Tolosan Cedex, F 31326, France

Professor Atle Mysterud
Centre for Ecological and Evolutionary Synthesis (CEES), Department of Biology, University of Oslo, PO Box 1066, Blindern, N-0316 Oslo, Norway

Dr Bostjan Pokorny
ERICo Velenje, Ecological Research and Industrial Cooperation, Koroška 58, 3320 Velenje, Slovenia

Professor Rory Putman
(School of Biological Sciences, Manchester Metropolitan University), *correspondence address:* Keil House, Ardgour, by Fort William, Inverness-shire, Scotland, PH33 7AH, UK

Professor Friedrich Reimoser
Research Institute of Wildlife Ecology, Vienna Veterinary University, Austria

Professor Bernt-Erik Sæther
Centre for Conservation Biology, Department of Biology, Norwegian University of Science and Technology, N-7491 Trondheim, Norway

Dr Ir Chris Smit
Community and Conservation Research Group, Centre for Evolutionary and Ecological Studies, University of Groningen, Postbus 11103, 9700 CC Groningen, The Netherlands

Dr Erling Solberg
Norwegian Institute for Nature Research, Tungasletta 2, NO-7485, Trondheim, Norway

Dr Joaquín Vicente
Instituto de Investigación en Recursos Cinegéticos (IREC), Ronda de Toledo 13071, Ciudad Real, Spain

Dr Bjørnar Ytrehus
National Veterinary Institute, Section for Wildlife Diseases, PO Box 750, Sentrum, NO-0106 Oslo, Norway

Dr Frank E. Zachos
Zoological Institute, Christian-Albrechts-University Kiel, Olshausenstrasse 40, 24118 Kiel, Germany

Scientific names of species referred to in this text

In this book, common names are used through the text. Species implied are:

Cervidae:	
Chinese muntjac	*Muntiacus reevesi*
Chinese water deer	*Hydropotes inermis*
moose	*Alces alces*
reindeer	*Rangifer tarandus*
roe deer	*Capreolus capreolus*
white-tailed deer	*Odocoileus virginianus*
red deer	*Cervus elaphus*
sika deer	*Cervus nippon*
wapiti	*Cervus canadensis*
fallow deer	*Dama dama*
axis (or chital)	*Axis axis* and their subspecies
Bovidae/Ovidae:	
European bison or wisent	*Bison bonasus*
musk ox	*Ovibos moschatus*
alpine chamois	*Rupicapra rupicapra*
Pyrenean chamois	*Rupicapra pyrenaica*
Barbary sheep (aoudad)	*Ammotragus lervia*
mouflon	*Ovis orientalis musimon*
alpine ibex	*Capra ibex*
Spanish ibex	*Capra pyrenaica*
wild goat	*Capra aegagrus* and their subspecies
Suidae:	
wild boar	*Sus scrofa*

Common names are used throughout the text in place of formal scientific names, except where the latter are used in text or in subheadings to identify a particular subspecies

1
Introduction

RORY PUTMAN, REIDAR ANDERSEN AND MARCO APOLLONIO

1.1 Introduction to this volume

In November 2004, a seminal meeting was held in Erice (Sicily) where representatives from a wide range of European countries were asked to come and offer presentations on ungulate populations and their management in their respective countries. The overall idea was to learn from each others' experiences (and each others' mistakes), in the hopes of developing improved management strategies for the future. Speakers were asked to review the status of populations of wild ungulates in their countries, describe current legislation and management philosophy, and review problems and actual practice with day-to-day management.

The meeting was an enormous success – and very revealing in highlighting the diversity of attitudes and approaches to management of wild ungulates in general, as well as the very different issues faced by wildlife managers in different countries. After the meeting was over the organisers decided to prepare a book to 'encapsulate' the material to make it more widely available – to academic researchers, wildlife managers and policy makers alike. At the meeting in Erice, however, presentations covered only some 12 countries from within Europe; and it was determined that the book should be extended in order to include contributions from as many European countries as possible. That book, including coverage from some 28 countries (all EU countries except Malta, plus Norway, Switzerland and Croatia) was published by Cambridge University Press in 2010 (Apollonio *et al.*, 2010a, *European Ungulates and their Management in the 21st Century*); we believe that this was the first time anyone had attempted to try and draw together information on wild ungulates and their management across Europe.

Ungulate Management in Europe: Problems and Practices, eds. Rory Putman, Marco Apollonio and Reidar Andersen. Published by Cambridge University Press.

To try to ensure that authors provided material on all relevant topics (and all covered the same ground), and to try to facilitate an analysis of similarities and differences between the different countries involved, we asked authors to prepare chapters to a common template.
All chapters thus considered:

(i) the ungulate species present in their country and the distribution and numbers of each species
(ii) genetic status of populations of each species (whether native or introduced; whether genetically 'pure' or affected by subsequent introductions of animals of different genetic origin)
(iii) legislation and administrative structure for management
(iv) actual management practice; hunting philosophy, hunting methods
(v) problems with ungulate management (expanding populations of pest species; need for conservation of endangered species or subspecies)
(vi) impacts of ungulates on agriculture, forestry or conservation habitats; extent of ungulate involvement in road traffic accidents
(vii) an analysis of the extent to which management is addressing problems effectively (or the extent to which the problems are exacerbated by inappropriate management!).

Such was the scale of the project that we did not at that time attempt to offer a detailed synthesis of this diverse body of material. Our main aim was, explicitly, to draw together in one place a convenient single source of reference for the primary information itself – and with the book already extending to some 30-odd chapters, an equivalent 'weight' of synthesis would render the work so large as to be virtually unusable. However, one of the main themes rehearsed again and again by authors in the first volume, and highlighted by the editors in conclusion, was a need for science-based management (Apollonio *et al.*, 2010b).

In different countries, and for different species, management may be directed variously towards

- control of population numbers, or control of damaging impacts
- exploitation of a sustainable resource, for meat or recreation
- conservation of endangered species or subspecies.

But whatever its primary aim, that management will only be effective if it is well informed.

The aim of this book

In this volume, therefore, we have invited experts in a number of different areas of wildlife biology or management to review the different management approaches adopted in each of the various different countries of Europe, in relation to each of a number of management issues identified and to prepare some sort of synthesis of that experience.

Of course, in a purely factual sense, this new volume draws on the information which is now summarised in that earlier book (Apollonio *et al.*, 2010a). In some senses therefore, this new book may be seen as a companion volume to that earlier work – but it is our hope that it may also be valid as a separate contribution in its own right. While inevitably each chapter is to an extent *informed* by the material summarised in that earlier volume, each author uses his or her own research experience and expertise to essay further development of the material in particular topical areas. Indeed the true nature of the relationship between the two books is perhaps that the first offers a convenient source of the information which underpins the analyses of this new volume (enabling any reader who may wish to seek more detail about individual examples, or about management practices in a particular given country, to return to the detailed descriptions of individual practices presented in that earlier work and find all the relevant background detail in one place).

As was the previous volume, this book is directed at practising wildlife managers and stalkers; policy makers in local regional or national administrations responsible for formulating policies affecting management of different wildlife species; and others who may be actively involved in research into improving methods of wildlife management.

In this book we review a number of issues which seem to crop up again and again as problems in management (or issues affecting management). Many of these were highlighted by Apollonio *et al.* (2010b), but are now developed in more detail. Topics include:

An overview of the basic resource and the administrative structures within which management is carried out:

- An overview of the ungulate species present: native species, problems associated with the introduction of non-native genotypes of native species, problems associated with the introduction of exotic species, and the implications for management of this varied genetic resource.
- Value systems for ungulates in Europe; management systems and the exploitation of ungulate populations for meat or sport.

- Management context: European legislation; an overview of different national and international legal constraints.
- Other constraints on management: hunting seasons in relation to biological breeding seasons and the implications for the control or regulation of ungulate populations.

A consideration of management issues:

- Impact of large ungulates on agriculture, forestry and conservation habitats in Europe.
- Road traffic accidents involving ungulates and available measures for mitigation.
- Large herbivores as 'environmental engineers' or agents of deliberate habitat change.
- Large carnivores and the impact of predation on populations of wild ungulates.
- The role of diseases in limiting or regulating large ungulate populations.
- Wild ungulates as vectors of disease.
- Climate change and implications for the future distribution and management of ungulates in Europe.

Through an exploration of the underlying biology and a comparison of the experience gained from different management approaches adopted in different places we then attempt to tease out what works (in what circumstances) and what does not.

One size does **not** *fit all!!!*

But inevitably, there are no 'holy grails' to be discovered, no 'universal' solutions. Different countries support different species of ungulates and different species-mixtures. Even with regard to the same species, management objectives may differ markedly in different places or in different contexts (whether directed towards control of populations and their impacts, management for exploitation, or a need for active conservation). Local circumstances may also affect what management options are actually available, or may affect the utility of any given method.

Superimposed on such variation is an equal variation in attitudes and cultural approaches to hunting and game management.

- In some countries with a long tradition of game management, hunting is positively celebrated.
- In other countries, while hunters are perhaps in the minority, there is no widespread 'objection' to the idea amongst the general public.

- In other countries again, the idea of hunting (taking life for pleasure) is widely considered repugnant, and hunting is only accepted by the general public if it is formally justified as necessary to maintain animal populations in balance with their wider environment, and to fulfil other management objectives.
- In such cases hunting is usually 're-branded' as 'game management' ... and often viewed with some reservations by society in general.
- Finally, in some countries hunting is actually illegal. All animal species are fully protected by law and permission to kill them for management purposes needs to be specifically applied for in every individual instance by seeking specific exemption under the law.

With such diversity even in cultural attitudes to hunting (together with profound differences between countries in legal and administrative regulation of hunting, it is not surprising that there is an equal diversity of hunting practice and, as above, we should not expect to find any single optimal solution. Rather, in this volume we focus on the issues in order to try and present an informed scientific basis on which any such solutions may be more soundly based.

But any such solution must 'fit' within the social attitudes and expectations which characterise any given culture, and it is perhaps instructive in this initial chapter to offer some overview of that same cultural diversity. We tend to become accustomed to what is common practice in our own country and assume that practice is similar elsewhere. Nothing could be further from the truth! And if we are better to understand the management systems and practices adopted in other countries it is helpful to understand their different social and cultural context.

1.2 Cultural attitudes to hunting

Attitudes to hunting and game management are likely to be influenced by a number of factors, amongst them: the legal status of game; the legal status of the right to hunt (and in both cases, therefore, whether hunting is seen as a socially divisive 'elitist' pursuit); the history of hunting and its place within cultural tradition; the proportion of the population as a whole who are engaged in hunting activities; and the level of 'urbanisation' of the human population (and thus their increasing detachment from the land).

This list is far from exclusive, and clearly many of these factors interact (or are different reflections of a common underlying basis). What is significant, however, is that cultural attitudes are changing in many countries – and changing quite rapidly as the result of a declining interest in hunting (as the percentage of the human population actively engaged in hunting or related activities) and with increasing urbanisation. To the extent that public attitudes and public perceptions may influence what is deemed 'acceptable', it is clear that changing attitudes may well influence management approaches, management systems and management legislation in the future. This in turn may have implications (or provide future conflicts) with achieving effective management of wildlife populations.

At the present time there is little formal data available on attitudes to hunting throughout Europe. A new project just started under the European Commission 7th Framework Programme aims to assess the social, cultural, economic and ecological functions and impacts of hunting across a range of contexts in eight European and African countries, and to understand what influences attitudes to hunting, how these attitudes influence and determine individual and societal behaviour in relation to hunting and, finally, how hunting behaviour influences biodiversity. This project has, however, only just commenced, and in advance of any results from this or similar projects any presentation of the range of cultural attitudes across Europe must necessarily be somewhat subjective.

In attempting to provide some overview of stakeholder perceptions and attitudes it is helpful (if somewhat reductionist) to try and 'group together' the various different cultural systems existing in different places into a number of different 'clusters'. In effect we may reduce the diversity of cultural systems that exist to four basic cultural 'models' recognised by Yves Lecocq, the Secretary-General of FACE (the European Fédération des Associations de Chasse) as the Scandinavian (North European) model, the Germanic (Central European) model, the Anglo-Saxon model and the southern European model (Lecocq, 2007). Such a device is clearly oversimplistic and perhaps something of a caricature, and not every country fits into its regional stereotype, but by the same token it emphasises distinct differences in perspective of both hunter and wider public to hunting, game management and related (welfare) issues. This is useful not only in reviewing attitudes and perceptions, but also in beginning to understand differences in legislative requirements.

Lecocq notes the following characteristics.

1. *North European (Scandinavian) model:*
[Norway, northern Sweden, Finland, Denmark]

- Hunting is recreational, but with a major focus still concerned with generating food.
- Hunting is popular and widespread – with the highest proportion of hunters per head of population in Europe.

Lecocq uses Denmark as his 'type species'. We shall here substitute Norway (after Andersen *et al.*, 2010): "The main objective for cervid management in Norway is based upon sport/recreational hunting, but with a focus on venison production. Most of the venison harvested is consumed by the hunter and his family/friends, and only small amounts are accessible for trade. Although there are more than 400 000 persons in the official register of hunters, a much smaller proportion are actually active hunters at any one time: thus only about half of these (a total of 195 200 persons) paid the hunting fee for small game or large game for the hunting year 2005/2006 (some 5% of the total population). Nonetheless, in general, hunting is widely accepted and there are no ethical objections raised to the exploitation or harvest of wildlife species. Legislation simply emphasises that the concept of *sustainable use* should underpin all wildlife management" (Norway: Andersen *et al.*, 2010). In Sweden: 'The only general and national objective for the management of game species in Sweden is that they should be preserved in viable populations, but not be allowed to seriously damage other vital interests of the society' (Liberg *et al.*, 2010).

2. *Central European (German) model:*
[Germany, Hungary, Austria, Poland, other countries of the former German, Polish or Austro-Hungarian Empires, such as Slovakia, Croatia, Romania, etc.]
Lecocq characterises this group as having:

- very closely regulated hunting, strongly circumscribed by administrative and regulatory requirements and constraints as well as traditional practices
- a very long tradition and very strict 'rules' or codes of conduct (e.g. St Hubert's)
- hunting more concerned with management of ungulate populations than exploitation (at least for venison), but trophy quality important
- hunting not carried out by individuals, but rather by members of well-established hunting groups or hunting associations with long traditions
- well-trained hunters – with training provided by those same hunters associations or (Slovenia) hunters' families.

Clearly the 'expression' of this system varies somewhat from country to country, but with Austria as a type example we may note that perhaps 1.5% of the population are involved in hunting; in Slovenia figures are similar with 1.1% of the population as active hunters.

Hunters clubs are long-established with a great strength of tradition. In many cases there is a traditional hunters 'uniform' or dress code – worn by forest managers or hunters. Hunters often serve a long 'apprenticeship' within the association before they are recognised as full hunting members; the whole concept of hunting is seen as an honoured and very honourable tradition.

3. *Anglo-Saxon model:*
[Typified perhaps by the UK and Ireland]
Lecocq suggests that here:

- Hunting is largely recreational.
- There is a relatively small number of participants – Lecocq suggests perhaps some 1 in 60 of the entire population (1.7%), similar to Austria, Germany etc.
- There is a high proportion of professional stalkers.

In the present analysis it is important to add that – perhaps because of the small number of active hunters and the long-standing association of the right to hunt with ownership of land, the wider general public (as a largely urban society) regards hunting with some disfavour, either simply because it involves the killing of animals or because they see it as the recreational pursuit of a land-owning elite. Such suspicion may also stem at least in some part from recognition that hunting in the UK is perhaps the least regulated of any country in Europe (i.e. least state intervention in management and management practice; Putman 2008a; see also this volume, Chapter 3).

4. *Southern European model:*
[Lecocq typifies this with Spain, but also includes Portugal, France, Italy, Greece and other Mediterranean countries]
This category is perhaps the most diverse and while it is adopted here for simplicity, it might not truly represent a single homogeneous group. Here hunting is relatively common and perhaps more widely accepted [perhaps 3% of the population are hunters] and would indeed appear to be a more social activity. However, attitudes are changing. In the last 50 years urbanisation led to a strong differentiation between the rural world and the urban society. In the former, hunting is still popular and widely accepted, while in the latter hunting is increasingly strongly opposed (see below).

As noted already, this characterisation of different 'national' attitudes is oversimplistic; hunting practices and attitudes in some countries fit uneasily within their 'type' while others really do not easily fit into any of Lecocq's categories (Netherlands, Belgium, Switzerland, perhaps Italy). Nonetheless the implications are clear ... that there is no single European 'model' and significantly that: *attitudes and expectations amongst stakeholders and the wider public in relation to hunting will be strongly coloured by the 'traditional' view of hunting within the national culture and the proportion of the human population who are themselves actively engaged in hunting.*

It is in fact quite hard to undertake any formal analysis of the factors influencing public attitudes and public perceptions and the ways these may indeed be changing with increased urbanisation of human societies, as there are few objective surveys available.

However, such formal surveys as have been undertaken confirm that one of the primary factors affecting individual attitudes to hunting is personal experience (as a hunter, or closely related to others who hunt). In a survey of 415 interviews in Louisiana (Floyd *et al.*, 1986), the major factor influencing attitudes to hunting was direct participation in hunting or having family members and friends who hunt. Similar results were found by Stokke (2004) in a survey of 1000 Norwegians, weighted by gender, age, place of residence, income and education, to suggest a representative sample of Norway's population.

Once again, attitudes were found to be significantly coloured by personal experience of hunting, and a generally positive attitude towards hunting is linked to the fact that a total of 60% of the Norwegian population has a direct relationship with hunting and hunters either because they are themselves hunters or have close relatives or friends who hunt (Stokke, 2004).

Other factors are likely to include (as above)

(i) the legal status of wildlife (whether the state or the private individual or whether they are in effect *res nullius*)
(ii) the degree of state intervention in (and thus state regulation of) management
(iii) the proportion of the human population actively involved in hunting
(iv) the cultural and historical traditions of hunting (as e.g. within countries of a more Germanic tradition)
(v) the degree to which human society is increasingly urbanised (although this last may in itself have an indirect effect simply through the reduction in the proportion of the overall population who have direct experience of hunting and simultaneous reduction of proportion of population having direct contact with animals at all whether wild or domestic).

Clearly however, there could be significant implication for future management in any country as attitudes change over time in response to changes in these driving variables. Aware of these changes in perception and attitude, management systems and legislation are currently under active review in a significant number of countries in Europe and North America, with increasing attention being paid to management systems, ways of improving welfare (e.g. Putman, 2008b, 2008c), and review of alternative non-lethal management approaches such as immunocontraception or translocation (Putman, 1997, 2004; Green, 2008).

1.3 Conclusions

We develop this theme of contrasting attitudes to hunting in some detail here since not only is it extremely striking, but it has a profound effect on many other aspects of game management and its administration (training, hunting practice, etc.). There is similar (and often related) variation in legislative systems and the administration of hunting (explored in more detail in Chapter 3). Taken in combination with differences in game species present and differences in objectives of management, we should then expect a diversity of solutions to management of game animals and their impacts.

Whatever the solution, however, the driving theme of this book is that solutions should be informed by a proper scientific appraisal of the management issues and problems to be resolved.

The chapters which follow aim to offer a review of some of the most relevant issues which must be considered by managers (and legislators).

References

Andersen, R., Lund, E., Solberg, E. and Sæther B-E. (2010) Ungulates and their management in Norway. In M. Apollonio, R. Andersen and R. Putman (eds.) *European Ungulates and their Management in the 21st Century*. Cambridge, UK: Cambridge University Press, pp. 14–36.

Apollonio, M., Andersen, R. and Putman, R. (eds.) (2010a) *European Ungulates and their Management in the 21st Century*, Cambridge, UK: Cambridge University Press. 604 pp.

Apollonio, M., Andersen, R. and Putman, R. (2010b) Present status and future challenges for European ungulate management. In M. Apollonio, R. Andersen and R. Putman (eds.) *European Ungulates and their Management in the 21st Century*. Cambridge, UK: Cambridge University Press, pp. 578–604.

Floyd, H.H., Bankston, W.B. and Burgesion, R.A. (1986) An examination of the effects of young adults' social experience on their attitudes toward hunting and hunters. *Journal of Sport Behavior* **9**(4), 116–130.

Green, P. (2008) *Can Contraception Control Deer Populations in the UK?* Report to the Deer Initiative, Wrexham, UK.

Lecocq, Y. (2007) Demonstrating competence: a European perspective. Address to an Open Seminar of the Deer Commission for Scotland; Drumossie, Inverness; 19 June 2007.

Liberg, O., Bergström, R., Kindberg, J. and von Essen, H. (2010) Ungulates and their management in Sweden. In M. Apollonio, R. Andersen and R. Putman (eds.) *European Ungulates and their Management in the 21st Century*. Cambridge, UK: Cambridge University Press, pp. 37–70.

Putman, R.J. (1997) *Chemical and Immunological Methods in the Control of Reproduction in Deer and other Wildlife: Potential for Population Control and Welfare Implications*. RSPCA Technical Bulletin. Horsham, UK: Royal Society for the Prevention of Cruelty to Animals, 50 pp.

Putman, R.J. (2004) *The Deer Manager's Companion: A Guide to Deer Management in the Wild and in Parks*. Shrewsbury, UK: Swan Hill Press.

Putman, R.J. (2008a) *Management Systems and the Administrative Framework for the Management of Ungulates in Different European Countries*. Report to the Norwegian Directorate for Nature Management, Trondheim, Norway.

Putman, R.J. (2008b) *A Report on Current Perception, Legal Status and Expectation with Respect to Deer Welfare in Other Countries*. Contract report for the Deer Commission for Scotland.

Putman, R.J. (2008c) *A Report on the Potential Responsibilities of Care for Upland Red Deer*. Contract report for the Deer Commission for Scotland.

Stokke, E. (2004) Nordsmenns Holdninger til jakt: Norwegians' attitudes to hunting. Masters thesis at the Department of Ecology and Natural Resource Management, Norwegian University of Life Sciences, Ås, Norway.

2

Status and distribution patterns of European ungulates: genetics, population history and conservation

JOHN D.C. LINNELL AND FRANK E. ZACHOS

2.1 Introduction

Within Europe as a whole, the distributional range, population size and the status of many species has been greatly influenced by human activity – not simply through the negative influences of humans on land-use patterns and in overexploitation, but also through active attempts to 'restore' and 'augment' species distributions.

A number of indigenous subspecies have been (or may be currently) threatened whether due to habitat loss, overexploitation or simply by lack of positive management to protect them. In addition, the genetic integrity of such endangered taxa may be compromised by the introduction to those populations of animals of different genetic background in misguided, although well-intentioned, attempts to bolster dwindling populations. Even within well-established populations apparently not under threat, introduction of animals of different genetic types may have been quite commonplace (usually in an attempt to try and improve the 'trophy quality' of antlers) – and thus the special genetic status of particular local populations has been greatly altered by the introgression of alien genes. Reintroduction of species to local areas from which they had previously become extinct has also often been undertaken without due regard to the genetic provenance of those individuals released, thus causing other discontinuities in genetic distributions.

For some whole species (such as fallow deer, whose range now extends far beyond its original post-glacial distribution in Turkey and the southern Mediterranean) the modern geographical distribution has largely resulted from deliberate introductions well beyond their native range (Chapman and Chapman, 1980). And, while fallow deer may at least have some claim to a

Ungulate Management in Europe: Problems and Practices, eds. Rory Putman, Marco Apollonio and Reidar Andersen. Published by Cambridge University Press.

European origin in the first place, during the last interglacial, introductions have also been made to the European fauna of a number of exotics (American wapiti, sika, axis, muntjac, white-tailed deer, Chinese water deer, etc.) which have no claim to any European past.

In some cases, such introductions have led to further compromise of genetic stocks because of hybridisation between exotic and native species: hybridisation between red and sika has been documented in both the United Kingdom and the Republic of Ireland (e.g. Harrington, 1973; Abernethy, 1994; Goodman *et al.*, 1999; Díaz *et al.*, 2006; Pemberton *et al.*, 2006; Senn and Pemberton, 2009) and in the Czech Republic (Bartoš *et al.*, 1981; Bartoš and Žirovnický, 1981; Zima *et al.*, 1990), and there is concern about the potential for hybridisation elsewhere (see Bartoš, 2009). But even without such problems from potential hybridisation, non-indigenous ungulates may have a profound impact on wider biodiversity through impacts on native vegetation or through competition with native ungulates (Spear and Chown, 2009).

In short, human activities have had a profound effect on both the distributions and genetics of ungulate populations throughout Europe. In this chapter we consider the natural phylogeography of ungulate species in Europe and the effect of anthropogenic interventions such as these on distribution and genetic status of our native ungulates. For the purposes of this chapter we limit ourselves largely to Western Europe, excluding Russia and Ukraine. Therefore Turkey and the Caucasus with their very diverse ungulate faunas are excluded. Taxonomically, we focus on select examples from the Cervidae and Bovidae.

2.2 A short history of ungulates in Europe: decline and recovery

Although there have been a number of extinctions from the European ungulate fauna since the end of the Pleistocene, these were much fewer than those experienced in North America over the same period (Kurtén, 1968; Kurtén and Anderson, 1980; Yalden, 1999). In practice, since that time, we have lost only five species: the mammoth, the woolly rhinoceros, the Irish elk, the wild horse, and the aurochs. And even some of these species persisted long after the ice age ended in some European locations (Stuart *et al.*, 2004). For example, the aurochs only became extinct some 400 years ago (the last one was killed on the Vistula River in 1627), and the last wild horses persisted until the nineteenth century. All other species present at the end of the Pleistocene have persisted in some form, although their numbers, distributions and ecological situations have changed – and there have been numerous other, non-native additions.

As human influence spread across Europe, gradually changing from a hunter-gathering to an agricultural lifestyle, the abundance and distribution of wild ungulates declined. The low point for most wild ungulates, in fact for large mammals in general in Europe, was the end of the nineteenth century and the early twentieth century. This was when human pressure on the ecosystem was at its greatest. Ungulates were hunted directly as food and their habitats were under great pressure from deforestation and competition with domestic livestock. Although few whole species were driven to extinction, many populations were extirpated, and some species only just survived. For example, a European endemic, the alpine ibex, only persisted in one royal hunting reserve in Italy (where because of its status, protection was more strongly enforced).

Human attitudes began changing just in time. The late nineteenth century and twentieth century have seen a growing awareness of the value of large ungulates. By far the most significant factor in this was the value placed on them as game species, and for most of recent history it has been hunters who have been the driving force for the restoration (and reintroduction) of ungulate populations. Modern-day thinking about the importance of biodiversity and concern with conservation is only 30 years old and has really only played a minor role in the recovery of these species.

As mentioned above, the alpine ibex only persisted in one royal hunting reserve in Italy and countless private estates provided crucial refuges for remnant populations of other species. For example, all roe deer in Scandinavia descend from a population that persisted in the environs of one Swedish hunting estate (Andersen *et al.*, 2004b). From such refuges an enormous number of translocations have been performed all across Europe, again largely by hunters, or motivated by a desire to extend the range of favoured game species. At the same time, while human activities were markedly affecting the distribution and status of native species, many landowners were also keen to introduce new species into Europe whose origins were well outside the normal European range – species from Asia or North America.

At the same time as these various efforts to restore or expand the distribution of ungulate species were being undertaken there has been a dramatic change in human land use. Forest cover has been restored over much of Europe, livestock numbers have decreased, human use of marginal lands (especially uplands) has decreased through intensification of agriculture in lowlands and legislation has improved dramatically in favour of conservation or at least sustainable use. The net result is that (although there are exceptions, as we will see later) both native and naturalised ungulates in general are now thoroughly reintegrated into the European landscape, although recovery has been stronger in western, central and northern Europe and weakest in the south. There are few areas where at least one species does not occur, and many areas are home to communities of up to five or six species. No species is at risk of extinction (although there are some threatened subspecies). However, in the scramble to restore the species to their former ranges (and even more widely), and in the introduction of new, exotic, species from elsewhere, very little attention was paid to population genetics or risks of 'corruption' of the genetic identity of local races or subspecies, as the following case studies illustrate.

It would not be practical to explore in these pages the changing fortunes of every native or introduced species of ungulate within Europe, and the way in which its distribution, population size or genetics have been affected by human activities. Indeed, many of the reintroductions and translocations that have occurred have not even been recorded. Instead we focus on a number of individual species as examples, selecting them to illustrate particular points or issues of more general application.

Red and roe deer are today among the most common European ungulates and are arguably, together with wild boar, the most important game species. Nonetheless, they are of conservation concern for a variety of reasons. Firstly, in both species, molecular analyses have identified genetically distinct populations (which are sometimes, but not always, taxonomically acknowledged as subspecies) in particular need of protection. Secondly, human influences – in particular selective hunting regimes, translocations and habitat fragmentation – have resulted in many challenges for the management and conservation of other local or regional populations. Red deer in particular also offer a well-studied example of hybridisation of an indigenous species with a closely related introduced exotic (in this case, sika deer). The other species covered in detail, the European bison and the ibex, exemplify ungulates that have recovered from near-extinction but may still be vulnerable due to this bottleneck and concomitant reduced genetic variation. The ibex further serves to illustrate the challenges of teasing apart taxonomic identity of species (also known from chamois/isard), where what was once regarded as being one species is now regarded as two (the alpine and the Spanish ibex).

2.3 Case study 1: Red deer

The red deer, *Cervus elaphus*, is the best studied and, together with the roe deer and the wild boar, the most widespread European ungulate species, occurring in most of continental Europe and the British Isles. It is absent from northern Scandinavia, Iceland and Finland (Mitchell-Jones *et al.*, 1999).

Globally, red deer are Holarctic in distribution, with a basal split between a western 'red deer type' group in Europe, North Africa and west Asia, and an eastern 'wapiti type' group in east Asia and North America (Ludt *et al.*, 2004), with intermediate forms in central Asia.

In Europe, six subspecies are usually listed (e.g. Dolan, 1988): Swedish red deer (*C. e. elaphus*), Norwegian red deer (*C. e. atlanticus*), central European red deer (*C. e. hippelaphus*), British red deer (*C. e. scoticus*), Spanish red deer (*C. e. hispanicus*), and Corsican red deer (*C. e. corsicanus*). The North African Barbary red deer (*C. e. barbarus*) shows close affinities to the Corsican red deer (e. g. Ludt *et al.*, 2004; Skog *et al.*, 2009), and it seems possible that the Barbary deer derive from animals introduced from Sardinia and/or Corsica in historical times (Hajji *et al.*, 2008; see also Dobson, 1998) or vice versa (Ludt *et al.*, 2004).

This simple taxonomic classification, however, does not reflect the complete diversity of genetic structure of the European red deer. The Quaternary history and phylogeography of the red deer in Europe has recently been analysed in detail (Skog *et al.*, 2009; see also Ludt *et al.*, 2004; Sommer *et al.*, 2008). Based on mitochondrial DNA (mtDN) data, two main lineages were detected (plus a third one comprising only Sardinian and African red deer) that hint at distinct western and eastern glacial refuges (in the Iberian Peninsula and southern France, and in the Balkans and Carpathians, respectively).

In the Holocene, western and northern parts of Europe (including the British Isles) as well as large parts of central Europe were recolonised by the western mtDNA lineage, while the eastern lineage remains still largely confined to eastern and east-central Europe (Skog *et al.*, 2009). How far into the east the western lineage extends and where the two lineages meet is not yet clear. Italy is peculiar in this regard in that present-day populations of red deer almost all derive from the eastern lineage; by contrast the only remaining native Italian population from Mesola (see below) shows a haplotype whose relationship to the two main lineages is ambiguous.

Re-establishment of red deer within their former range has been in part due to natural recolonisation. After their extinction in the seventeenth century red deer recolonised the area of eastern Switzerland in the late nineteenth century and genetic analyses suggest that these deer very probably came from Liechtenstein (Kuehn *et al.*, 2004). In a further example: the Italian Alps were recolonised after extinction by red deer immigrating from Austria, Switzerland and Slovenia (Mattioli, 1990).

But recovery of red deer has, in addition, undoubtedly been facilitated by an immense number of translocations: many of these are recorded in documentary records (e.g. Niethammer, 1963), although these themselves probably only reflect a small proportion of the translocations undertaken.

In some cases red deer were reintroduced to areas where they had previously been eradicated, e.g. in Italy or parts of Germany. One German stock north of Hamburg was founded with animals from Poland, Hungary and Austria, and these origins are still evident in the genetic make-up of this population which is fundamentally different from other north-German populations (F. E. Zachos *et al.*, unpublished data). Such extraneous introductions might thus introduce a marked disruption of the natural genetic patterns observed across the continent, by interpolating populations whose genetic origins lie somewhere else completely.

In other regions deer were introduced into areas where there were already existing populations of native red deer present, which thus had a marked effect on local stocks. Hmwe *et al.* (2006b), for example, presented evidence that Scottish red deer populations free from (recorded) introductions showed a more natural pattern of genetic differentiation with respect to geographic distance than did an analysis of a wider sample of Scottish populations overall – a possible consequence of introduced genotypes blurring an underlying natural pattern of increased isolation by distance. This suggests that while populations free from introductions show relatively similar genetic character, other populations such as those on the islands with clear evidence of past introductions still bear the legacy of that genetic admixture. In the same study, rather low mtDNA variability was found in those populations analysed from offshore islands (Islay and Arran), localities where current populations of red deer are known to have been heavily affected by past human management (Hmwe *et al.*, 2006b). Red deer genetic variability has also been assessed on the island of Rum, where a low number of mtDNA haplotypes and the existence of a divergent haplotype closely related to Corsican red deer (*Cervus elaphus corsicanus*) give further evidence of the effects of past human activities on Scottish red deer island populations (Nussey *et al.*, 2006).

Curiously, despite such evidence for effects of introduction on this more local scale and although we know that red deer, as a prestigious game species, have been translocated throughout Europe for centuries (Niethammer, 1963; Hartl *et al.*, 2003), in the phylogeographic analysis of Skog *et al.* (2009) few individual red deer specimens out of a sample of nearly 600 were found to show discordances between their geographical location and their overall genetic lineage (some Spanish individuals with a Sardinian/African haplotype and some individuals from Sardinia, Romania and Italy with a western haplotype). To our knowledge evidence for only one more between-lineage translocation has been found so far: Nussey *et al.* (2006) found a mitochondrial haplotype in the introduced population on the Isle of Rum, Scotland, that only differed by one mutational step from a Sardinian haplotype. This is perhaps suggestive that (at least as far as females are concerned; see below), the majority of translocations were carried out within the main lineages rather than among them and have not, after all, blurred the large-scale phylogeographical pattern of the species.

However, it must be stressed that since the most widely used molecular marker in phylogeographical analyses is the maternally inherited mitochondrial DNA these studies only reveal the female side of the story, rather than reflecting the effects of (possibly more frequent) translocations of male deer. Such focus on maternally transmitted genetic markers (coupled perhaps with inevitably limited sampling) may explain why Skog *et al.* found so few discontinuities in distributional pattern – despite evidence for the effects of introduction on a more local scale, as in the Hmwe *et al.* example above. While the polygynous mating system in this species ensures that not all introduced males will contribute to the next generations, single successful males may have a disproportionately large impact. Studies on paternally transmitted markers are much needed to shed light on the eventual consequences of these translocations. Unfortunately, Y chromosome markers have turned out not to be of equal variability to mtDNA in red deer (Barbosa and Carranza, 2010).

Similar limitations in analysis may explain the failure of Feulner *et al.* (2004) to detect 'inconsistent' genetic markers among Carpathian red deer. These deer have long been considered distinct from other European mainland red deer based on visual appearance (greyish colour, lack of mane and absence of a dark mark on the rump patch) (Dobroruka, 1960; Groves and Grubb, 1987). In line with this, they were classified as *C. e. montanus* (Botezat, 1903). Based on mtDNA control region sequences and nuclear microsatellite loci, Feulner *et al.* (2004) suggested that the red deer from the Romanian Carpathian Mountains indeed were a genetic unit different from surrounding populations and might represent one of the few remaining natural populations of the species in Europe. This analysis was, however, based on some 40 deer from the Romanian Carpathians, a sample perhaps not representative of the whole Carpathian population. In practice it is known that Austrian red deer with smaller antlers than indigenous stags (but with multi-tine crowns) were introduced into the Carpathians, especially in the western regions, in the nineteenth and early twentieth centuries, leading to genetic admixture (Micu *et al.*, 2010). Local hunters distinguish morphologically between 'Austrian deer' (individuals whose antlers have many tines) and 'Carpathian deer' (those individuals that count 12 or fewer tines; Comşia, 1961).

Despite all these translocations and introductions there do remain populations of red deer that are phylogenetically distinct and deserve special attention from a conservation point of view as they represent unique genetic lineages that are, or were, threatened with extinction.

In the western Palaearctic, this particularly applies to the red deer from the Tyrrhenian islands (*C. e. corsicanus*, Corsica, Sardinia) and to those from North Africa (*C. e. barbarus*) and Mesola (Po delta, Italy; not acknowledged taxonomically). These three have recently been analysed in detail and there is growing evidence that they are related phylogenetically (Zachos *et al.*, 2003; Ludt *et al.*, 2004; Hmwe *et al.*, 2006a; Hajji *et al.*, 2008; Skog *et al.*, 2009).

The Tyrrhenian red deer of Corsica and Sardinia[1] have often been considered to be descendants of introduced North African Barbary deer (e.g. Ludt *et al.*, 2004). In practice, other more comprehensive evidence suggests that it is perhaps more likely that it was the other way around (Dobson, 1998; Hmwe *et al.*, 2006a; Hajji *et al.*, 2007, 2008 and references therein). In fact, genetic analyses not only showed close relationships between *C. e. corsicanus* and *C. e. barbarus* but also between *C. e. corsicanus* and the indigenous red deer from Mesola in the Po delta region on mainland Italy. This relict population is the only surviving stock of native Italian red deer (see Hmwe *et al.*, 2006a and references therein), and its relationship with Sardinian and Corsican red deer is indicative that *C. e. corsicanus* was probably of Italian origin, that relict populations survived on Sardinia and around Mesola, and that some of these deer were used in introductions to North Africa.

The genetically distinct status of the red deer population from Mesola in the Po delta region, Italy, is not acknowledged taxonomically (it is classified as part of the European mainland subspecies *C. e. hippelaphus*). However, it is the last remaining native Italian red deer population (Hmwe *et al.*, 2006a; Zachos *et al.*, 2009 and references therein) and harbours a single but unique mitochondrial haplotype somewhat intermediate between those of the two major European lineages (Skog *et al.*, 2009). As a consequence of severe bottlenecks and an estimated long-term effective population size of a mere 15 individuals (Lorenzini *et al.*, 1998), its genetic variability is similar to that of *C. e. corsicanus*, among the lowest found in this species (e.g. Zachos *et al.*, 2009). The Mesola population is therefore considered to be of significant

[1] This subspecies is usually called Corsican red deer. In order to avoid misunderstandings with respect to which island population is meant, we use the term Tyrrhenian red deer.

conservation interest. At the moment, plans are being developed for splitting the population in order to reduce its susceptibility to environmental stochasticity, a measure also encouraged by a recent population viability analysis (PVA) which identified catastrophic impacts (and inbreeding depression) as the main threats to the population's survival (Zachos *et al.*, 2009). Although the population's general prospects are rather bleak, this PVA and the fact that management measures taken from 1994 (reduction of competition with fallow deer, habitat improvement, winter feeding) have resulted in an increase in reproduction and a decrease in mortality (Mattioli *et al.*, 2003) show that conservation actions are promising.

The Tyrrhenian red deer similarly underwent a severe bottleneck with only 100–150 animals remaining in the 1970s – the animals became extinct on Corsica in 1970 and were subsequently re-established with 13 animals introduced from Sardinia in the 1980s and 1990s (Kidjo *et al.*, 2007). Now, after strict protection and breeding programmes, the combined population comprises at least 2700 animals according to Lovari *et al.* (2007). Kidjo *et al.* (2007) give a more conservative estimate of 'slightly more than 1000' but independent estimates for the population of Sardinia alone in 2005 were more optimistic, suggesting populations in excess of 6000 individuals (M. Apollonio, pers. comm), and Puddu *et al.* (2009) reported more than 1000 rutting males from Sardinia.

The subspecies is still classified as 'Endangered' but its recent recovery may warrant a reclassification as 'Near Threatened' (Kidjo *et al.*, 2007). Interestingly, a recent study revealed significant genetic differentiation between the source Sardinian population and its descendant population on Corsica based on nuclear microsatellite loci, which is indicative of a founder effect (Hajji *et al.*, 2008). The pairwise F_{ST}-value (fixation index in Wright's F-statistics) between Sardinia and Corsica was 0.151, which is considered to be moderate or even great differentiation (see Balloux and Lugon-Moulin, 2002 and references therein). Differentiation between the two subpopulations of *C. e. corsicanus* was further substantiated by other methods such as Bayesian clustering, assignment tests and individual-based trees. Thus, a further exchange between the two islands seems desirable in order to keep the effective population size as large as possible.

While the main anthropogenic impacts relevant to the conservation of genetic diversity of red deer in Europe were translocations and selective hunting in the past, probably the most important for the future will be habitat fragmentation. Especially in Central Europe, expanding human infrastructure (settlements, roads etc.) causes populations to become more and more isolated from each other. As a consequence, gene flow is reduced, effective population size decreases and genetic drift and inbreeding in small and discontinuously distributed populations result in the loss of genetic variability.

For Bavarian red deer populations, Kuehn *et al.* (2003) were able to show that genetic structure was determined in the past by equilibrium of genetic drift and gene flow, whereas today it is influenced primarily by genetic drift alone due to a disruption of gene flow. In general, in polygynous species with male-biased dispersal spatial genetic structuring is expected to be much more pronounced in the female sex than in males. This has been confirmed in studies on red deer (Nussey *et al.*, 2005; Frantz *et al.*, 2008) with an extreme example of fine-scale structuring (< 100 m) in female red deer in the introduced population on the Isle of Rum off Scotland (Nussey *et al.*, 2005). A recent study on red deer in the Scottish Highlands has nicely demonstrated that landscape features significantly affect gene flow by imposing differential costs of migration on the animals (Pérez-Espona *et al.*, 2008). In particular, it was found that sea lochs, mountain slopes and roads, but also forests, were barriers to gene flow, whereas inland lochs and rivers rather facilitated gene flow. Simple geographical distances between populations thus only explained a fraction of the genetic differentiation, and more detailed habitat analyses are needed when differentiation patterns are addressed in conservation and management issues.

At present, a considerable amount of research into migration corridors and habitat connectivity is being carried out. In northern Germany, the red deer has been included in the regional Red List as 'Near Threatened' due to its being confined to small and isolated populations in danger of genetic depletion and inbreeding depression (Zachos *et al.*, 2007). Indeed, in northern Germany strong evidence has been found for inbreeding depression: in an isolated and genetically depauperate population of about 50 animals no less than eight cases of brachygnathia inferior (shortened lower jaw) have been recorded so far, a malformation known to be a corollary of inbreeding (Zachos *et al.*, 2007). Inbreeding depression has also been confirmed for the red deer population on the Scottish Isle of Rum where inbreeding has been shown to reduce lifetime breeding success (Slate *et al.*, 2000), a trait directly associated with overall fitness. To address such problems considerable attention is being offered in many countries to the development of habitat corridors linking isolated populations and re-establishing gene flow. In northern Germany, an ongoing project is evaluating genetic isolation of red deer populations in fragmented landscapes combined with telemetry-based studies on dispersal behaviour in order to determine the need and optimal location of green bridges and/or translocations (F.E. Zachos *et al.* unpublished data).

Similar efforts are being made in the Netherlands to provide habitat corridors between isolated populations of red deer in the Oostvaardersplassen, the Horstervold and the Veluwe in the central Netherlands (ICMO, 2006).

Selective hunting also has the potential to significantly influence the genetic composition and fitness parameters of natural populations (Milner *et al.*, 2007). It has been shown that selective harvesting in red deer may drive adaptive responses from adult survival and growth to early and lightweight reproduction (Proaktor *et al.*, 2007). For the red deer in particular, analyses in Spain have shown that different hunting practices result in different selection regimes with respect to body and antler size, which is of relevance to conservation and management (Torres-Porras *et al.*, 2009). A general problem with selective hunting is that not only are the desired traits selected for but also those that are linked with the traits under direct selection. A detailed study on French red deer populations (Hartl *et al.*, 1991a, 1995b) revealed that selective hunting in favour of a large number of antler points and against small spikes in young males has caused a detectable change in the frequencies of some alleles at certain allozyme loci. These allozyme loci, however, are also known to be associated with viability in females (Pemberton *et al.*, 1988) and, thus, selective hunting for antler traits may have an indirect and unintended effect on fitness components. As the genetic basis of the quantitative traits selected for is generally unknown (as is the amount of linkage with loci governing components of fitness), this is a general caveat of selective harvesting regimes.

2.4 Case study 2: Introgressive hybridisation of red and sika deer in Europe

Apart from highlighting issues of intraspecific taxonomy, habitat fragmentation, translocations and selective hunting, there is one more very important aspect relevant to red deer conservation in Europe: risk of losing genetic identity through hybridisation with introduced sika deer, *Cervus nippon*. Red and sika deer are very closely related. Genetic analyses based on mitochondrial cytochrome b sequences have suggested that sika deer are actually closer to Asian and North American red deer taxa (wapiti) than the latter are to European red deer (Pitra *et al.*, 2004). As long as these analyses are based on single or few molecular loci, however, taxonomic conclusions are rash due to the potential discordance between gene and species trees (see Zachos, 2009 and references therein).

Sika are native to eastern Asia but, just like the red deer, have been widely introduced elsewhere. In Europe, the largest populations of sika occur in the British Isles, but the species is also widespread in the Czech Republic and can be found in some 10 countries on the European mainland (Bartoš, 2009; Apollonio *et al.*, 2010a). The extant sika deer in both the British Isles and the European mainland are of Japanese origin (Goodman *et al.*, 2001; Pitra *et al.*, 2005), with the exception of the Czech Republic where sika deer derived from mainland Asia may also be present (Barancekova *et al.*, 2007).

Where they occur sympatrically within their native range (eastern Asia), red and sika deer are known to occasionally hybridise, producing fertile offspring (Bartoš *et al.*, 1981; Geist, 1998). Introgression of sika genes into European red deer gene pools is therefore a matter of great interest for conservationists and game managers and illustrates yet another problem of introduction and translocation, albeit in this case introduction of an exotic species. There is some evidence that hybridisation occurs in Germany (Gehle and Herzog, 1998), Austria (introgression of red deer alleles into sika gene pools, F. Suchentrunk, pers. comm. based on unpublished data) and the Czech Republic (Bartoš and Žirovnický, 1981; Bartoš *et al.*, 1981; Barancekova *et al.*, 2007), but by far the most extensive hybridisation is known from the British Isles (Harrington, 1973; Lowe and Gardiner, 1975; Abernethy, 1994; Goodman *et al.*, 1999; Pérez-Espona *et al.*, 2009). Based on a study on red and sika deer in Argyll, Scotland, Abernethy (1994) even considered the integrity of Scottish mainland red deer at risk from introgression of sika. In a reanalysis, the strength of this conclusion was softened but nonetheless, in areas where the two taxa overlap, about 40% of the deer were found to show introgressed alleles (Goodman *et al.*, 1999).

Sika deer were introduced to the British Isles in 1860, and first reports on hybridisation with red deer (from Powerscourt Park, Wicklow, Ireland) date to the late nineteenth century (see Pérez-Espona *et al.*, 2009 and references therein). Normally, assortative mating strategies prevail and hybridisation events are rare even in sympatry, but sometimes assortative mating breaks down and hybrid swarms emerge. At present, there are mainly three areas where such hybrid swarms occur: Wicklow in south-eastern Ireland; Cumbria in north-west England; and West Loch Awe/Loch Avich on the Kintyre peninsula in Argyll, western Scotland (McDevitt *et al.*, 2009; Pérez-Espona *et al.*, 2009; Senn and Pemberton, 2009).

What exactly leads to a breakdown of assortative mating in sympatrically occurring red and sika deer is unknown, but it might be surmised that initially a few very successful sika stags triggered the process of hybridisation by generating multiple hybrids simultaneously (Pérez-Espona *et al.*, 2009; Senn and Pemberton, 2009). Interestingly, although sika are smaller, it is mainly sika stags mating with red deer hinds when hybridisation occurs because early-generation hybrids frequently have red deer mitochondrial DNA

(Goodman *et al.*, 1999; Senn and Pemberton, 2009). This may be due to the sika stags' exceptional aggressiveness in the rut and also due to red deer harems being dispersed over rather large areas, especially in forests (Pérez-Espona *et al.*, 2009). As it now stands, hybridisation in the wild between sika and red deer is not as common as previously feared but it nonetheless does occur, at least locally, in frequencies high enough to threaten the genetic integrity of red deer populations. Without appropriate management measures, sika are likely to spread further in the British Isles, and thus the danger of introgressive hybridisation will increase even more. These findings are also relevant for areas in mainland Europe, e.g. northern Germany, where due to conservation actions red deer are expanding their present distribution range into areas where previously only sika deer occurred.

2.5 Case study 3: European roe deer

The European roe deer, *Capreolus capreolus*, is by far the most numerous ungulate species of the continent, with an estimated 10 million roe deer occuring in Europe (Linnell *et al.*, 1998; Apollonio *et al.*, 2010a). Accordingly it is one of the most important game species. It is also the species most implicated in accidents with traffic (Groot Bruinderink and Hazebroek, 1996; Chapter 8, this volume) and increasingly is considered of key significance (with red deer and wild boar) in programmes for the conservation or reintroduction of large carnivores, as potential prey for these large predators.

Roe deer occur throughout most of the continent, from southern Spain to northern Norway and from the Black Sea to the Atlantic. They are absent only in southern Greece, some parts of mainland Italy and the Iberian Peninsula and Sicily (where they were present but were subsequently exterminated), Corsica, Sardinia, Ireland and Iceland (where they never occurred) (Mitchell-Jones *et al.*, 1999; Apollonio *et al.*, 2010a). During the late nineteenth and early twentieth century there was a serious decline in roe deer numbers and distribution in Europe, due to hunting, and in some countries the species was on the brink of extinction – as for example in England (Hewison and Staines, 2008; Putman, 2010) or Sweden, (Liberg *et al.*, 2010), but its distribution range has been restored and even extended northward, particularly in Scandinavia (Sempéré *et al.*, 1996; Andersen *et al.*, 2004a; Thulin, 2006).

Roe deer have perhaps been less studied genetically than red deer (for a review of studies until the 1990s see Hartl *et al.*, 1998), but they, too, have undergone similar human influences such as selective hunting, habitat fragmentation and translocations, and the genetic studies carried out have, just like in red deer, revealed intraspecific substructuring relevant to conservation (and possibly to taxonomy). Generally, genetic variability in the roe deer is high compared with other deer species (e.g. Hartl *et al.*, 1991b) as would be expected in a widely distributed species with high effective population sizes (due to a weakly polygynous mating system!), but it has been shown that variability measures may differ widely depending on the marker system studied (Zachos *et al.*, 2006).

Recently, detailed research has been carried out into the species' phylogeography (Vernesi *et al.*, 2002; Randi *et al.*, 2004; Lorenzini and Lovari, 2006; Royo *et al.*, 2007; reviewed and discussed in Sommer *et al.*, 2009). Combined patterns of genetic and palaeontological data suggest several regions in Iberia, southern France, Italy and the Balkans as well as the Carpathians and/or Eastern Europe as Pleistocene glacial refuge areas (Sommer *et al.*, 2009). It would appear that, in contrast to the red deer (see above), roe deer seem to have recolonised Europe from the east, not out of one of the traditional Mediterranean or Balkan refuges (Sommer *et al.*, 2009).

With regard to intraspecific systematics, the European roe deer is most often regarded as a monotypic species throughout Europe but recently attention has been focused on the Iberian and Italian roe deer where different studies yielded unequivocal evidence of substantial genetic substructuring (Lorenzini *et al.* 2002, 2003; Randi *et al.*, 2004; Lorenzini and Lovari, 2006; Royo *et al.*, 2007). In Italy, the central and southern indigenous populations are genetically distinct irrespective of the molecular marker analysed, which is in line with their sometimes being classified as *C. c. italicus*. In Iberia, genetic analyses have also shown two distinct lineages of roe deer, a central-southern and a north-western one, which also differ with regard to skull shape (Aragón *et al.*, 1998) and coat pattern (Geist, 1998), again in line with the presence of a south-Iberian subspecies *C. c. garganta*. If the *italicus* and *garganta* roe deer lineages are indeed unique, then they should be managed accordingly; that is, admixture with other European roe deer should be avoided (although sadly, this may be too late for some *italicus* populations, due to introductions of roe deer from continental populations; Gentile *et al.*, 2009).

Roe deer have adapted very well to the human-dominated cultural landscape in Europe, much more so than red deer, and this ecological plasticity has prevented the roe deer from being reduced to small isolated populations due to habitat fragmentation as is so often the case in red deer. Nevertheless, landscape connectivity does have an influence on genetic structure in roe deer as well. Coulon *et al.* (2006), in a comprehensive analysis of roe deer that recently recolonised a fragmented area of 2200 km^2 in south-western France, showed that highways, rivers and canals did not act as absolute barriers to movement but were able to produce subtle population differentiation. Interestingly, similar to results found in red deer (see above), inter-individual genetic distances were better correlated to pairwise geographical distances when the latter were not calculated as Euclidean distances (straight lines) but as least-cost distances maximising the use of forest corridors (Coulon *et al.*, 2004). In a similar approach, based however on populations rather than individuals, Wang and Schreiber (2001) found a correlation between genetic distances and differentiation between adjacent roe deer populations and the density of human settlements (measured as the percentage of land surface covered by villages and roads). Kuehn *et al.* (2007) directly analysed the influence of a fenced motorway on roe deer populations in Switzerland and found that the barrier induced genetic differentiation between, but no reduction in genetic diversity within, populations, the latter being due probably to high effective population sizes and gene flow among populations on the same side of the motorway.

In line with studies on antler characteristics and allozyme polymorphisms in red deer (see above), Hartl *et al.* (1995a) also found a correlation between certain genotypes at two allozyme loci and antler size in yearling roe deer. In adult males the results were similar, yet not statistically significant. The polymorphisms at these loci were found in more or less all roe deer populations analysed, which is a strong argument in favour of their being maintained by natural selection. Again, this result implies that selective trophy hunting may change allele frequencies at particular loci (and those linked to them, functionally or physically), with unknown consequences on the population level in the long term.

As indicated above, neither red nor roe deer are threatened as species. Nonetheless, there is considerable interest among conservationists because they exemplify the general problems faced by large mammals in Europe – living in intimate proximity to a dense human population and, concomitantly, habitat fragmentation and isolated populations experiencing high amounts of genetic drift and inbreeding. Red and roe deer, being among the most important European game species, additionally serve to show the consequences of deliberate human impacts such as translocations and selective hunting.

Conservation issues for these two species tend to be regional or local in nature, affecting local populations or races (rather than being of global concern as for example in the case of bison and ibex, below). But the extensive genetic studies available in these two species have uncovered evolutionary subtaxa worthy of protection measures as they represent unique and very distinct genetic lineages. There is a priori no reason to think that such units are not present in other, poorly studied species, too, and the fact that for many species such lineages are as yet unknown may be due to our ignorance rather than to genetic homogeneity.

2.6 Case study 4: European bison

The European bison, *Bison bonasus*, is without doubt one of the most charismatic flagship species in Europe. It is also one of *the* textbook examples of a species that became extinct in the wild and was saved by a captive breeding programme. It is currently classified as 'Vulnerable' as a species by the IUCN, but the two breeding lines (see below) are classified differently: the Lowland line as 'Vulnerable' and the Lowland–Caucasian line as 'Endangered' (Olech, 2008). It is important to note that neither the Bern Convention nor the European Commission's Habitats Directive mentions the subspecies and therefore the legal mandate for conservation in Europe focuses simply on the species as a whole.

By the end of the nineteenth century, after a dramatic decline, the European bison was confined to only two populations, one in the Bialowieza primeval forest straddling the Polish–Belarusian border, belonging to the Lowland subspecies *B. b. bonasus*, and one in the western Caucasus belonging to the Caucasian subspecies *B. b. caucasicus*. By 1919 *B. b. bonasus* was extinct in the wild, and in 1927 the same held true for *B. b. caucasicus*. The whole present population, free-living and captive, goes back to a founder population of merely 12 captive animals, seven females and five males. All but one of the males belonged to the Lowland subspecies, and today there are two breeding lineages: a pure lowland line going back to seven animals from Bialowieza (three females and four males) and a mixed Lowland–Caucasian line going back to all 12 founders; the distinct Caucasian subspecies is therefore extinct (Olech and Perzanowski, 2002; Pucek, 2004; Perzanowski and Olech, 2007; Olech, 2008; K. Perzanowski, pers. comm.). Free-ranging (reintroduced) Lowland populations now occur in Poland, Belarus, northern Ukraine, Lithuania and Russia; Lowland–Caucasian bison can be found in Russia and were introduced to the Polish Carpathians (Bieszczady), and at present further populations are being established along the Carpathian mountain chain in Slovakia, Ukraine and Romania aimed at establishing a viable metapopulation of interconnected stocks (Olech and Perzanowski, 2002; Perzanowski *et al.*, 2004; Perzanowski and Olech, 2007; K. Perzanowski, pers. comm.).

The total European bison population is currently about 3100 individuals, with some 1900 living in the wild and 1200 individuals in zoological gardens and breeding centres. About 65% of the free-living population belong to the

Lowland line (Perzanowski and Olech, 2007). This line is, after a decline in the 1990s, now increasing, whereas the Lowland–Caucasian line has been decreasing over the last 20 or so years (Olech, 2008). The Bialowieza population is the largest free-ranging population comprising almost 600 individuals, more than half of which live on the Polish side of the border (Flisikowski *et al.*, 2007; Perzanowski and Olech, 2007). Unfortunately, the Polish–Belarusian border is fenced, limiting the genetic exchange between bison from the two countries. Thus, there are really two isolated Bialowieza populations rather than one. Recent simulation analyses (Daleszczyk and Bunevich, 2009) have shown that particularly the Belarusian population would greatly benefit from the creation of passages by partial removal of the border fence as the Polish population, although derived from a smaller founder group, had more favourable genetic parameters in terms of balanced contribution of the founding individuals (which increases the effective population size and thus reduces inbreeding effects).

In line with expectations after a severe bottleneck, molecular analyses repeatedly yielded low levels of genetic variation irrespective of the markers chosen (allozymes: Hartl and Pucek, 1994; mtDNA: Tiedemann *et al.*, 1998; Burzynska *et al.*, 1999; Wójcik *et al.*, 2009; microsatellites: Gralak *et al.*, 2004; Luenser *et al.*, 2005; MHC genes: Radwan *et al.*, 2007). Comparison of extant bison mtDNA control region sequences with those from medieval bison remains found in Lithuania reflected the loss of genetic variation: three out of four medieval bison analysed yielded a haplotype not found in the extant bison population (Anderung *et al.*, 2006). Loss of mitochondrial variability is a priori expected to be very high in extant bison because (1) the genetically effective population size with respect to mtDNA is only one fourth of the nuclear effective size (due to mtDNA being haploid and transmitted only maternally), and (2) the only *B. b. caucasicus* founder was a bull that contributed only nuclear alleles from the Caucasian lineage to the founding gene pool of the present bison population (although this is not relevant for the results from Lithuania as these bison were from the Lowland line).

The most immediate threat to small populations having experienced a bottleneck or founder event and concomitant loss of genetic variability is inbreeding depression. The average inbreeding coefficient in bison with full pedigree data is 44% in the Lowland line and 26% in the Lowland–Caucasian line (Olech, 2008), the difference probably reflecting the different number of founder genomes in the two breeding lines (7 versus 12). These values are very high when compared with outbred populations of large mammals. Indeed, inbreeding has been shown to be correlated with shorter lifespan, higher juvenile mortality, skeletal malformations, and the mean period between calving bison (Olech, 1987, 2008). Although on average less inbred, the

Lowland–Caucasian line shows a more pronounced decline in reproduction rate than the pure Lowland line (Olech, 2008). The low variability at major histocompatibility complex (MHC) loci (see above) also suggests a potentially increased susceptibility to pathogens, and there have been repeated reports of balanoposthitis (an inflammation of the penis and prepuce) in bison bulls (Olech, 2008). A breeding programme aimed at minimising inbreeding and further loss of genetic variability through drift, an international bison breeding centre and the establishment of a gene resource bank for semen have explicitly been recommended as appropriate conservation measures in an IUCN action plan (Pucek, 2004). Apart from the potential hybridisation between the two European breeding lines, hybridisation with American bison (*Bison bison*) is also a matter of great concern. In the Caucasus mountains two free-living herds of European–American bison hybrids have been established in proximity to herds of the Lowland–Caucasian line, and there are hybrid herds elsewhere, too (Olech, 2008). Introgression of domestic cattle has also been verified, as one out of four analysed European bison from the Puschino Research Station in Russia (Ward *et al.*, 1999) carried a cattle mtDNA haplotype. Although introgression of cattle DNA is much more serious in American bison where more than 5% of the bison analysed yielded cattle haplotypes (Ward *et al.*, 1999), measures to avoid introgression should also be considered for the European species.

While in former times habitat degradation and excessive hunting or poaching were the most critical factors in the decline of the European bison, at present the most important issues in bison conservation are politics and genetics. Political instability has resulted in the decline and even extinction of reintroduced bison herds in the Caucasus (Pucek, 2004), and the fenced border between Poland and Belarus severely reduces the effective size of the Bialowieza population. Connectivity and gene flow among populations is critical to the viability of large mammals in regions like Europe where anthropogenic impacts generally do not allow for the existence of single populations large enough to be genetically self-sustaining in the long term. A possible exception is the Carpathian mountain range, and apart from the Bialowieza population a future interconnected meta-population of European bison in the Carpathians seems most important in the preservation of the species' genetic potential.

2.7 Case study 5: Alpine and Spanish ibex

Systematics, and particularly species delineation, within the Caprinae is still uncertain and an area of very active research. In Europe, apart from introduced species, this group is mainly represented by ibex and chamois both of which are believed to comprise two species: *Capra ibex* (alpine ibex), *Capra pyrenaica* (Spanish ibex), *Rupicapra rupicapra* (alpine or northern chamois) and *Rupicapra pyrenaica* (Pyrenean or southern chamois, often called isard). Together with the European bison, the alpine ibex is Europe's best-known near-extinct ungulate species and a model of a mammalian species saved by human conservation efforts. Unfortunately, two of the four acknowledged subspecies of Spanish ibex were not so lucky and died out within the last centuries. Ibex also highlight the danger of hybridisation between wild and domestic species (goats in this case), an issue that is of similar relevance to other European large mammals such as bison (see above) or wolves.

Driven to extinction throughout its whole distribution area except for the Gran Paradiso massif in Italy near the French and Swiss borders in the nineteenth century, all extant populations of alpine ibex (*Capra ibex*) go back to about 100 animals from Gran Paradiso (Shackleton, 1997), either through natural recolonisation or through translocations. At present, the total population size is estimated at >35 000 (Apollonio *et al.*, 2010a). Alpine ibex are again widely distributed throughout the Alps and have also been introduced to Germany, Austria and Bulgaria (Mitchell-Jones *et al.*, 1999; see also Maudet *et al.*, 2002; Aulagnier *et al.*, 2008). In line with expectations after such a severe bottleneck, genetic diversity in the alpine ibex is very low, and there are as yet no significant differences between the source, Gran Paradiso, population and its descendants (Maudet *et al.*, 2002).

Even after the reintroductions, the species' distribution remains fragmented, and often populations are small and thus susceptible to environmental and demographic stochasticity, inbreeding depression and diseases, particularly sarcoptic mange or 'scabies' (Aulagnier *et al.*, 2008). While in the past genetic aspects were largely neglected in the choice of founder individuals for newly established populations, there are now data at hand based on genetic analyses and simulations that can help choose appropriate founders in order to mitigate as much as possible the negative corollaries of small population sizes and bottlenecks (Maudet *et al.*, 2002). Apart from this the main goals in alpine ibex conservation are to create viable (meta-) populations and to reduce the negative impacts of domestic sheep and goats when these occur sympatrically with ibex (Aulagnier *et al.*, 2008). Sheep and goats are competitors for food, potential transmitters of parasites and diseases, and in the case of goats there is also the danger of hybridisation with alpine ibex, as already detected in the Bregaglia valley in southern Switzerland where morphological as well as genetic analyses have shown that extensive hybridisation occurred between 1989 and 2001 (all goats and hybrids were subsequently culled; Giacometti *et al.*, 2004).

The Spanish ibex (*Capra pyrenaica*) historically occurred in south-western France, Andorra, Spain and Portugal. It is usually divided into four subspecies, two of which, however, are now extinct. *C. p. lusitanica*, formerly endemic to northern Portugal and southern Galicia, died out at the end of the nineteenth century, and the last specimen of the Pyrenean subspecies *C. p. pyrenaica* was killed in 2000, ironically, by a falling tree (Manceau *et al.*, 1999; Pérez *et al.*, 2002; Herrero and Pérez, 2008). The two remaining subspecies are *C. p. victoriae* in the Sierra de Gredos–Batuecas area of central Spain, and *C. p. hispanica*, the most widespread of the four, occurring in the south and east of the Iberian Peninsula. Present overall population size is about 50 000, distributed in over 50 subpopulations (Pérez *et al.*, 2002), but numbers have been increasing, mainly as a result of habitat changes due to widespread rural abandonment (Herrero and Pérez, 2008), from only about 7900 individuals in the early 1990s (Shackleton, 1997), although there are issues of competition with domestic goats and the exotic aoudad (*Ammotragus lervia*; Acevedo *et al.*, 2006, 2007).

Currently the Spanish ibex is expanding also into Portugal where in 2003 there were no less than 75 animals and where it is classified as 'Critically Endangered' on a national level (Moco *et al.*, 2006; Herrero and Pérez, 2008). This demographic history of long-standing decline and subsequent increase is still mirrored in a very low genetic diversity even at otherwise highly polymorphic loci such as MHC genes and microsatellites (Amills *et al.*, 2004). These results refer to the two remaining subspecies; the extinct ones certainly exhibited even lower values as reflected by the fact that the last representative of *C. p. pyrenaica* was homozygous at all 13 microsatellite loci analysed (Amills *et al.*, 2004) – unequivocal evidence of a very high degree of inbreeding in this now extinct population.

The delineation of the four subspecies of Spanish ibex was based solely on coat colour and horn morphology and, as is often the case with subspecies, it is doubtful whether they correctly reflect evolutionary units. In a study based on mtDNA, Manceau *et al.* (1999) found that *C. p. pyrenaica* was genetically distinct (nearly as distinct from the other Spanish ibex subspecies as the level of divergence between alpine and Spanish ibex), but the recognition of *C. p. victoriae* and *C. p. hispanica* was not supported. Rather, they found two genetic clusters combining the northern and southern populations, respectively, from both alleged subspecies. An important consequence of this latter finding pertaining to conservation is that from the viewpoint of (mitochondrial) genetics there is no reason for keeping *victoriae* and *hispanica* populations apart. It has to be kept in mind, though, that single-locus analyses only show single gene trees that need not be in accordance with organism-level phylogeny (Zachos, 2009). Incomplete lineage sorting ('hemiplasy', Avise and Robinson, 2008) or recent hybridisation between *C. p. victoriae* and *C. p. hispanica* may have blurred potential adaptive differences between them. Reciprocal monophyly is only expected after $4N_e$ generations (N_e being the effective population size), and mtDNA markers might simply be too conservative to correctly mirror recent evolutionary dynamics at the intraspecific level. Therefore, further data should be at hand before a definitive decision on the conservation status of Spanish ibex populations can be made.

The Spanish ibex is, at present, not threatened, but competition with introduced Barbary sheep (*Ammotragus lervia*) and perhaps fallow deer, mouflon and domestic ungulates may cause problems in the future (Pérez *et al.*, 2002; Herrero and Pérez, 2008). Human-caused habitat deterioration and selective hunting may have impacts on the species' viability but the most serious problem is sarcoptic mange (scabies) which is known to have occurred in several Spanish ibex populations, sometimes with mortality rates of more than 95% (Pérez *et al.*, 2002).

2.8 Reprise

As we have noted in our introduction, anthropogenic factors have had a pronounced influence on the status, genetic status and distribution of ungulates right across Europe.
It is clear, for example, that:

- A number of indigenous subspecies and other distinct genetic units have been (or may be currently) threatened by overexploitation or simply by a lack of positive management to conserve them.

While some subspecies or geographical races (or even in some cases, whole species) have been under considerable threat in the past (for example alpine ibex, Pyrenean chamois, Tyrrhenian red deer, moose and bison), active conservation measures have supported the recovery of many (see, for example, review by Apollonio *et al.*, 2010b).

Roe deer too, while considerably reduced in numbers in the past, have widely recovered numbers and distribution to become one of the most numerous ungulates present in Europe. There is still concern expressed about the endemic races (putative subspecies) of Italian roe deer (*C. c. italicus*) and Iberian roe deer (*C. c. garganta*), but at least awareness is high and thus some form of active positive management may be anticipated.

- The genetic integrity of such endangered taxa may be compromised by introduction to those populations of animals of different genetic background in well-intentioned attempts to augment dwindling populations or in order to improve trophy quality.

The red deer is perhaps the species that has undergone the most extensive translocations (Niethammer, 1963; Hartl *et al.*, 2003), mostly in attempts at improving trophy quality or establishing hunting grounds. In particular in central Europe it is doubtful if there are really non-affected indigenous populations left as red deer have great dispersal capacities so that introduced animals or their offspring may disperse into other populations of the region. Northern Germany, apart from indigenous red deer, also harbours animals from Poland, Hungary and Austria, and even the Carpathians are not free from introductions (see above). Whether this eventually leads to ecological problems due to disruption of adapted genotypes is unclear and perhaps not very likely, but the natural genetic pattern of red deer, at least at a regional scale, has been blurred or even destroyed.

Roe deer, too, have extensively been translocated, including attempts at introducing Siberian roe deer (*Capreolus pygargus*) into the former Czechoslovakia; their genetic legacy is allegedly still visible although *capreolus* females sired by *pygargus* males often died due to the foetuses being too large for them (Niethammer, 1963). Perhaps the greatest risk to the conservation of the two local roe deer races or subspecies in Italy and Iberia is the introgression of genes from nearby populations of roe of different genetic provenance. This is a potentially fatal risk to evolutionarily distinct intraspecific units (whether acknowledged as subspecies or not).

Since genetic diversity is one of three levels of biodiversity acknowledged by the IUCN (apart from ecosystem and species diversity), the preservation of genetically unique populations is one of the core issues of ungulate conservation. Therefore, whenever it is clear from genetic or other (morphological, ecological) data that a taxon represents a unique fraction of a species' overall diversity the introduction of non-indigenous animals to augment dwindling populations should definitely be the last resort when, for example, inbreeding depression proves to be fatal (as in the famous case of the Florida panther which was finally saved by the introduction of mountain lions from Texas). In these cases, however, genetic analyses should be carried out beforehand to inform conservationists about which populations should be chosen for introduction. The examples of bison and red deer from Mesola and Sardinia have shown that *in situ* measures or breeding programmes can be powerful tools that should be applied before introduction dilutes the genetic integrity of unique lineages.

- Reintroduction of species to areas from which they have actually become extinct has equally been undertaken without due regard to the genetic provenance of those individuals released, thus causing other discontinuities in genetic distributions.

As noted earlier, there have been many documented introductions and reintroductions of red deer, both to subsidise existing local populations (or with the intention of 'improving' their genetic stock), but also to re-establish them in areas from which they have become locally extinct. Red deer have been reintroduced in this way to various parts of the Netherlands (Veluwe, Oostvaardersplassen) and Germany, amongst other countries. Unless considerable care is taken in selecting animals for introduction from an appropriate genetic source, such introductions may end up blurring natural patterns of genetic distribution and evolutionary phylogeography. As we have described, one German stock north of Hamburg was founded with animals from Poland, Hungary and Austria, and these origins are still evident in the genetic make-up of this population, which is fundamentally different from other north-German populations (F.E. Zachos *et al.*, unpublished data).

Despite such 'interpolations' of alien genetic types within the natural genetic distribution of red deer across Europe, in the phylogeographic analysis of Skog *et al.* (2009) only few individual red deer specimens out of a sample of nearly 600 were found to show discordances between their geographical location and their genetic lineage, with one additional discrepancy noted by Nussey *et al.* (2006) on the Isle of Rum, Scotland. While this is perhaps suggestive that the majority of translocations were carried out within the main lineages rather than among them and have not, after all, blurred the large-scale phylogeographical pattern of the species, we note that such conclusions are currently based primarily on evidence from maternally inherited mitochondrial DNA and may underestimate this problem.

Similar translocations of alien genetic stock have been made in roe deer, with populations in England almost all derived from a number of discrete reintroductions from small, refuge populations in Scotland, from Germany (Chapman *et al.*, 1985), Austria (Lowe, 1979) or elsewhere (Prior, 1968). Recent analyses of isozymes and skull morphometrics (Hewison, 1995, 1997), or more recently, of DNA (Baker, 2009) confirm distinct genetic lineages in different areas within England (see also Hewison and Staines, 2008).

Introductions of non-native stocks of European roe to 'empty' areas in central Italy, now threaten (by expansion of range) the genetic integrity of isolated populations of *C. c. italicus*. Taken to even greater extremes records also exist of attempts to introduce Siberian roe deer to Estonia and areas within the former Czechoslovakia (see Niethammer, 1963), although it seems probable that the releases failed to establish.

- For some whole species, the modern geographical distribution has in practice largely resulted from deliberate introductions well beyond their native range.

The most obvious example of this is the fallow deer, whose current distributional range extends far beyond its immediate post-glacial distribution in Turkey and the eastern Mediterranean (Chapman and Chapman, 1980). But we should note that other species too such as chamois and mouflon have seen significant 'expansion' of their distributions thanks to human agency.

In addition there have been widespread introductions of a number of exotics (American wapiti, American white-tailed deer, sika, axis, muntjac and Chinese water deer). The only caprid successfully introduced was the Barbary sheep (*Ammotragus lervia*) that is currently well established in Spain (Cassinello *et al.*, 2004) and, after an escape from a zoo in 1976, has also established a small population in the Czech Republic (Mitchell-Jones *et al.*, 1999). These various introductions and their current geographical distribution within Europe are summarised in Table 2.1.

These exotic introductions were part of a deliberate policy popular in the nineteenth century to enrich the fauna of European countries. Such introductions were conducted by well-organised 'acclimatisation societies' such as the French *La Société Zoologique d'Acclimatation* founded in 1854. In many cases, however, because these species have been introduced into novel environments with which they have not coevolved, they may have strong negative impacts on those environments and on native species (see, for example, Spear and Chown, 2009).

For example, apart from the possible problems arising from hybridisation with red deer, already described, where sika reach high densities, they can also be implicated in significant economic damage to commercial forestry (e.g. Ratcliffe, 1989; Lowe, 1994; Chadwick *et al.*, 1996; Abernethy, 1998). Damage may be caused through browsing of both lateral and leading shoots, and also by bark stripping. The economic significance of such damage may be considerable at a local scale (see for example Pérez-Espona *et al.*, 2009). At high densities, sika may potentially also have negative impacts on habitats of natural heritage importance, for example preventing the regeneration of native woodlands or impacting on wetland sites where they may cause damage to reedbeds and saltmarsh (e.g. Diaz *et al.*, 2005).

Table 2.1 *Distribution of 'exotic' ungulate species in various European countries*

The table includes species that are native to Europe but have been introduced to areas outside their post-glacial distribution. If a species is present in a country, but is confined to an island or fenced hunting reserve it is listed under the appropriate category; however, if a species is free-ranging on the mainland no note is made of its additional presence on islands or fenced areas.

	Free-ranging on mainland	Physically restricted range	
		Fenced areas only	Offshore islands only
White-tailed deer *Odocoileus virgianus*	CZ, FIN	SRB	
Fallow deer *Dama dama*	CZ, HR, BL, DK, A, D, FIN, F, UK, IRL, H, I, NL, P, ROM, SLVK, SLV, S, PL, ESP	GR, SRB, MK	GR
Sika deer *Cervus nippon*	LITH, CZ, DK, A, D, F, UK, IRL, CH, PL	H	
Axis deer *Axis axis*			HR
Muntjac *Muntiacus reevesi*	UK, NL		
Chinese water deer *Hydropotes inermis*	UK, F		
Mouflon *Ovis orientalis*	CZ, HR, BL, A, D, F, H, I, SLVK, SLV, S, CH, PL, ESP	DK, GR, NL, P, ROM, SRB, MK	DK, FIN, GR
Barbary sheep *Ammotragus lervia*	CZ, ESP		
Muskox *Ovibos moschatus*	N, S		
Wild goat *Capra aegagrus*	CZ		
Ibex *Capra ibex*	D, BULG		
Reindeer *Rangifer tarandus*		UK	

A, Austria; BL, Belgium; BULG, Bulgaria; CH, Switzerland; CZ, Czech Republic; D, Germany; DK, Denmark; ESP, Spain; F, France; FIN, Finland; GR, Greece; H, Hungary; HR, Croatia; I, Italy; IRL, Ireland; LITH, Lithuania; MK, Macedonia; N, Norway; NL, Netherlands; P, Portugal; PL, Poland; ROM, Romania; S, Sweden; SLV, Slovenia; SLVK, Slovakia; SRB, Serbia; UK, United Kingdom

Muntjac too, may cause significant environmental problems, suppressing regeneration in native woodlands and causing extensive damage in coppice woodlands (Cooke and Lakhani, 1996; Cooke and Farrell, 2001; Cooke, 1998, 2004, 2005, 2006). Arnold Cooke has also reported comprehensively on the effects of muntjac at high densities on other elements of the ground flora (primroses, *Primula vulgaris*; bluebells, *Hyacynthoides nonscripta*; dog's mercury, *Mercurialis perennis*; and common spotted orchid *Dactylorhiza fuchsii*), within Monk's Wood National Nature Reserve in the UK (summarised, for example, in Cooke, 1994, 1995, 2005, 2006).

There is considerable niche overlap between muntjac and native roe deer, and correlational data (with declines in density of roe correlated with increases in muntjac density) suggest muntjac may displace roe from areas of sympatry (Forde, 1989; Wray, 1994; Hemami *et al.*, 2005).

Like sika, North American wapiti (*Cervus canadensis*), where introduced, have the potential for compromising genetic integrity of local red deer stocks through hybridisation. In addition, where wapiti have been introduced into Europe, they brought with them the trematode *Fascioloides magna*, highly pathogenic in red deer and roe deer (Novobilsky *et al.*, 2006); in Italy this has resulted in virtual extermination of these two species from the region into which wapiti were introduced.

- In some cases, such introductions have led to further modification of genetic stocks because of hybridisation between exotic and native species

As well as their direct impact on forestry and natural habitats, we have discussed in some detail the additional threat now posed by sika in their ability to hybridise with native populations of red deer. This now poses a real concern in a number of European countries, such as the United Kingdom (e.g. Lowe and Gardiner, 1975; Ratcliffe *et al.*, 1992; Putman and Hunt, 1994; Goodman *et al.*, 1999; Pemberton *et al.*, 2006; Pérez-Espona *et al.*, 2009), the Republic of Ireland (Harrington, 1973, 1982; McDevitt *et al.*, 2009) and is of increasing concern elsewhere, for example in the Czech Republic (Bartoš and Žirovnický, 1981; Bartoš *et al.*, 1981; Barancekova *et al.*, 2007), in Austria (F. Suchentrunk, pers. comm.) and in Germany (Gehle and Herzog, 1998).

Similar problems may occur where other *Cervus* taxa are introduced, such as wapiti (*Cervus canadensis*), rusa deer (*C. timorensis*) or sambar (*C. unicolor*), all of which are known at least to have the capacity to hybridise with red deer in captivity. It seems likely that this genus of closely related species provides the 'arena' of greatest risk, although, as above, there are circumstantial reports of introgression of Siberian roe deer into European roe deer stocks.

- Hybridisation of wild and domestic forms and the problems of establishment of primitive domesticates as wild species.

In similar context we may review the issue of introgression of genes from domestic livestock, subject to artificial selection for many generations, back into wild populations (e.g. wild goats) or the active naturalisation of primitive domestic breeds (as, for example, introductions of mouflon to many countries).

The distinction between wild and domestic is further blurred by the current trend to use domestic species as ecological or functional replacements of species that have become extinct. This includes primitive or specially designed breeds of cattle (e.g. Heck cattle, a phenotypic re-creation of the aurochs) and horse (e.g. Polish ponies or konik: the nearest we seem to have to the extinct forest tarpan). Furthermore, feral or free-ranging livestock, especially sheep and goats, are very common in the European ecosystems. In some cases these feral populations are present because of unintentional escapes, but in other cases they are grazed as part of extensive livestock production systems or to achieve specific conservation objectives for maintaining grazing-dependent cultural landscapes associated with high species diversity and/or strong aesthetic appeal (see Chapter 9).

As mentioned above, hybridisation between feral goats and alpine ibex as well as between cattle and bison is of potential conservation concern. In caprines, insufficient knowledge about species delineation complicates an evaluation of which populations are wild and which are domestic or of hybrid origin.

Of relevance for this review is the often disputed taxonomic status of two species, the mouflon (*Ovis orientalis*, but often also referred to as *O. gmelini, O. ammon* and *O. mousimon*) and the wild goat (*Capra aegagrus*). Both species occur on Mediterranean islands which have an ancient history of human influence. Mouflon presence on Corsica, Sardinia and Cyprus and wild goat presence on Majorca and Crete extends back many thousand years. However, there is much debate about whether they should be regarded as feral domestic sheep and goats (albeit from very early forms of the modern breeds) or if they represent naturalised populations of wild ancestors of these domestic breeds (Bar-Gal *et al.*, 2002). The latter view is highly complex because of widespread controversies about taxonomy within the Caprinae and because of recent evidence about the complexity of the domestication process that has led to many of our modern breeds. The so-called 'natural' populations (Mediterranean islands) of mouflon and wild goat are strictly protected under the EC Habitats Directive and the wild goats also receive strict protection under the Bern Convention. However, the IUCN did not consider them in the recent European Mammal Assessment: 'as they are considered to be feral descendants of early domestic stock, they are classed as Not Applicable' (Temple and Terry, 2007). The debate about their status is as much a discussion about their cultural historical value as about their genetic identity, and the fact that they are listed in the appendices of both pan-European bodies of legislation implies that their conservation is legally mandated despite not being included in the IUCN Red List of Threatened Species.

2.9 A summary of conservation status and legislation

While reviewing, in general, the distribution and population genetics of ungulate species in Europe and the major impacts of human activity, one explicit emphasis of this chapter has been to highlight issues which may be of conservation concern. Table 2.2 summarises the current conservation status of the different species under the Habitats Directive and the Bern Convention.

Of all the ungulates present in Europe there is only one full species, the European bison, which makes its way onto the international IUCN red lists where it is classed as 'Vulnerable'; there is no other ungulate species that presently occurs in Europe with a global threat status. Species like moose, reindeer, roe deer and red deer all have very wide distributions in the Palaearctic or Holarctic, and even those that are confined to Europe exist in large populations that number in the many thousands, tens of thousands or even hundreds of thousands. Accordingly all are listed as being of 'Least Concern' on a global basis.

However, if we look at lower levels of taxonomic organisation such as distinct regional populations, the picture is far more varied. Traditionally, taxonomists have recognised very many subspecies of ungulate in Europe based on morphological characteristics. These classifications have been supported to varying degrees by modern genetical approaches. Although the subspecies concept has been the subject of much criticism, it is useful as it focuses attention on the importance of intraspecies levels of diversity.

For the sake of consistency with common usage and European policy documents we retain the common forms and use the classification adopted by the IUCN European Mammal Assessment (Temple and Terry, 2007). In those instances where this assessment did not make a formal classification of subspecies threat status we have followed the earlier assessment of Shackleton (1997) for caprinids and Wemmer (1998) for cervids. This should not be taken

Table 2.2 *Listing of European ungulates under the Habitats Directive and the Bern Convention*

	Habitats Directive				Bern Convention	
	Priority species	II[1]	IV[2]	V[3]	II[4]	III[5]
Caprinae						
Capra aegagrus (natural populations)		✓	✓		✓	
*Capra pyrenaica (*except *C. p. pyrenaica)*				✓		✓
Capra pyrenaica pyrenaica	✓	✓	✓		✓	
Capra ibex				✓		✓
Ovis gmelini musimon/Ovis ammon musimon (Corsica and Sardinia)		✓	✓			✓
Ovis gmelini ophion/Ovis orientalis ophion		✓	✓			
Rupicapra rupicapra (all subspecies)						✓
Rupicapra rupicapra (except *R. r. balcanica, ornata, tatrica*)				✓		
Rupicapra rupicapra balcanica		✓	✓			
Rupicapra rupicapra ornata	✓	✓	✓		✓	
Rupicapra rupicapra tatrica	✓	✓	✓			
Ovibos moschatus					✓	
Cervidae						
All species						✓
Cervus elaphus corsicanus	✓	✓	✓		✓	
Rangifer tarandus fennicus		✓				
Bovinae						
Bison bonasus	✓	✓	✓			✓

[1]Annex II: Animal and plant species of community interest whose conservation requires the designation of special areas of conservation.
[2]Annex IV: Animal and plant species of community interest in need of strict protection.
[3]Annex V: Animal and plant species of community interest whose taking in the wild and exploitation may be subject to management measures.
[4]Appendix II: Strictly protected fauna.
[5]Appendix III: Protected fauna species.

as any type of formal endorsement of classification. Rather classifications should be viewed as handy labels to place on local forms whose conservation is needed to maintain the full diversity of genetical, morphological and ecological variation present within the species.

When looking at the subspecies levels there is a far greater variation in status and threat assessment, with some individual subspecies qualifying for the highest threat levels. Again this is reflected in the Habitats Directive and the Bern Convention with many individual subspecies being given stricter protection than others, although all species of cervids, caprids and bovids in Europe are afforded some degree of conservation protection by one or both of these bodies of legislation.

Among the caprinids the major conservation threats lie with three of the chamois subspecies, the Tatra chamois, the Apennine chamois and the Chartreuse chamois. These three subspecies all occur as small, single populations, although they do have access to more unoccupied habitat. A fourth chamois subspecies, the Balkan chamois, is also recognised as being under threat. Despite a wide distribution from Bosnia through Montenegro, Macedonia, Albania to Greece and Bulgaria, most of the populations are small and isolated. The now extinct Pyrenean subspecies of Spanish ibex was also afforded high conservation status. Among cervids it is the Tyrrhenian subspecies of red deer that has greatest priority followed by the forest reindeer in Finland.

Although not formally identified as priorities for conservation on red lists or European legislation there are a number of other subspecies or populations that deserve some focus. For example, wild mountain reindeer are only represented in Europe by the populations in south Norway. While the number of animals is in the tens of thousands, these populations remain under threat from habitat encroachment and fragmentation – mainly by transport, energy and recreational infrastructure. The two putative subspecies of roe deer found in southern Iberia and southern Italy are also in need of conservation focus. In addition, there are some populations without subspecific status that genetic research has identified as being carriers of specific haplotypes or alleles that are in urgent need of conservation activity. These include the Mesola red deer in Italy's Po delta and fallow deer on the island of Rhodes. Finally, there remain many native and unmixed populations that have not even been examined genetically that may or may not contain unique genetic characteristics, such as the population of native red deer in south-western Ireland.

2.10 European ungulates: the future

A wide range of threats exist for the different species/subspecies of conservation concern. For a few cases these are directly linked to small population size which represents both demographic and genetic threats. Many European ungulates have been forced through very small bottlenecks (e.g. alpine ibex, bison) which could potentially have long-term effects on viability, although the experience so far shows that these populations are remarkably robust.

For many others the immediate threats come from issues such as poaching or poorly regulated harvests, habitat loss to infrastructure development, disturbance from human activities, hybridisation with domestic derivatives, and disease transfer between wild and domestic species.

There is also a wide range of conflicts between humans and the more abundant species that pose challenges for their future conservation. These include problems arising from impacts on agricultural crops and commercial forestry (Chapter 6) as well as from collisions with vehicles (Chapter 8).

But overall, the picture for European ungulates in the twenty-first century is one of hope. This optimism stems from the historical trends that have led to the recovery of many species, subspecies and populations. The fact that it is possible to restore such large mammals to a crowded continent like Europe is a testimony to both the adaptable ecology and life histories of these species and to the fact that they appeal to a wide range of human interest groups, from environmentalists to hunters, which are willing to work for their conservation. Many of these species are now back in an increasingly diverse ecosystem, influencing the vegetation on which they feed and in turn providing prey for recovering populations of large carnivores like wolf and lynx across Europe.

There are, however, some cautionary lessons to be learnt here. Some subspecies and populations are still under threat and in need of serious conservation attention to improve their status and safeguard their genetic integrity. In addition, in many parts of Europe there is a need to reform wildlife management systems to ensure that wild ungulate management is sustainable (see also Chapter 13). It should be a goal to maintain as much of the genetic diversity as possible, and potential local adaptation, which remains in Europe today to ensure that wild ungulates are equipped to meet the challenges of future global change. There is therefore a need to come up with a definitive mapping of genetic diversity and taxonomic clarification at the intraspecific level for all species to ensure that legal and operational priorities are focused on real evolutionary units rather than on subjective classifications. However, the issue of local genetic identity also needs to be considered, as there may be a trade-off between maximising diversity while conserving uniqueness.

There are also many issues concerning trophic interactions. At one end of the scale there is a need to address issues of low density populations, especially in areas where predators like wolves exist. Across many of the Mediterranean countries there is a need to restore wild ungulates as a crucial prey species for wolves and to restore some semblance of a natural mammalian community. However, there are also areas where predator density may be threatening the survival of wild ungulate populations of conservation interest; examples include Finnish forest reindeer and some of the small chamois populations (Kojola *et al.*, 2009). As large predator populations expand (Hetherington, 2006; Linnell *et al.*, 2009) these issues are going to increase in relevance. At the other end of the density scale are the many areas where very high density (often referred to as overabundant) populations of wild ungulates may be exerting major effects on vegetation in a manner that may conflict with the conservation of other species (plants, insects, etc.) of conservation or aesthetic importance.

Dealing with some of these challenges will depend on access to good ecological knowledge about ecosystem functionality, but it also requires a focus on social and political issues concerning societal goals for wild ungulates and the ecosystems in which they live (Chapter 13). Society needs to debate the relative importance of different management objectives that may conflict and to make decisions concerning the type of ecosystems that we want to have in the future.

Questions like 'how many ungulates do we want to have?' need to be answered. Such questions then lead to discussions about animal welfare where the debate about what management measures are acceptable becomes central. Recreational hunting is the main, and probably only, practical tool available for large-scale population regulation. In many areas of Europe, hunting remains a widely accepted activity and hunter numbers remain stable; however, in other areas hunting is becoming increasingly controversial and hunter numbers are in decline. Therefore, the debate about how we want our ecosystems to look in the future must also be balanced by a discussion about what goals are possible given the tools available to us and which are accepted by society.

Hunting is also a complex issue from the point of view of ecology, as we are only just beginning to understand the complex ways in which hunting can influence the demographics and genetics of ungulate populations (Milner *et al.*, 2007). Such debates often include a discussion about the potential role of large predators in regulating wild ungulate populations. There is no doubt that predator influence can be dramatic in some situations, but there is a need to be cautious about what general impact they can have on prey populations, especially those that occur at high densities (Andersen *et al.*, 2006; Melis *et al.*, 2009), and we should be cautious about making predictions concerning how predator–prey relationships will function in our heavily modified European ecosystems (Linnell *et al.*, 2005).

In conclusion, we are entering a challenging time. While some focus will remain on classical conservation of endangered populations, the majority of the focus will switch to the issue of living with success. This will require that the toolkits of the ecologist be joined with those of the social scientist and social economist in an effort to build multidisciplinary approaches to address complex issues on the conceptual borderlines between human societies and the ecosystems which they share with wild ungulates. However, the fact that we can still experience the thrill of standing face to face with a wild bison in the forest, of watching wild ibex clash their horns together in the mountains or hearing the rutting call of red deer in the morning mists of the twenty-first century gives us all reason to be optimistic.

References

Abernethy, K. (1994) The establishment of a hybrid zone between red and sika deer (genus *Cervus*). *Molecular Ecology* **3**, 551–562.

Abernethy, K. (1998) *Sika Deer in Scotland.* Deer Commission for Scotland. Edinburgh, UK: The Stationery Office.

Acevedo, P., Cassinello, J. and Gortázar, C. (2006) The Iberian ibex is under an expansion trend but displaced to suboptimal habitats by the presence of extensive goat livestock in central Spain. *Biodiversity and Conservation* **16**, 3361–3376.

Acevedo, P., Cassinello, J., Hortal, J. and Gortázar, C. (2007) Invasive exotic aoudad (*Ammotragus lervia*) as a major threat to native Iberian ibex (*Capra pyrenaica*): a habitat suitability model approach. *Diversity and Distributions* **13**, 587–597.

Amills, M., Jiménez, N., Jordana, J., *et al.* (2004) Low diversity in the major histocompatibility complex class II *DRB1* gene of the Spanish ibex, *Capra pyrenaica. Heredity* **93**, 266–272.

Andersen, R., Herfindal, I., Linnell, J.D.C., *et al.* (2004a) When range expansion rate is faster in marginal habitats. *Oikos* **107**, 210–214.

Andersen, R., Linnell, J.D.C., Hustad, H. and Brainerd, S. (2004b) Large carnivores and human communities: a guide to coexistence in the 21st century. *Norwegian Institute for Nature Research Temahefte* **25**, 1–48.

Andersen, R., Linnell, J.D.C. and Solberg, E.J. (2006) The future role of large carnivores on terrestrial trophic interactions: the northern temperate view. In K. Danell, R. Bergström, P. Duncan and J. Pastor (eds.) *Large Herbivore Ecology, Ecosystem Dynamics and Conservation.* Cambridge, UK: Cambridge University Press, pp. 413–448.

Anderung, C., Baubliene, J., Daugnora, L. and Götherström, A. (2006) Medieval remains from Lithuania indicate loss of a mitochondrial haplotype in *Bison bonasus. Molecular Ecology* **15**, 3083.

Apollonio, M., Andersen, R. and Putman, R. (2010a) *European Ungulates and their Management in the 21st Century.* Cambridge, UK: Cambridge University Press. 604 pp.

Apollonio, M., Andersen, R. and Putman, R. (2010b) Present status and future challenges for European ungulate management. In M. Apollonio, R. Andersen, and R. Putman (eds) *European Ungulates and their Management in the 21st Century.* Cambridge, UK: Cambridge University Press, pp. 578–604.

Aragón, S., Braza, F., San José, C. and Fandos, P. (1998) Variation in skull morphology of roe deer (*Capreolus capreolus*) in western and central Europe. *Journal of Mammalogy* **79**, 131–140.

Aulagnier, S., Kranz, A., Lovari, S., *et al.* (2008) *Capra ibex*. In *IUCN Red List of Threatened Species*. Online: www.iucnredlist.org.

Avise, J.C. and Robinson, T.J. (2008) Hemiplasy: a new term in the lexicon of phylogenetics. *Systematic Biology* **57**, 503–507.

Baker, K. (2009) An investigation into the population genetic history of the British roe deer (*Capreolus capreolus*). Abstract for 9th European Roe Deer Congress held in Edinburgh, July 2009.

Balloux, F. and Lugon-Moulin, N. (2002) The estimation of population differentiation with microsatellite markers. *Molecular Ecology* **11**, 155–165.

Barancekova, M., Vallo, P., Prokesova, J., Koubek, P. (2007) Genetic roots of sika deer in the Czech Republic inferred from mitochondrial gene for cytochrome b. In 1st International Conference on Genus *Cervus*, held at Primiero, Trentino, Italy. Abstracts, p. 68.

Barbosa, A.M. and Carranza, J. (2010) Lack of geographical variation in Y-chromosomal introns of red deer (*Cervus elaphus*). *Journal of Negative Results* **7**, 1–4.

Bar-Gal, G.K., Smith, P., Tchernov, E., *et al.* (2002) Genetic evidence for the origin of the agrimi goat (*Capra aegagrus cretica*). *Journal of Zoology, London* **256**, 369–377.

Bartoš, L. (2009) Sika deer in Continental Europe. In D.R. McCullough, S. Takatsuki and K. Kaji (eds.) *Sika Deer: Biology, Conservation and Management of Native and Introduced Populations*, Tokyo: Springer, pp. 573–594.

Bartoš, L. and Žirovnický, J. (1981) Hybridisation between red and sika deer. II. Phenotype analysis. *Zoologischer Anzeiger, Jena* **207**, 271–287.

Bartoš, L., Hyanek, J. and Žirovnický, J. (1981) Hybridisation between red and sika deer. I. Craniological analysis. *Zoologischer Anzeiger, Jena* **207**, 260–270.

Botezat, E. (1903) Gestaltung und Klassifikation der Geweihe des Edelhirsches, nebst einem Anhange über die Stärke der Karpathenhirsche und die zwei Rassen derselben. *Gegenbaurs Morphologisches Jahrbuch* **32**(1), 104–158.

Burzynska, B., Olech, W. and Topczewski, J. (1999) Phylogeny and genetic variation of the European bison *Bison bonasus* based on mitochondrial DNA D-loop sequences. *Acta Theriologica* **44**, 253–262.

Cassinello, J., Serrano, E., Calabuig, G. and Pérez, J.M. (2004) Range expansion of an exotic ungulate (*Ammotragus lervia*) in southern Spain: ecological and conservation concerns. *Biodiversity and Conservation* **13**, 851–866.

Chadwick, A.H., Ratcliffe, P.R. and Abernethy, K. (1996) Sika deer in Scotland: density, population size, habitat use and fertility some comparisons with red deer. *Scottish Forestry* **50**, 8–16.

Chapman, N.G. and Chapman, D.I. (1980) The distribution of fallow deer: a world wide review. *Mammal Review* **10**, 61–138.

Chapman, N.G., Claydon, K., Claydon, M. and Harris, S. (1985) Distribution and habitat selection by muntjac and other species of deer in a coniferous forest. *Acta Theriologica* **30**, 287–303.

Comşia, A.M. (1961) *Biologia şi Principiile Culturii Vânatului*. Bucharest: Editura Academiei.

Cooke, A.S. (1994) Colonisation by muntjac deer *Muntiacus reevesi* and their impact on vegetation. In M.E. Massey and R.C. Welch (eds.) *Monks Wood National Nature Reserve. The Experience of 40 Years: 1953–93*. Peterborough, UK: Natural England, pp. 45–61.

Cooke, A.S. (1995) Muntjac damage in woodland. *Enact* **3**, 12–14.

Cooke, A.S. (1998) Survival and regrowth performance of coppiced ash (*Fraxinus excelsior*) in relation to browsing damage by muntjac deer (*Muntiacus reevesi*). *Quarterly Journal of Forestry* **92**, 286–290.

Cooke, A.S. (2004) Muntjac and conservation woodland. In C.P. Quine, R.F. Shore and R.C. Trout (eds.) *Managing Woodlands and their Mammals*. Proceedings of a joint Mammal Society/Forestry Commission Symposium. Edinburgh, UK: Forestry Commission, pp. 65–69.

Cooke, A.S. (2005) Muntjac deer *Muntiacus reevesi* in Monks Wood NNR: their management and changing impact. In C. Gardiner and T. Sparks (eds.) *Ten Years of Change: Woodland Research at Monks Wood NNR, 1993–2003*. Research Report 613, Peterborough, UK: English Nature, pp. 65–74.

Cooke, A.S. (2006) *Monitoring Muntjac Deer* Muntiacus reevesi *and their Impacts in Monks Wood National Nature Reserve*. Research Report 681. Peterborough, UK: English Nature.

Cooke, A.S. and Farrell, L. (2001) Impact of muntjac deer (*Muntiacus reevesi*) at Monks Wood National Nature Reserve, Cambridgeshire, Eastern England. *Forestry* **74**, 241–250.

Cooke, A.S. and Lakhani, K. (1996) Damage to coppice regrowth by muntjac deer *Muntiacus reevesi* and protection with electric fencing. *Biological Conservation* **75**, 231–238.

Coulon, A., Cosson, J.F., Angibault, J.M., *et al.* (2004) Landscape connectivity influences gene flow in a roe deer population inhabiting a fragmented landscape: an individual-based approach. *Molecular Ecology* **13**, 2841–2850.

Coulon, A., Guillot, G., Cosson, J-F., *et al.* (2006) Genetic structure is influenced by landscape features: empirical evidence from a roe deer population. *Molecular Ecology* **15**, 1669–1679.

Daleszczyk, K. and Bunevich, A.N. (2009) Population viability analysis of European bison populations in Polish and Belarusian parts of Bialowieza Forest with and without gene exchange. *Biological Conservation* **142**, 3068–3075.

Diaz, A., Pinn, E.H. and Hannaford, J. (2005) Ecological impacts of sika deer on Poole Harbour saltmarshes. In J. Humphreys and V. May (eds.) *The Ecology of Poole Harbour*. Amsterdam, Netherlands: Elsevier.

Díaz, A., Hughes, S., Putman, R., Mogg, R. and Bond, J.M. (2006) A genetic study of sika (*Cervus nippon*) in the New Forest and in the Purbeck region, southern England: is there evidence of recent or past hybridization with red deer (*Cervus elaphus*)? *Journal of Zoology, London* **207**, 227–235.

Dobroruka, L.J. (1960) Der Karpatenhirsch, *Cervus elaphus montanus* Botezat 1903. *Zoologischer Anzeiger* **165**, 481–483.

Dobson, M. (1998) Mammal distributions in the western Mediterranean: the role of human intervention. *Mammal Review* **28**, 77–88.

Dolan, J.M. (1988) A deer of many lands: a guide to the subspecies of the red deer *Cervus elaphus* L. *Zoonooz*, **62**(10), 4–34.

Feulner, P.G.D., Bielfeldt, W., Zachos, F.E., *et al.* (2004) Mitochondrial DNA and microsatellite analyses of the genetic status of the presumed subspecies *Cervus elaphus montanus* (Carpathian red deer). *Heredity* **93**, 299–306.

Flisikowski, K., Krasinska, M., Maj, A., *et al.* (2007) Genetic polymorphism in selected gene loci in a sample of Bialowieza population of European bison (*Bison bonasus*). *Animal Science Papers and Reports* **25**, 221–230.

Forde, P. (1989) Comparative ecology of muntjac *Muntiacus reevesi* and roe deer *Capreolus capreolus* in a commercial coniferous forest. PhD thesis, University of Bristol, UK.

Frantz, A.C., Hamann, J-L. and Klein, F. (2008) Fine-scale genetic structure of red deer (*Cervus elaphus*) in a French temperate forest. *European Journal of Wildlife Research* **54**, 44–52.

Gehle, T. and Herzog, S. (1998): Is there evidence for hybridization between red deer and sika deer in Germany? In Z. Zomborszky (ed.) *Advances in Deer Biology*. Proceedings of the 4th International Deer Biology Congress. Kaposvár, Hungary: Pannon University of Agriculture, pp. 121–123.

Geist, V. (1998) *Deer of the World. Their Evolution, Behavior, and Ecology*. Mechanicsburg, PA: Stackpole Books.

Gentile, G., Vernesi, C., Vicario, S., *et al.* (2009) Mitochondrial DNA variation in roe deer (*Capreolus capreolus*) from Italy: evidence of admixture in one of the last *C. c. italicus* pure populations from central-southern Italy. *Italian Journal of Zoology* **76**, 16–27.

Giacometti, M., Roganti, R., De Tann, D., Stahlberger-Saitbekova, N. and Obexer-Ruff, G. (2004) Alpine ibex *Capra ibex ibex* × domestic goat *C. aegagrus domestica* hybrids in a restricted area of southern Switzerland. *Wildlife Biology* **10**, 137–143.

Goodman, S., Barton, N., Swanson, G., Abernethy, K. and Pemberton, J. (1999) Introgression through rare hybridization: a genetic study of a hybrid zone between red and sika deer (genus *Cervus*) in Argyll, Scotland. *Genetics* **152**, 355–371.

Goodman, S., Tamate, H., Wilson, R., *et al.* (2001) Bottlenecks, drift and differentiation: the population genetic structure and demographic history of sika deer (*Cervus nippon*) in the Japanese archipelago. *Molecular Ecology* **10**, 1357–1370.

Gralak, B., Krasinska, M., Niemczewski, C., Krasinski, Z.A. and Zurkowski, M. (2004) Polymorphism of bovine microsatellite DNA sequences in the lowland European bison. *Acta Theriologica* **49**, 449–456.

Groot Bruinderink, G.W.T.A. and Hazebroek, E. (1996) Ungulate traffic collisions in Europe. *Conservation Biology* **10**, 1059–1067.

Groves, C.P. and Grubb, P. (1987) Relationships of living deer. In C.M. Wemmer (ed.) *Biology and Management of the Cervidae*. Washington, DC: Smithsonian Institution Press, pp. 21–59.

Hajji, G.M., Zachos, F.E., Charfi-Cheikrouha, F. and Hartl, G.B. (2007) Conservation genetics of the imperilled Barbary red deer in Tunisia. *Animal Conservation* **10**, 229–235.

Hajji, G.M., Charfi-Cheikrouha, F., Lorenzini, R., *et al.* (2008) Phylogeography and founder effect of the endangered Corsican red deer (*Cervus elaphus corsicanus*). *Biodiversity and Conservation* **17**, 659–673.

Harrington, R. (1973) Hybridisation among deer and its implications for conservation. *Irish Forestry Journal* **30**, 64–78.

Harrington, R. (1982) The hybridisation of red deer (*Cervus elaphus* L. 1758) and Japanese sika deer (*C. nippon* Temminck, 1838). *International Congress of Game Biologists* **14**, 559–571.

Hartl, G.B. and Pucek, Z. (1994) Genetic depletion in the European bison (*Bison bonasus*) and the significance of electrophoretic heterozygosity for conservation. *Conservation Biology* **8**, 167–174.

Hartl, G.B., Lang, G., Klein, F. and Willing, R. (1991a) Relationships between allozymes, heterozygosity and morphological characters in red deer (*Cervus elaphus*), and the influence of selective hunting on allele frequency distributions. *Heredity* **66**, 343–350.

Hartl, G.B., Reimoser, F., Willing, R. and Köller, J. (1991b) Genetic variability and differentiation in roe deer (*Capreolus capreolus* L) of Central Europe. *Genetics, Selection, Evolution* **23**, 281–299.

Hartl, G.B., Apollonio, M. and Mattioli, L. (1995a) Genetic determination of cervid antlers in relation to their significance in social interactions. *Acta Theriologica, Suppl.* **3**, 199–205.

Hartl, G.B., Klein, F., Willing, R., Apollonio, M. and Lang, G. (1995b) Allozymes and the genetics of antler development in red deer (*Cervus elaphus*). *Journal of Zoology, London* **237**, 83–100.

Hartl, G.B., Hewison, A.J.M., Apollonio, M., Kurt, F. and Wiehler, J. (1998) Genetics of European roe deer. In R. Andersen, P. Duncan and J.D.C. Linnell (eds.) *The European Roe Deer: The Biology of Success.* Oslo: Scandinavian University Press, pp. 71–90.

Hartl, G.B., Zachos, F. and Nadlinger, K. (2003) Genetic diversity in European red deer (*Cervus elaphus* L.): anthropogenic influences on natural populations. *Comptes Rendus Biologies* **326**, S37–S42.

Hemami, M.R., Watkinson, A.R. and Dolman, P.M. (2005) Population densities and habitat associations of introduced muntjac *Muntiacus reevesi* and native roe deer *Capreolus capreolus* in a lowland pine forest. *Forest Ecology and Management* **215**, 224–238.

Herrero, J. and Pérez, J.M. (2008) *Capra pyrenaica*. In *IUCN Red List of Threatened Species*. Online: www.iucnredlist.org

Hetherington, D.A. (2006) The lynx in Britain's past, present and future. *Ecos* **27**, 66–70.

Hewison, A.J.M. (1995) Isozyme variation in roe deer in relation to their population history in Britain. *Journal of Zoology, London* **235**, 279–288.

Hewison, A.J.M. (1997) Evidence for a genetic component of female fecundity in British roe deer from studies of cranial morphometrics. *Functional Ecology* **11**, 508–517.

Hewison, A.J.M. and Staines, B.W. (2008) Roe deer. In S. Harris and D.W. Yalden (eds.) *Mammals of the British Isles: Handbook*, 4th edn. London: Mammal Society, pp. 605–617.

Hmwe, S.S., Zachos, F.E., Eckert, I., *et al.* (2006a) Conservation genetics of the endangered red deer from Sardinia and Mesola with further remarks on the phylogeography of *Cervus elaphus corsicanus*. *Biological Journal of the Linnean Society* **88**, 691–701.

Hmwe, S.S., Zachos, F.E., Sale, J.B., Rose, H.R. and Hartl, G.B. (2006b) Genetic variability and differentiation in red deer (*Cervus elaphus*) from Scotland and England. *Journal of Zoology, London* **270**, 479–487.

International Committee on the Management of Large Herbivores in the Oostvaardersplassen (ICMO) (2006) *Reconciling Nature and Human Interest.* Report to the Dutch Ministry of Agriculture, Nature and Food Quality; The Hague ISBN 9032703528.

Kidjo, N., Feracci, G., Bideau, E., e*t al.* (2007) Extirpation and reintroduction of the Corsican red deer *Cervus elaphus corsicanus* in Corsica. *Oryx* **41**, 488–494.

Kojola, I., Tuomivaara, J., Heikkinen, S., *et al.* (2009) European wild forest reindeer and wolves: endangered prey and predators. *Annales Zoologica Fennici* **46**, 416–422.

Kuehn, R., Schroeder, W., Pirchner, F. and Rottmann, O. (2003) Genetic diversity, gene flow and drift in Bavarian red deer populations (*Cervus elaphus*). *Conservation Genetics* **4**, 157–166.

Kuehn, R., Haller, H., Schroeder, W. and Rottmann, O. (2004) Genetic roots of the red deer (*Cervus elaphus*) population in eastern Switzerland. *Journal of Heredity* **95**, 136–143.

Kuehn, R., Hindenlang, K.E., Holzgang, O., *et al.* (2007) Genetic effect of transportation infrastructure on roe deer populations (*Capreolus capreolus*). *Journal of Heredity* **98**, 13–22.

Kurtén, B. (1968) *Pleistocene Mammals of Europe*. London: Weidenfeld and Nicolson.

Kurtén, B. and Anderson, E. (1980) *Pleistocene Mammals of North America*. New York: Columbia University Press.

Liberg, O., Bergström, R., Kindberg, J. and von Essen, H. (2010) Ungulate management in Sweden. In M. Apollonio, R. Andersen, R. and R. Putman (eds.) *European Ungulates and their Management in the 21st Century*. Cambridge, UK: Cambridge University Press, pp. 37–70.

Linnell, J.D.C., Duncan, P. and Andersen, R. (1998) The European roe deer: a portrait of a successful species. In R. Andersen, P. Duncan and J.D.C. Linnell (eds.) *European Roe Deer: The Biology of Success*. Oslo: Scandinavian University Press, pp. 1–22.

Linnell, J.D.C., Promberger, C., Boitani, L., *et al.* (2005) The linkage between conservation strategies for large carnivores and biodiversity: the view from the 'half-full' forests of Europe. In J.C. Ray, K.H. Redford, R.S. Steneck and J. Berger (eds.) *Carnivorous Animals and Biodiversity: Does Conserving One Save the Other?* Washington, DC: Island Press, pp. 381–398.

Linnell, J.D.C., Breitenmoser, U., Breitenmoser-Würsten, C., Odden, J. and von Arx, M. (2009) Recovery of Eurasian lynx in Europe: what part has reintroduction played? In M.W. Hayward and M.J. Somers (eds.) *Reintroduction of Top-order Predators*. Oxford, UK: Wiley-Blackwell, pp. 72–91.

Lorenzini, R. and Lovari, S. (2006) Genetic diversity and phylogeography of the European roe deer: the refuge area theory revisited. *Biological Journal of the Linnean Society* **88**, 85–100.

Lorenzini, R., Mattioli, S. and Fico, R. (1998): Allozyme variation in native red deer *Cervus elaphus* of Mesola Wood, northern Italy: implications for conservation. *Acta Theriologica, Suppl.* **5**, 63–74.

Lorenzini, R., Lovari, S. and Masseti, M. (2002) The rediscovery of the Italian roe deer: genetic differentiation and management implications. *Italian Journal of Zoology* **69**, 367–379.

Lorenzini, R., San José, C., Braza, F. and Aragón, S. (2003) Genetic differentiation and phylogeography of roe deer in Spain, as suggested by mitochondrial DNA and microsatellite analysis. *Italian Journal of Zoology* **70**, 89–99.

Lovari, S., Cuccus, P., Murgia, A., *et al.* (2007) Space use, habitat selection and browsing effects of red deer in Sardinia. *Italian Journal of Zoology* **74**, 179–189.

Lowe, R. (1994) Deer management: developing the requirements for the establishment of diverse coniferous and broadleaf forests. Unpublished report, Coillte (Irish Forestry Board), Bray, Ireland.

Lowe, V.P.W. (1979) Wild and feral deer in Great Britain. Unpublished NERC contract report HF3/05/43, Institute of Terrestrial Ecology, Merlewood Research Station, 77 pp.

Lowe, V.P.W. and Gardiner, A.S. (1975) Hybridisation between red deer and sika deer, with reference to stocks in north-west England. *Journal of Zoology, London* **177**, 553–566.

Ludt, C.J., Shroeder, W., Rottmann, O and Kuehn, R. (2004) Mitochondrial DNA phylogeography of red deer (*Cervus elaphus*). *Molecular Phylogeny and Evolution* **31**, 1064–1083.

Luenser, K., Fickel, J., Lehnen, A., Speck, S. and Ludwig, A. (2005) Low level of genetic variability in European bisons (*Bison bonasus*) from the Bialowieza National Park in Poland. *European Journal of Wildlife Research* **51**, 84–87.

Manceau, V., Crampe, J-P., Boursot, P. and Taberlet, P. (1999) Identification of evolutionary significant units in the Spanish wild goat, *Capra pyrenaica* (Mammalia, Artiodactyla). *Animal Conservation* **2**, 33–39.

Mattioli, S. (1990) Red deer in the Italian peninsula with particular reference to the Po delta population. *Deer* **8**, 95–98.

Mattioli, S., Fico, R., Lorenzini, R. and Nobili, G. (2003) Mesola red deer: physical characteristics, population dynamics and conservation perspectives. *Hystrix, Italian Journal of Mammalogy* **14**, 87–94.

Maudet, C., Miller, C., Bassano, B., *et al.* (2002) Microsatellite DNA and recent statistical methods in wildlife conservation management: applications in Alpine ibex *Capra ibex* (*ibex*). *Molecular Ecology* **11**, 421–436.

McDevitt, A.D., Edwards, C.J., O'Toole, P., *et al.* (2009) Genetic structure of, and hybridisation between, red (*Cervus elaphus*) and sika (*Cervus nippon*) deer in Ireland. *Mammalian Biology* **74**, 263–273.

Melis, C., Jedrzejewska, B., Apollonio, M., *et al.* (2009) Predation has a greater impact in less productive environments: variation in roe deer, *Capreolus capreolus*, population density across Europe. *Global Ecology and Biogeography* **18**, 724–734.

Micu, I., Náhlik, A., Neguş, S., Mihalache, I. and Szabó, I. (2010) Ungulates and their management in Romania. In M. Apollonio, R. Andersen and R. Putman (eds.) *European Ungulates and their Management in the 21st Century*. Cambridge, UK: Cambridge University Press, pp. 319–337.

Milner, J.M., Nilsen, E.B. and Andreassen, H.P. (2007) Demographic side effects of selective hunting in ungulates and carnivores. *Conservation Biology* **21**, 36–47.

Mitchell-Jones, A.J., Amori, G., Bogdanowicz, W., *et al.* (1999) *The Atlas of European Mammals*. London: T. and A. D. Poyser.

Moco, G., Guerreiro, M., Ferreira, A.F., *et al.* (2006) The ibex *Capra pyrenaica* returns to its former Portugese range. *Oryx* **40**, 351–354.

Niethammer, G. (1963) *Die Einbürgerung von Säugetieren und Vögeln in Europa*. Hamburg, Germany: Paul Parey.

Novobilsky, A., Horackova, E., Hirtova, L., Modry, D. and Koudela, B. (2006) The giant liver fluke *Fascioloides magna* (Bassi 1875) in cervids in the Czech Republic and potential of its spreading to Germany. *Parasitological Research* **100**, 549–553.

Nussey, D.H., Coltman, D.W., Coulson, T., *et al.* (2005) Rapidly declining fine-scale spatial genetic structure in female red deer. *Molecular Ecology* **14**, 3395–3405.

Nussey, D.H., Pemberton, J., Donald, A. and Kruuk, L.E.B. (2006) Genetic consequences of human management in an introduced island population of red deer (*Cervus elaphus*). *Heredity* **97**, 56–65.

Olech, W. (1987) Analysis of inbreeding in European bison. *Acta Theriologica* **32**, 373–387.

Olech, W. (2008) *Bison bonasus*. In *IUCN Red List of Threatened Species*. Online: www.iucnredlist.org

Olech, W. and Perzanowski, K. (2002): A genetic background for reintroduction program of the European bison (*Bison bonasus*) in the Carpathians. *Biological Conservation* **108**, 221–228.

Pemberton, J.M., Albon, S.D., Guinness, F.E. and Clutton-Brock, T.H. (1988) Genetic variation and juvenile survival in red deer. *Evolution* **42**, 921–934.
Pemberton, J., Swanson, G., Barton, N., Livingstone, S. and Senn, H. (2006) Hybridisation between red and sika deer in Scotland. *Deer* **13**, 22–26.
Pérez, J.M., Granados, J.E., Soriguer, R.C., *et al.* (2002) Distribution, status and conservation problems of the Spanish ibex, *Capra pyrenaica* (Mammalia: Artiodactyla). *Mammal Review* **32**, 26–39.
Pérez-Espona, S., Pérez-Barbería, F.J., McLeod, J.E., *et al.* (2008) Landscape features affect gene flow of Scottish Highland red deer (*Cervus elaphus*). *Molecular Ecology* **17**, 981–996.
Pérez-Espona, S., Pemberton, J.M. and Putman, R.J. (2009) Red and sika deer in the British Isles, current management issues and management policy. *Mammalian Biology* **74**, 247–262.
Perzanowski, K. and Olech, W. (2007) A future for European bison *Bison bonasus* in the Carpathian ecoregion? *Wildlife Biology* **13**, 108–112.
Perzanowski, K., Olech, W. and Kozak, I. (2004) Constraints for re-establishing a meta-population of the European bison in Ukraine. *Biological Conservation* **120**, 345–353.
Pitra, C., Fickel, J, Meijaard, E. and Groves, C. (2004) Evolution and phylogeny of old world deer. *Molecular Phylogenetics and Evolution* **33**, 880–895.
Pitra, C., Rehbein, S. and Lutz, W. (2005) Tracing the genetic roots of the sika deer *Cervus nippon* naturalized in Germany and Austria. *European Journal of Wildlife Research* **51**, 237–241.
Prior, R. (1968) *The Roe Deer of Cranborne Chase*. Oxford, UK: Oxford University Press.
Proaktor, G., Coulson, T. and Milner-Gulland, E.J. (2007) Evolutionary responses to harvesting in ungulates. *Journal of Animal Ecology* **76**, 669–678.
Pucek, Z. (ed) (2004) *European Bison. Status Survey and Conservation Action Plan*. Gland, Switzerland and Cambridge, UK: IUCN/SSC Bison Specialist Group, IUCN.
Puddu, G., Maiorano, L., Falcucci, A., Corsi, F. and Boitani, L. (2009) Spatial-explicit assessment of current and future conservation options for the endangered Corsican red deer (*Cervus elaphus corsicanus*) in Sardinia. *Biodiversity and Conservation* **18**, 2001–2016.
Putman, R.J. (2010) Ungulates and their management in Great Britain and Ireland. In M. Apollonio, R. Andersen and R. Putman (eds.) *European Ungulates and their Management in the 21st Century*. Cambridge, UK: Cambridge University Press, pp. 129–164.
Putman, R.J. and Hunt, E. (1994) Hybridisation between red and sika deer in Britain. *Deer*, **9**, 104–110.
Radwan, J., Kawalko, A., Wojcik, J.M. and Babik, W. (2007) MHC-DRB3 variation in a free-living population of the European bison, *Bison bonasus*. *Molecular Ecology* **16**, 531–540.
Randi, E., Alves, P.C., Carranza, J., *et al.* (2004) Phylogeography of roe deer (*Capreolus capreolus*) populations: the effects of historical genetic subdivisions and recent nonequilibrium dynamics. *Molecular Ecology* **13**, 3071–3083.
Ratcliffe, P.R. (1989) The control of red and sika deer populations in commercial forests. In R.J. Putman (ed.) *Mammals as Pests*. London: Chapman and Hall, pp. 98–115.

Ratcliffe, P.R., Peace, A.J., Hewison, A.J.M., Hunt, E.J. and Chadwick, A.H. (1992) The origins and characterization of Japanese sika deer populations in Great Britain. In N. Maruyama, B. Bobek, Y. Ono, *et al.* (eds.) *International Symposium on Wildlife Conservation: Present Trends and Perspectives for the 21st Century*. Tokyo: Japan Wildlife Research Center, pp. 185–190.

Royo, L.J., Pajares, G., Alvarez, I., Fernández, I. and Goyache, F. (2007) Genetic variability and differentiation in Spanish roe deer (*Capreolus capreolus*): a phylogeographic reassessment within the European framework. *Molecular Phylogenetics and Evolution* **42**, 47–61.

Sempéré, A.J., Sokolov, V.E. and Danilkin, A.A. (1996) *Capreolus capreolus*. In *Mammalian Species* No. 538. American Society of Mammalogists. Online: www.mammalsociety.org.

Senn, H.V. and Pemberton, J.M. (2009) Variable extent of hybridisation between invasive sika (*Cervus nippon*) and native red deer (*Cervus elaphus*) in a small geographic area. *Molecular Ecology* **18**, 862–876.

Shackleton, D.M. (ed.) (1997) *Wild Sheep and Goats and their Relatives. Status Survey and Conservation Action Plan for Caprinae*. Gland, Switzerland and Cambridge, UK: IUCN/SSC Caprinae Specialist Group, IUCN.

Skog, A., Zachos, F.E., Rueness, E.K., *et al.* (2009) Phylogeography of red deer (*Cervus elaphus*) in Europe. *Journal of Biogeography* **36**(1), 66–77.

Slate, J., Kruuk, L.E.B., Marshall, T.C., Pemberton, J.M. and Clutton-Brock, T.H. (2000) Inbreeding depression influences lifetime breeding success in a wild population of red deer (*Cervus elaphus*). *Proceedings of the Royal Society B–Biological Sciences* **267**, 1657–1662.

Sommer, R.S., Zachos, F.E., Street, M., *et al.* (2008) Late Quaternary distribution dynamics and phylogeography of the red deer (*Cervus elaphus*) in Europe. *Quaternary Science Reviews* **27**, 714–733.

Sommer, R.S., Fahlke, J.M., Schmölcke, U., Benecke, N. and Zachos, F.E. (2009) Quaternary history of the European roe deer *Capreolus capreolus*. *Mammal Review* **39**, 1–16.

Spear, D. and Chown, S.L. (2009) Non-indigenous ungulates as a threat to biodiversity. *Journal of Zoology* **279**, 1–17.

Stuart, A.J., Kosinysev, P.A., Higham, T.F.G. and Lister, A.M. (2004) Pleistocene to Holocene extinction dynamics in giant deer and woolly mammoth. *Nature* **431**, 684–690.

Temple, H.J. and Terry, A. (compilers) (2007) *The Status and Distribution of European Mammals*. Luxembourg: Office for Official Publications of the European Communities.

Thulin, C.-G. (2006) Microsatellite investigation of roe deer (*Capreolus capreolus*) in Scandinavia reveals genetic differentiation of a Baltic Sea Island population. *European Journal of Wildlife Research* **52**, 228–235.

Tiedemann, R., Nadlinger, K. and Pucek Z (1998) Mitochondrial DNA: RFLP analysis reveals low levels of genetic variation in European bison *Bison bonasus*. *Acta Theriologica, Suppl.* **5**, 83–87.

Torres-Porras, J., Carranza, J. and Pérez-González, J. (2009) Selective culling of Iberian red deer stags (*Cervus elaphus hispanicus*) by selective montería in Spain. *European Journal of Wildlife Research* **55**, 117–123.

Vernesi, C., Pecchioli, E., Caramelli, D., *et al.* (2002) The genetic structure of natural and reintroduced roe deer (*Capreolus capreolus*) populations in the Alps and central Italy, with reference to the mitochondrial DNA phylogeography of Europe. *Molecular Ecology* **11**, 1285–1297.

Wang, M. and Schreiber, A. (2001) The impact of habitat fragmentation and social structure on the population genetics of roe deer (*Capreolus capreolus* L.) in Central Europe. *Heredity* **86**, 703–715.

Ward, T.J., Bielawski, J.P., Davis, S.K., Templeton, J.W. and Derr, J.N. (1999) Identification of domestic cattle hybrids in wild cattle and bison species: a general approach using mtDNA markers and the parametric bootstrap. *Animal Conservation* **2**, 51–57.

Wemmer, C. (1998) *Deer: Status Survey and Conservation Action Plan*. Gland, Switzerland and Cambridge, UK: IUCN/SSC Deer Specialist Group, IUCN.

Wójcik, J.M., Kawalko, A., Tokarska, M., *et al.* (2009) Post-bottleneck mtDNA diversity in a free-living population of European bison: implications for conservation. *Journal of Zoology, London* **277**, 81–87.

Wray, S. (1994) Competition between muntjac and other herbivores in a commercial coniferous forest. *Deer* **9**, 237–245.

Yalden, D.W. (1999) *The History of British Mammals*. London: T. and A.D. Poyser.

Zachos, F.E. (2009) Gene trees and species trees: mutual influences and interdependences of population genetics and systematics. *Journal of Zoological Systematics and Evolutionary Research* **47**, 209–218.

Zachos, F., Hartl, G.B., Apollonio, M. and Reutershan, T. (2003) On the phylogeographic origin of the Corsican red deer (*Cervus elaphus corsicanus*): evidence from microsatellites and mitochondrial DNA. *Mammalian Biology* **68**, 284–298.

Zachos, F.E., Hmwe, S.S. and Hartl, G.B. (2006) Biochemical and DNA markers yield strikingly different results regarding variability and differentiation of roe deer (*Capreolus capreolus*, Artiodactyla: Cervidae) populations from northern Germany. *Journal of Zoological Systematics and Evolutionary Research* **44**, 167–174.

Zachos, F.E., Althoff, C, von Steynitz, Y., Eckert, I. and Hartl, G.B. (2007) Genetic analysis of an isolated red deer (*Cervus elaphus*) population showing signs of inbreeding depression. *European Journal of Wildlife Research* **53**, 61–67.

Zachos, F.E., Hajji, G.M., Hmwe, S.S., *et al.* (2009) Population viability analysis and genetic diversity of the endangered red deer population from Mesola, Italy. *Wildlife Biology* **15**, 175–186.

Zima, J., Kožená, I. and Hubálek, Z. (1990) Non-metrical cranial divergence between *Cervus elaphus*, *C. nippon nippon* and *C. nippon hortulorum*. *Acta Scientarum Naturalium, Brno* **24**, 1–41.

3

A review of the various legal and administrative systems governing management of large herbivores in Europe

RORY PUTMAN

This chapter attempts to review the legal constraints on hunting and game management in a range of different countries in Europe. It draws heavily on data presented in Apollonio *et al.* (2010) as a primary source – and focuses principally on the control of hunting of ungulates.

Different management systems in different countries largely reflect the legal status of game (whether they are considered to belong to the state, to the private landowner, to everyone, or to no one!) and cultural attitudes to hunting – which relate to the role that hunting plays within the wider culture of each country and the historical status of hunting in that country (reviewed briefly in Chapter 1). Within the legal framework established, the second part of this chapter thus considers the different administrative systems operating in different countries to regulate and control hunting.

3.1 General legislative framework

No attempt is made to list the relevant legislative instruments or regulations concerned with the management of game animals in the different countries within Europe – partly because in any one country, relevant regulations are contained in a large number of different laws on hunting, game management, conservation of wildlife, handling and sale of game meat, control of use of firearms, etc. There is an additional complication in the fact that in some countries, while a part of the legislation may be operative at national level, some elements of the legal provision are delegated to regulation at a provincial or regional level. For detail, therefore, the reader is referred back to that primary source (Apollonio *et al.*, 2010).

Ungulate Management in Europe: Problems and Practices, eds. Rory Putman, Marco Apollonio and Reidar Andersen. Published by Cambridge University Press.

Even then, it should be noted that legislation is never a fixed reality but is subject to continuous change: in the UK, for example, the Deer Act of 1991 was recently revised by a Regulatory Reform Order in 2007. In the same way, the current Deer (Scotland) Act 1996, which largely summarises legislation affecting deer and their management in Scotland, is also currently under review, with the expectation that it will shortly be replaced.

Finally, we should emphasise that the intention of this review is to attempt a synthesis, offering analysis of similarities and differences in the general 'shape' of the legislation; any such attempt to offer an overview inevitably risks oversimplification – even if the relevant regulations were unchanged. **It is not the intention of this review to offer comprehensive and definitive coverage of the exact regulations currently operative in any given individual territory. Any reader requiring specific detail of regulations applying in one or another individual country is advised to ensure they consult the most recent sources in that country**.

For the most part, legislation (wherever the particular regulation may be found, and whether it applies at national or provincial level) controls hunting seasons, permitted weapons and ammunition and permitted hunting methods, as well as the handling and disposal of venison or other game products. In most cases it also defines training requirements for hunters and responsibilities imposed on game management, in preparation of game management plans and in reporting details of cull (and sometimes non-cull) mortality. In some cases primary legislation may define the objective of hunting or management of game populations, outlining its role within a wider context of sustainable land-use objectives. Fundamentally, however, legislation first defines ownership of game and determines who has the right to hunt.

3.1.1 Who owns game animals?

Perhaps the first most important generality is that in no country in Europe does game belong to the owner of the land on which that game may occur. Legal definitions of game ownership define them in all cases as belonging to everyone, or to no one (*res communis* or *res nullius*). The distinction is a subtle one but the difference may be significant and in large part relates to the extent to which the state may determine or intervene in management structure or practice. (Thus where game is defined as *res communis*, the state may elect to sell licences to hunt, or allocate management of game management districts (GMDs), to individuals or hunting groups without reference to the landowner; where game is regarded as *res nullius*, the right to shoot that game more generally involves some sort of 'contract' with the landowner or his agent.)

Table 3.1 *Legal status of game ownership*

Game considered *res communis*	Game considered *res nullius*
Finland	Austria
Netherlands	United Kingdom
Poland	Norway
Lithuania	Sweden
Croatia	Belgium
Slovenia	Germany
Switzerland	Estonia
Hungary	Latvia
Romania	Czech Republic
Portugal	Spain
Italy	

Countries in which game is regarded as *res communis* include: Croatia, Finland, Hungary, Italy, Lithuania, the Netherlands, Poland, Portugal, Romania, Slovenia, Switzerland. Countries where game is clearly considered *res nullius* include Austria, Belgium, Czech Republic, Estonia, Germany, Latvia, Norway, Spain, Sweden, UK (Table 3.1).

3.1.2 The right to kill game

As noted, this question of ownership of game may be of significance in determining who has the right to kill that game.

In parts of Switzerland, management is entirely vested in the state under a system of '*Patentjagd*' (a licensing system practised in 16 cantons by about 24 000 hunters on 70% of the national territory). Under this system each hunter buys an annual licence from the cantonal hunting department, and the licence is valid throughout the canton.

More generally where the state (at national or regional level) maintains 'ownership' of the game, management is delegated by the state to hunters associations or hunters clubs, who take over the management of game in a prescribed GMD (even though the land may well be owned by other, third parties). Such a system, for example, operates in the remainder of Switzerland ('*Revierjagd*'), Croatia, Poland, Portugal, Romania and Slovenia. As an alternative, hunting is delegated to provinces, which define the hunting districts and manage them (Italy).

In all other areas, the rights to hunt technically belong to the owner of the land (which may of course itself be in state ownership). This does not necessarily imply that the landowner has free exercise of that right, but generally implies that he or she must be consulted and there is some 'partnership' in management between state and private owner.

Thus, for example, in Austria, Belgium, the Czech Republic, Estonia, Latvia, Germany and Hungary hunting rights belong to the landowner but there is defined a minimum area within which hunting or game management may be carried out; where individual landholdings are below that minimum viable area (defined as the limits of an effective biological population, and sometimes determined separately for different ungulate species), then owners must group themselves into syndicates or hunting associations to reach the required threshold area. Thereafter they themselves have the right to hunt within that area or may lease the hunting to other individuals (or, once again, to an independent hunting association or hunting club).

Minimum areas for approval vary from 75 ha (0.75 km^2) in parts of Germany, 115 ha in Austria to 500 ha in the Czech Republic (50 ha in enclosed hunting parks), 1000 ha in Belgium (Flanders; roe deer) or Lithuania, and 3000 ha in Hungary. In Latvia, separate thresholds are set for roe deer (200 ha), wild boar (1000 ha), red deer (2000 ha for stags; 1000 ha for hinds and calves) and moose (2500 ha). There is clearly considerable variation between countries and while in part this may reflect species present, with minimum area of GMDs set in relation to the population range of the largest or more mobile of the species present, there remains variation even in relation to a single species (200 ha for roe deer in Latvia, 1000 ha for roe deer in Flanders).

In France and Finland, a slightly different model applies in that hunting rights once again belong to the landowner, but the state (at regional or department level) retains the right to determine what may be killed and issues individual licences for the number (by age and sex) of each species which may/must be harvested. Once again the landowner may exercise the right to hunt himself, or may pass this right to others. Elsewhere in Scandinavia (in Norway, Sweden, Denmark) and in Spain and the United Kingdom, hunting rights belong exclusively to the landowner – who may, again, exercise these rights himself or sell/lease these to another. In the majority of cases (outside the UK), however, numbers to be harvested must be agreed within a mandatory management plan for the holding which must be approved by regional state authorities (see Section 3.4). Production of a management plan is in fact mandatory in all countries apart from those where the state assumes entire responsibility for control of hunting/game management (Finland, France, Slovenia and Switzerland) or in the UK and Sweden (for species other than moose).

3.1.3 Management objectives

In a number of countries, primary legislation defines not only ownership of game animals and who may have the right to kill game, but also defines what shall be the objectives of management. Increasingly frequently, management of ungulates is now being set within a wider context – and one primarily determined by general conservation objectives.

In the Netherlands, for example, all species of wildlife, including roe deer, red deer, wild boar and fallow deer are all fully protected and can only be culled

- for reasons of public health and public safety
- in the interest of safety of air traffic
- to prevent damage to crops, cattle, forests, fishery and waters
- to prevent damage to flora and fauna or
- other reasons, to be defined by specific order in Government.

And formal application must be made in every case.

Elsewhere, where national objectives are declared, legislative systems declare as the major objective either conservation per se or define as that objective the integration of conservation with sustainable exploitation. In Switzerland, the main goals of the national hunting laws are maintenance of biodiversity, protection of threatened species, limitation of damage caused by free-ranging ungulates, and sustainable culling of populations. In Scandinavia, legislation emphasises that 'the concept of *sustainable use* should underpin all wildlife management' (Norway: Andersen *et al.*, 2010). In Sweden: 'The only general and national objective for the management of games species in Sweden is that they should be preserved in viable populations, but not be allowed to seriously damage other vital interests of the society' (Liberg *et al.*, 2010). In Hungary the same principle of balance between conservation of biodiversity and sustainable use is clearly expressed in the preamble of the Game Act:

> Recognizing, that all wild animal species form an irreplaceable part of the renewable natural resources of Earth and of ecosystems, being aware, that wild animals carry aesthetic, scientific, cultural, economical and genetic values, and therefore – as a treasure of humankind as a whole and of our nation – they should be conserved in their natural state for the future generations also, Parliament creates the following act in the interest of conserving nature and rationally utilizing the populations of game species... (Act LV/1996 on game conservation, game management and hunting).

This concept that hunting should be sustainable, and that populations of game animals should be maintained in balance with other land-use objectives (with control of damage to agriculture, forestry or other conservation objectives as a major consideration in management of ungulate populations) is increasingly accepted in the majority of European states.

3.1.4 National or regional legislation

In a number of countries game laws are established by national government and apply to the whole country. In others, while certain regulations may be established at national level (firearms laws, training requirements, etc.) details of management may be devolved and controlled by regional or provincial bye-laws. This is particularly the case in relation to the setting of game seasons, or requirements for returning data on animals harvested (Table 3.2).

3.2 Specific legislative provisions

Having explored the general legislative framework for hunting and game management and the legal status of game, we may now pay attention to specific provisions. In the following sections I will briefly review the different requirements for obtaining a hunting licence (licence to kill game), requirements for training, legal restrictions on permitted weapons and ammunition, permitted hunting methods, hunting seasons and a variety of other specific requirements (recording and return of information on animals harvested, inspection and tagging of carcases, handling and disposal (sale) of venison and game products etc.).

Table 3.2 *Summary to show those countries where regulation and administration are largely at national or national* and *regional levels*

Regulation/administration largely at national level	Regulation/administration at both national and regional levels
Finland	Belgium (Flanders/Wallonia)
Norway	United Kingdom
Sweden	Republic of Ireland
Denmark	Germany (states)
Netherlands	Austria (provinces)
Poland	Italy (provinces)
Lithuania	
Croatia	
Slovenia	
Switzerland	
Hungary	
Romania	

3.2.1 Legal requirements for obtaining a licence to shoot game

In every country considered, formal application must be made to hold a firearms (or shotgun) licence; in the majority of cases an additional licence is required in order to be permitted to shoot large ungulates (whether this is a general licence or a licence for a particular number of animals of given age and sex). This 'secondary' licence to shoot ungulates is commonly dependent on proof of the right to shoot over a particular area of land and in many cases it is also dependent on membership of a regional hunting association/ cooperative ('game management group') or recognised hunters club (e.g. Hungary, Romania, Poland, Slovakia, Slovenia, Croatia).

3.2.2 Training

In the vast majority of European countries, issue of a licence is dependent on some degree of formal training which is usually assessed by formal examination; this is (less commonly) offered by national or regional administrations, more commonly (whether formally or informally) through hunters associations. Training in many instances involves a period of supervised training in the field (e.g. Poland, Lithuania, etc.) and in some cases even after obtaining a hunter's licence the individual still 'suffers' a period of probation where he/she may only hunt under supervision (e.g. Romania). In Germany a hunter can only assume responsibility for his/her own hunting ground (revier) after he/she has held a full licence for 3 years.

The level and intensity of training (and the 'ease' of the examination) varies enormously from country to country, as does the 'syllabus'. Further, while in most cases the examination does require a practical test of shooting skills and accuracy (whether this is a requirement simply for gaining a firearms licence in the first place, or required more specifically towards a licence to hunt ungulates) the level of expectation is notably variable.

Specifically

Countries where no formal training is required after issue of a straightforward gun licence:

Netherlands, Belgium, UK, Spain

Countries where fairly minimal training is required before issue of a hunting licence for ungulates:

Finland (12 hours of voluntary lectures), Norway (mandatory 30 hour course and theory exam), Sweden, Denmark, Latvia, Switzerland, Portugal, France, Italy, Greece (no practical examination)

Countries with more extensive expectation of training:
Germany (120 hours mandatory plus exam), Hungary, Poland (1 year training in a hunting club), Estonia (56 hour mandatory course), Lithuania (1 year training in a hunting club), Czech Republic, Slovakia, Croatia (hunting exam and 1 year practical training), Austria, Slovenia, Romania.
The most comprehensive training is, not unexpectedly, required in German-speaking (and related) countries of central Europe, with a long-established tradition of game management through established (and respected) hunters' associations.

3.2.3 Legal restrictions on firearm calibre and muzzle energy/velocity

Most European countries do have explicit legislation prescribing weapons, calibres and projectile energy to be used when hunting different species of game animals.

Rather than rehearse in each case the particular detailed regulations, I attempt below to extract

- those countries where it is permissible to use a shotgun to pursue and kill ungulates of any species
- those countries where it is permissible to use small calibre rifles (and for what species)
- those countries/species where rifles of large calibre and providing a minimum projectile energy at 100 m (E100) are required.

Shotgun permitted for small species

Finland (roe; also fallow if single slug); ***Sweden*** (roe); ***Norway*** (roe, except males 10 August – 25 September); ***Denmark*** (roe, except males in summer, as Norway); ***UK:*** all species, but only where deer are causing damage to agricultural crops, growing timber or other property, and only where the shotgun is 12 bore or larger and loaded with AAA shot or a single slug of at least 350 grains (at least 380 grains in Scotland); [***Croatia*** wild boar only]; ***Austria*** (permitted for roe deer in one province only but not used in practice); ***Switzerland*** (roe, but primarily only in central plateau); [***Italy*** wild boar only, but the shotgun must be loaded with a single slug]; ***Portugal*** (all ungulate species but single slug only); ***France*** (all ungulate species).

Small calibre rifles

Sweden (roe); ***Norway*** (roe; restriction is that expanding bullets with E100 > 980 joules must be used), ***Denmark*** (roe; bullet weight must be > 50 grains

(3.2 g) and E100 > 800 J); ***Scotland*** (roe; bullet weight must be > 50 grains (3.2 g), muzzle velocity > 746.76 m/s and muzzle energy > 1356 J); ***England/Wales*** (rifle calibre of .220 inches or more, bullet weight > 50 grains (3.2 g) and muzzle energy > 1356 J: muntjac and Chinese water deer, but *not* roe); ***Netherlands*** (roe: E100 > 980 J); ***Belgium*** (roe: E100 > 980 J); ***Czech Republic*** (roe: muzzle energy > 1000 J); ***Croatia*** (roe: bullet weight must be > 50 grains (3.2 g) and E100 > 1000 J); ***Slovenia*** (roe: bullet weight must be > 50 grains (3.2 g) and E100 > 1000 J); ***Switzerland*** (roe); ***Hungary*** (roe: rifle with muzzle energy > 1000 J); ***Italy*** (all species with calibre 5.6 mm central percussion).

Larger game

In almost all cases, where shooting larger animals (i.e. species other than roe), all countries specify the use of rifles and ammunition capable of delivery of a bullet of 9 g or 10 g (140 or 154 grain) with a minimum E100 of respectively 2700 J or of the order of 2500 J.

Exceptions to this general rule are:

Denmark, Finland: E100 for 10 gram bullet 2000 J; ***Norway:*** 2200 J; ***Czech Republic:*** muzzle energy > 1500 J; ***England/Wales:*** (for roe and larger) a rifle of calibre .240 inches or more, and muzzle energy equal to or greater than 1700 footpounds (2305 J) which must deliver a soft-nosed or hollow-pointed bullet, but no defined weight; ***Scotland:*** a rifle of a muzzle energy not less than 1750 foot pounds (2373 J) with a muzzle velocity of not less than 2450 feet per second (746.76 m/s) delivering a single bullet with a weight of not less than 100 grain (6.48 g).

In ***Portugal***, ***France***, ***Spain*** no specific legal restrictions appear to be in force. [In Portugal, 'recommendations' are made in a technical manual: Victorino, J.A. (2001) Armas de fogo e munições de caça (weapons and hunting ammunitions). In *Carta de Caçador: Manual Para Exame*. Lisbon: Direcção Geral das Florestas – but these are recommendations only and not legally binding; C. Fonseca, pers. comm.]

3.2.4 Other permitted weapons

Use of shotguns (usually, but not always presuming these are loaded with a single ball or slug) is permitted for all ungulate species in Portugal and for smaller species (roe) in Finland, Sweden, Norway, Denmark (except in both cases males in summer), Austria (one province only), France and Switzerland.

Shotguns can currently also be used in the UK for all species, under restricted circumstances as outlined above. Shotguns loaded with a single slug may be used for wild boar in Italy.

Traps and snares are not permitted in the majority of countries, but bow hunting is permitted for roe deer in Finland, Denmark and for other species in France and Italy (bow), Portugal (bow, crossbow, spear). Bow hunting is also increasingly permitted in a number of States of the USA. In some cases there may be a restriction on bow power (as for example in Finland where the bow must have a minimum power of 180 N).

3.2.5 Hunting practice and the use of dogs

In most European and North American countries/states, hunting of ungulates is either carried out from a high seat, or by stalking. In some cases groups of hunters 'drive' an area by walking through it in line abreast (e.g. Denmark) or lines of human beaters drive game to stationary guns (e.g. Sweden, Germany, Hungary, Poland (silent drives only), Baltic Countries, Portugal), but this is relatively uncommon compared with situations where game may be driven by dogs.

Dogs are frequently used by moose hunters throughout Scandinavia to seek out and hold moose at bay; in many other countries single dogs or packs may be used (with or without human beaters) to flush and drive deer to waiting guns (Finland, Sweden, Norway, Denmark, Germany, Austria, the Baltics, Romania, Portugal, France and Spain). In Poland, Hungary and Slovakia, as well as in Italy, dogs may be used for hunting wild boar, but not for deer. Use of dogs for driving deer is specifically forbidden by law in the Netherlands, Belgium, UK and Republic of Ireland, while in the Czech Republic it is illegal to use dogs > 55 cm at the withers.

The other side of this particular coin is that in many countries it is mandatory to have, or to have access to a specially trained dog for following a blood trail and finding injured animals which have run to cover. We may note that access to a trained dog for location of injured animals is mandatory in Sweden, Norway, Slovakia, Germany and Austria; it is common also in Poland, Slovenia, Hungary and Italy.

3.2.6 Hunting seasons

Almost all European countries operate with a principle of restricted hunting periods for some or all species (exception Portugal, where technically the season lasts from 1 June in any year to 31 May of the following year, although most hunting activity is carried out between September and February).

Seasons are primarily determined for welfare purposes (to avoid orphaning of dependent juveniles), to restrict harvests of species present in relatively low numbers, or to provide equitability in harvest access between landholdings which may be in different parts of the range of species with distinct seasonal ranges. There is, however, an enormous diversity in length (and actual time of year) of the permitted season, which shows little consistency between countries and often little relation to actual biological breeding seasons (rut, parturition, period of dependency of young).

Further, we should note that in many instances it is possible to apply for an extension to the permitted season (France for wild boar; Italy) or an exemption from the regulations (permission to kill animals outside the legal season – e.g. United Kingdom countries and others) where ungulates may be shown to be causing damage to agriculture, forestry or more rarely (UK and Austria only) conservation habitats.

The mismatch between hunting and biological seasons, and the implications of this for welfare, social dynamics, and the ability (or failure) of hunters to regulate prey populations, is the subject of a separate review in this volume (Chapter 4) and thus this matter of seasons will not be addressed in detail here. However, it is clear that there is enormous variation in both timing and duration of seasons (compare, for example, seasons in the three adjacent countries of the Baltics: Latvia, Lithuania and Estonia, or the different parts of Belgium). Seasons also vary (often quite markedly) between different regions or provinces of one country (e.g. Italy, Austria, Germany).

3.2.7 Reporting of cull and cull statistics

Cull statistics and the requirement to make a statutory cull return

Table 3.3 offers a summary to show in which countries it is mandatory to submit to the competent authorities some form of cull return (whether this merely details numbers of animals shot of each species and sex, or provides more detailed information). In most cases statistics and cull records must be returned to the licensing authority.

It is clear that in the vast majority of European countries some statistical return is compulsory (although the detail required may vary enormously (see below) and many commentators question the accuracy of some of these returns). In England and Wales there is no such requirement. In Scotland there is no actual fundamental requirement to supply such information, but

Table 3.3 *A summary of the main legal and administrative requirements governing hunting of ungulates in different European countries*

Country	Regulation/administration national	Regulation/administration regional	GMDs compulsory	GMDs voluntary	Management plan mandatory	Management plan voluntary	Quota set by regional authority	Quota must be approved by authority	No regulatory control of quotas	Legal requirement for cull records	Other *national* scheme	No central cull records	Carcase tagging required
Finland	✓	✓	✓				✓			✓			
Sweden	✓			✓		✓				[a]	✓		×
Norway		✓				✓	✓	✓		✓			×[b]
Denmark	✓						✓			✓			×
Scotland						✓			✓	[c]			✓
England etc.									✓	×		✓	✓
Netherlands	✓		✓		✓			✓		?			✓
Belgium				✓	✓			✓		✓			✓
Germany		✓	✓		✓			✓		✓			
Poland	✓	✓	✓		✓		✓	✓		✓			×
Baltics	✓		✓		✓			✓		?			
Czech Rep.	✓		✓		✓			✓		✓			×
Slovakia		✓	✓		✓		✓			✓			×
Croatia	✓		✓		✓			✓		✓			✓
Austria		✓	✓		✓			✓		✓			✓
Slovenia		✓	✓				✓			✓			✓
Switzerland		✓					✓			✓			✓
Hungary		✓	✓		✓			✓		✓			✓
Romania	✓		✓		✓			✓		✓			✓
Portugal	✓	✓	✓		✓		✓			✓			✓
Spain		✓			✓			✓		×			[d]
France		✓					✓			✓			✓
Italy		✓	✓		✓		✓			×			✓

[a]Sweden: statutory return required for moose only. However, voluntary scheme coordinated by the Swedish Association for Hunting and Wildlife Management.
[b]Norway: tagging required for wild reindeer only.
[c]Scotland: Deer Commission for Scotland may demand cull records from individual landholdings or deer management groups; UK: carcase tagging compulsory for animals passed to game dealer/game meat handler.
[d]Spain: compulsory but only in some regions.
GMDs, game management districts.

the competent authority (the Deer Commission for Scotland) may request an annual cull return from anyone known to be culling deer and it is an offence not to comply with their request.

Types of data collected

Once again there is enormous variation in the detail of information collected in cull returns. At the lowest level, licence holders, or leaseholders of GMDs, may be required to submit no more than a simple numerical return for each species of numbers of animals culled. In some cases this is not even subdivided by sex or age.

More commonly, and especially where there is a requirement for each carcase to be identified by a unique, numbered, tag, whose number is cross-referenced to the cull record, cull record sheets ask for more detail (as, for example, France: the record form asks for tag number, date, number of hunters, shotgun or rifle, lead shot or bullet, weight (weighed or estimated), dead or clean, male/female, antlered or not, in velvet or not, antler height, female lactating or not). In Belgium, the form asks for:date and place where the animal was shot, person who shot the animal, sex and age class of the animal and eviscerated weight. Additional information such as the jaw length, number of embryos, kidney fat index and details on the antlers can be added but is not mandatory.

In Finland, after the end of the hunting season, each harvested animal has to be reported to the local GMD. This report must include information on the ungulate sex, age (adult/calf), site of the kill and antler tine number. By contrast, in Norway hunters are obliged only to submit reports of the number and sex of animals, and the annual number of cervids and small game harvested.

Perhaps the greatest level of detail is asked for in those countries adhering to a more Germanic system. In Austria, Hungary, Slovenia, Croatia (for example), and also Norway, leaseholders of each GMD are required to submit to the regional authorities not only a report on numbers of animals culled for each species (by sex and age) but also a return of natural mortality, numbers of animals killed in road traffic accidents, even in some cases to estimate losses to predation. It should be noted, however, that there is little 'control' for the accuracy of such reports (Putman, 2008).

Where individual licences are issued by state or regional authorities to individual hunters, a record for all animals culled (up to the quota allocated under the licence) is supposed to be returned to the body issuing such licence. But it is very difficult to assess the accuracy of such a return since hunters will of course ensure that returns match the quota allocated under the licence granted (and there is no obvious way of 'checking'). Thus:

- If a hunter has shot more than his/her permitted quota this will not be admitted; numbers by age and sex on any return will be presented to match the quota allocated and more detailed statistics, if required (for weight, reproductive condition, antlers, etc.) will simply be presented for the 'correct' number of animals declared.
- If a hunter has shot less than his/her permitted quota, this may be honestly represented if no penalties attach (and simply the licensing authority subsequently 're-lets' the shortfall under another licence to ensure that the total cull level is maintained). However, in many situations, the licensing authority applies penalties if the required cull is not taken and in this case, cull returns may be inflated to avoid such penalty.

[Even where carcase tagging is mandatory, this does not wholly resolve the issue of accuracy. Simply, animals taken in excess of quota will not be tagged or declared, while in the case of a shortfall on quota, surplus tags, where issued, can simply be disposed of. Only where some crosscheck of all tags is maintained from issue through to game dealer or final outlet would such anomalies be detected, and I know of no country where such systematic checking is undertaken.]

In a small number of countries, allocation of hunting rights within a GMD to landowners or to a hunting association is conditional on appointment of a paid game warden or gamekeeper to each GMD. These professional staff are responsible (usually) for preparation of a management plan for official approval, for organising hunting activities within the GMD and for preparation of annual statistics at the end of the season. While once again this may simplify the 'chain of command' and ensure that one particular individual is responsible for submission of an annual return, it does not of course increase the certainty of accuracy of that return.

This problem is stressed by many of those consulted in preparation of this review – and interestingly enough is perhaps one downside of management or administrative systems which tend to 'over-regulate'. If the granting of the lease or concession of a hunting district is dependent on meeting certain targets, then statistical returns from the leaseholder are likely to be manipulated where necessary to match targets set. Even within those administrative models where individual licences are issued for particular quotas in state-controlled Départements (e.g. France), one informant notes: 'from personal experience, I know that the information is not worth the paper it is written on; I know of hunters making up extra roe deer as they didn't want to shoot any more and they have financial penalties if they don't'.

3.2.8 Venison handling and marketing; trained hunter status

In most countries a separate section of legislation covers the handling and distribution of game meat – in most cases as an individual response to the European Union's Game Meat Regulations (EC 178/2002; EC 852/2004; EC 853/2004).

While ownership of the live animal and entitlement to kill game are variously defined, in most cases venison belongs to the hunter or landowner. Under European legislation, however, there is a requirement for countries within the EU that game meat is inspected by a competent person (someone regarded as 'of trained hunter status') before game may be passed on for human consumption. In addition there are commonly strict hygiene rules for handling of the carcase itself, which must be processed in approved premises with minimum specifications of lardering. There are, however, exemptions from these regulations for those handling only small quantities of game meat, or where venison is intended simply for domestic consumption by the hunter.

In addition, it is a mandatory requirement under EU legislation to retain records which allow full traceability of any game meat, from hunter and hunting ground, through to point of sale. In most cases it is also a legal requirement to attach an identifying tag to the carcase which accompanies that carcase through every stage of processing (Table 3.3).

3.2.9 Compensation for game damage and the provision of supplementary feed

Finally: In a number of countries, especially those where hunting clubs or hunting associations take on the lease of a defined management area (*revier*, or other), they are presumed to have taken on a legal responsibility for management of the area, with obligations to manage the animals in balance with the capacity of the environment.

In many instances they themselves are severally or collectively responsible for damage caused by deer or other ungulates to agriculture, forestry or other land-use interests and may be required to pay compensation. In other instances they are charged with the responsibility of 'management of a population in good condition' although in no case does there seem to be any specific or objective mechanism whereby this may be assessed or judged, and loss of a lease, or failure to have a lease renewed, is more commonly associated with failure to meet required or agreed quotas and cull targets, or complaints about excessive environmental impacts, than with actual assessment of population or individual condition.

Perhaps the only area in which specific action may be taken to try and address actual animal condition is in

- deliberate efforts made by managers to improve forage resources by habitat management or planting of game crops, or
- through provision of supplementary foodstuffs during periods of food shortage.

In many cases such manipulations are carried out at the instigation of individual or collective managers (individuals or hunting associations responsible for ungulate populations in a particular area), largely through self-interest, in that such measures are directed towards

- attempting to improve the condition or 'quality' of animals for harvest
- attempting to hold animals on their own ground (or own leased ground) and counteracting any potential drift of animals to neighbouring reviers, or hunting grounds, or
- as a simple diversionary tactic in attempting to sustain high densities on their own ground for hunting purposes, while reducing the damage caused to agriculture, forestry, etc.

The actual effectiveness and/or cost-effectiveness of such manipulations were considered by Putman and Staines (2003, 2004), together with a detailed consideration of the likely effects of (in particular) winter feeding on body condition and welfare. This analysis concluded that there is little objective evidence from studies throughout Europe and North America that winter feeding causes (at the population level) any increase in average body weight, any increase (at the population, rather than the individual, level) in antler size of males or fecundity of females, or any increase in rates of overwinter survival. By contrast, provision of concentrated or bulk foods in a limited number of feed sites, combined with a tendency for animals to become reliant on the artificial supplement in place of more extensive foraging, promotes aggregation of animals, leading to an increased probability of transmission of disease, increased competition for (usually) limited resources and may thus lead to actual loss of condition in less dominant males and especially among females (Putman and Staines, 2003, 2004).

Such conclusions are widely echoed by others (e.g. Arnold, 2002; Meile, 2006; Adamic and Jerina, 2010; Bartoš *et al.*, 2010; Findo and Skuban, 2010; Imesch-Bebié *et al.*, 2010). However, despite such argument, we should conclude this review by noting that winter supplementation remains compulsory in some countries and common practice in others. There is, however, no simple consensus and there are also countries in which such supplementation is explicitly forbidden by law (Table 3.4).

Table 3.4 *Provision of supplementary feed overwinter is in some countries obligatory, in other countries common but not mandatory and in yet other cases forbidden by law*

Feeding obligatory	Feeding common	Feeding sporadic	Feeding forbidden
Germany (part)	Germany (part)	Denmark (roe)	Netherlands
Austria	Hungary (esp. red)	Belgium (roe)	
Slovakia	Slovenia (esp. red)	Portugal	
Poland	Baltics	Spain	
Czech Republic	Switzerland	UK	
Croatia	Italy (wild boar)	Norway (to keep moose away from roads/railroads)	
	France (wild boar)		
	Finland (roe)		
	Norway (roe)		
	Sweden (roe)		

Compensation for damage

As noted above, where management of a game management district is leased or devolved in its entirety to a third party (or more commonly a hunting club or hunters association), this may also include an obligation to control damaging impacts and in some cases to provide compensation for damage to agriculture or forestry.

Thus compensation for damage to agriculture or private forestry is the legal responsibility of the hunter or hunters' association in the Baltic States, Belgium, Hungary, Poland and the Czech Republic. We might note, however, that in practice, financial compensation for the damage caused to forest and crop production is relatively low. According to the Report on Management in Forestry submitted by the Ministry of Agriculture of the Czech Republic in 2005, compensation reached more than €2 million in 2002, but only €1 million in 2003, while true damage levels in state forests are estimated to reach up to 1.5 billion euros every year (Bartoš *et al.*, 2010).

Compensation for game damage is also the responsibility of the appropriate hunting leaseholder in Switzerland, in relation to agricultural damage, but not damage caused in forestry. The government (whether at national or regional/provincial level) provides compensation to private landowners for crops and forest damage here, as also in Finland and Italy (but in this last case only for crops). Compensation for agricultural damage is also offered by the state in the Netherlands, although amounts concerned are generally very small (van Wieren and Groot Bruinderink, 2010).

3.3 Legislative control; an overview

It will be apparent throughout this that although there are variations between countries in, for example, permitted hunting seasons, whether or not bow hunting is permitted, whether or not dogs may be used to hunt or drive game, these differences are, in general, slight and there is for the most part remarkable consistency between countries in relation to many elements of the legal situation. In part this may reflect the response by individual countries to EU legislation (for example, in relation to handling of game meat, tagging of carcases, etc.), but many similarities pre-date such pan-European legislation.

There is considerable consistency, for example, in what firearms and ammunition may be used to take different species (muzzle energy and type of ammunition) – although only one country (Croatia) has in recent legislation taken the step of imposing a maximum distance in law over which animals may be shot. This would appear to be a useful innovation.

There is similarly general consistency in a requirement for formal training in order to be granted a licence to hunt, although the detail and requirements of that training do vary (and are at their most demanding amongst those countries with a more Germanic tradition and a well-established structure of hunting clubs or hunting associations).

Greatest differences relate to the fundamental status of game (whether considered *res communis* or *res nullius*) and thereafter the rights to kill game. This in itself has profound implications for subsidiary legislation and the degree to which the state may determine or dictate hunting quotas, as well as the administrative system through which hunting is licensed.

3.4 Administration of ungulate hunting

As noted above, different management systems in different countries largely reflect the legal status of the ownership of game, and cultural attitudes to hunting – which relate to the role that hunting plays within the wider culture of each country and the historical status of hunting in that country. A fuller treatment of these different cultural attitudes is presented in Chapter 1, but we may offer a convenient summary here. Thus:

- In some countries with a long tradition of game management, hunting is positively celebrated.
- In other countries, while hunters are perhaps in the minority, there is no widespread 'objection' to the idea amongst the general public.
- In other countries again, the idea of hunting (taking life for pleasure) is widely considered repugnant, and hunting is only accepted by the general public if it is formally justified as necessary to maintain animal populations in balance with their wider environment, and to fulfil other management objectives. In such cases hunting is usually 're-branded' as 'game management', and often viewed with some reservations by society in general.
- In some countries, such as the Netherlands, hunting is actually illegal. All animal species are fully protected by law and permission to kill them for management purposes needs to be specifically applied for in every individual instance by seeking specific exemption under the law.

Cultural attitudes (and political history) also affect what level of state intervention may be permitted in law, and what extent of intervention may indeed be acceptable to the people. And these two factors have a strong influence on administrative systems which are developed in different countries. It will become clear that systems vary enormously from almost total state control (the state defines objectives of management, state determines management plan and determines how many animals of each species shall be killed each year; state determines quotas and either leases quotas to hunting associations or sells individual licences to individual hunters for one animal of specified age and sex) to systems where management (or lack of it) is left entirely to the individual landowner.

In reviewing the situation in 30 different countries, it becomes quickly apparent just how much diversity there is in management objectives, management priorities, management practice – and level of state regulation. It would not be appropriate here to consider each individual system in detail, but much in the same way as in Chapter 1 where we reduced the diversity of cultural attitudes to hunting to four different 'models' or groups, we may also cluster the different administrative systems of different countries into a number of discrete models or broad categories.

Model A

- The state (either at national level, or where there is a strong provincial structure, at the level of semi-autonomous regions) sets clear objectives for management of game species.

- The appropriate Administrative Authority (national or regional authority, but essentially: the state) establishes within each region well-defined hunting units (or game management districts; GMDs) and determines for each a clear management plan.
- Within the context of its declared objectives and its assessment of current population trends, the administrative authority concerned sets harvest quotas for each GMD.
- In some cases management is directly by state-appointed hunters, although more generally this applies only to a proportion of the total land area (such as in national parks); more usually, the national or regional authority issues (sells) hunting licences (to individuals or groups) with one hunting licence entitling the hunter to shoot one animal of predetermined species age and sex.

Such 'state' control of hunting is typified by (for example): Finland: (15 defined GMDs, each with its own, state-determined management plan); Denmark; parts of Switzerland: (cantons develop annual management plans and quotas; thereafter issue individual hunting licences each with a defined quota); France.

Model B

- The state (again, either at national or regional level) defines hunting units (GMDs) and determines a clear management plan.
- The state then
- either manages directly as in Model A (and, where this occurs at all, it usually applies only to part of the land area, or a proportion of GMDs) or
- devolves management of individual GMDs to individual landowners, groups of landowners, an approved hunting association or other leaseholder. After such allocation, the licence issued to the leaseholder is a 'global' licence up to the permitted or required quota established in the state-produced management plan.

This model effectively describes the situation in the Baltics, Romania, Slovenia.

Model C

- The state, at national or (now more commonly) regional level determines GMDs but management is subsequently devolved to landowners, hunting association or other leaseholder.
- The manager then produces a management plan (and suggests annual quotas), *but production of such a plan is mandatory and the plan itself*

(and annual quota) must be approved by state administration. Once the submitted plan is approved, a 'global' licence is issued up to permitted quota. State regulation is still very apparent and in some cases devolution of management responsibility to the leaseholder is dependent on (compulsory) appointment of a professional game warden or game manager within each hunting district.

This summarises the situation in Lithuania, Hungary, Poland, Austria and the Czech Republic. In both Austria and the Czech Republic the GMD must employ a professional gamekeeper.

Model D

- The integration of individual landholdings into cooperatively managed hunting districts or game management areas is initiated by groups of landowners/ hunters associations (or equivalent) who propose GMDs for approval.
- Hunting, however, may only be carried out within approved GMDs, and such GMDs remain a mandatory requirement. Once a GMD has been approved, the proposers must then submit a (mandatory) management plan for approval by administrative authorities and must submit proposals for harvest quotas (usually annually).
- Proposals are scrutinised by the responsible state authority (usually Forest District staff, or regional Wildlife Boards) who may alter the proposals as they consider necessary; approval of the (annual) plan then implies approval of a 'global' licence up to the agreed quota (and may imply penalties if that quota is not met).
- Once again, in some countries it is mandatory for the 'syndicate' managing any given GMD to appoint a professional gamekeeper or game warden, who has responsibility for coordinating hunts and for collating harvest statistics/cull records and reporting them to the responsible authorities.

This model describes the situation in Germany, Slovakia, Belgium, Spain, Italy.

Model E

Throughout all this we are seeing a relaxation of state involvement or state control of hunting activities. At the extreme, there is virtually no state regulation or intervention.

In this 'cluster' there is no mandatory requirement for organisation of landholdings into GMDs or requirement for a state-approved management plan. Where unified management districts are created, these are maintained on a voluntary basis (or not at all). Game licences are issued to individual hunters (simply as licence to hunt) but no quota limit is set.

This is the model which applies largely to the UK (Scotland, England and Wales, Northern Ireland) and the Republic of Ireland. To the best of my understanding, outside the British Isles, only Sweden operates a similar system (and there only in relation to red deer, roe deer and fallow; far tighter control is exercised in relation to hunting of moose).

Mixed models

In conclusion we should note that a number of countries operate what we might best describe as 'mixed models' – with part of the land area administered by one method, part by another, even within regions.

We have already noted that management of ungulates in Switzerland embraces alternative systems in different, autonomous, cantons. Thus in some cantons the authorities operate a system of '*Patentjagd*' (where management responsibilities are retained entirely by the state and each hunter buys individual licences from the cantonal hunting department, and the licence is valid throughout the canton), whereas other cantons operate a system of '*Revierjagd*', where, while the state maintains 'ownership' of the game, management is delegated by the state to hunters associations or hunters clubs, who take over the management of game in a given area (see Section 3.1.2). Switzerland, in effect, thus operates a mixture between Models A and B – although the system operated within any canton is consistent.

In other countries, elements of two 'models' may operate side by side even within the same administrative area (though obviously not within any single GMD). In Norway, for example, the default system is that municipalities develop management plans for their area and issue licences for defined quota to individuals or hunter groups; but, in practice, increasing emphasis is put upon encouraging landowners to develop local population management plans which may be submitted for approval to the municipality. If these plans are adopted/ approved, the municipality then endorses the plan and confirms quota for the next 5-year period without further intervention. In this case (in contrast to the situation in Switzerland) a combination of Models A and C may run side by side even within the same regional area. In reality, this is very possibly simply a transitional phase as Norway moves increasingly towards widespread adoption of Model C.

In a similar way Portugal also operates, throughout, an integrated mixture of Models B and C.

Table 3.5 *The key features of different management systems*
The degree of state intervention in management and regulation decline as one moves from left to right across the table

Model A/B

	Imposed/Determined by State [National or Regional Authority]	Proposed by Landowners Association/ Hunters' Association but must be approved by State Authorities	Proposed by Landowners Association/ Hunters' Association or equivalent Voluntary
Game Management District/Group	X		
Management Objectives	X		
Management Plan	X		
Quota/Cull Targets	X		
	Cull carried out by State hunters	Individual licences allocated (per animal)	Global Quota allocated to leaseholders
Global Quota/ Individual licences	A	A	B

e.g. A: Finland, Denmark, Switzerland, France
B: Latvia, Romania; ?Slovenia?

Model C

	Imposed/Determined by State [National or Regional Authority]	Proposed by Landowners Association/ Hunters' Association but must be approved by State Authorities	Proposed by Landowners Association/ Hunters' Association or equivalent Voluntary
Game Management District/Group	X		
Management Objectives	X	X	
Management Plan		X	
Quota/Cull Targets		X	
	Cull carried out by State hunters	Individual licences allocated (per animal)	Global Quota allocated to leaseholders
Global Quota/ Individual licences			X

e.g. Lithuania, Hungary, Poland, Austria*, Czech Republic*
[* GMD must employ a professional gamekeeper]

Model D

	Imposed/Determined by State [National or Regional Authority]	Proposed by Landowners Association/ Hunters' Association but must be approved by State Authorities	Proposed by Landowners Association/ Hunters' Association or equivalent Voluntary
Game Management District/Group		X	
Management Objectives		X	
Management Plan		X	
Quota/Cull Targets		X	
	Cull carried out by State hunters	Individual licences allocated (per animal)	Global Quota allocated to leaseholders
Global Quota/ Individual licences			X

e.g. Germany*, Slovakia*, Belgium, Spain, Italy
[* GMD must employ a professional gamekeeper]

Model E

	Imposed/Determined by State [National or Regional Authority]	Proposed by Landowners Association/ Hunters' Association but must be approved by State Authorities	Proposed by Landowners Association/ Hunters' Association or equivalent Voluntary
Game Management District/Group			X
Management Objectives			X
Management Plan			X
Quota/Cull Targets			X
	Cull carried out by State hunters	Individual licences allocated (per animal)	Global Quota allocated to leaseholders
Global Quota/ Individual licences			

e.g. UK and Sweden (for all species except moose)

Table 3.5 presents an oversimplified summary of these different 'models', based on a standard template and with a shift from strong state control to complete devolution to the individual landowner represented by a shift across the table from left to right.

3.5 In conclusion

In reviewing these different administrative systems I have myself been impressed by the diversity of attitude and management practice in the various different countries. We tend to become accustomed to what is common practice in our own country and assume that practice is similar elsewhere. Nothing could be further from the truth! Indeed we should expect that different countries will differ in both management objectives and management practice.

- Different countries support different species of ungulates and different species mixtures.
- Even with regard to the same species, management objectives may differ markedly in different places or in different contexts (whether directed towards a need for control of populations and their impacts, management for exploitation, or a need for active conservation). Local circumstances may also affect what management options are actually available, or may affect the utility of any given method.
- Superimposed on such variation is an equal variation in attitudes and cultural approaches to hunting and game management, cultural attitudes to the degree of 'regulation' (intervention) which may be tolerated by hunters themselves. All of these differences facilitate or constrain management in particular contexts and ensure that there is no simple single solution to any given problem.

As a result, it is no surprise that there are marked differences in the way in which management of game animals (or hunting) is administered and regulated in different countries, with the degree of state control/intervention ranging from total to virtually none at all. There is enormous variation in the mechanisms in place which determine the number of animals which may be shot (or indeed in many cases, must be shot) and whether this harvest quota, or required cull, is determined by state authorities or by the hunters themselves. Because of this variation in objective and attitude, we should not expect to find a single 'perfect' solution. It is, however, very instructive to see what does go on in other countries if only so that we may review our own management systems in some wider context.

References

Adamic, M. and Jerina, K. (2010) Ungulates and their management in Slovenia. In M. Apollonio, R. Andersen and R. Putman (eds.) *European Ungulates and their Management in the 21st Century*. Cambridge, UK: Cambridge University Press, pp. 507–526.

Andersen, R., Lund, E., Solberg, E. and Sæther B.-E. (2010) Ungulates and their management in Norway. In M. Apollonio, R. Andersen and R. Putman (eds.) *European Ungulates and their Management in the 21st Century*. Cambridge, UK: Cambridge University Press, pp. 14–36.

Apollonio, M., Andersen, R., and Putman, R.J. (eds.) (2010) *European Ungulates and their Management in the 21st Century*, Cambridge, UK: Cambridge University Press, 604 pp.

Arnold, W. (2002) Der verborgene Winterschlaf des Rotwildes. *Der Anblick* (Graz) **2**, 28–33.

Bartoš, L., Kotrba, R. and Pintíř, J. (2010) Ungulates and their management in the Czech Republic. In M. Apollonio, R. Andersen and R. Putman (eds.) *European Ungulates and their Management in the 21st Century*. Cambridge, UK: Cambridge University Press, pp. 243–261.

Findo, S. and Skuban, M. (2010) Ungulates and their management in Slovakia. In M. Apollonio, R. Andersen and R. Putman (eds.) *European Ungulates and their Management in the 21st Century*. Cambridge, UK: Cambridge University Press, pp. 262–290.

Imesch-Bebié, N., Gander, H. and Schnidrig-Petrig, R. (2010) Ungulates and their management in Switzerland. In M. Apollonio, R. Andersen and R. Putman (eds.) *European Ungulates and their Management in the 21st Century*. Cambridge, uk: Cambridge University Press, pp. 357–391.

Liberg, O., Bergström, R., Kindberg, J. and von Essen, H. (2010) Ungulates and their management in Sweden. In M. Apollonio, R. Andersen and R. Putman (eds.) *European Ungulates and their Management in the 21st Century*. Cambridge, UK: Cambridge University Press, pp. 37–70.

Meile, P. (2006) Wildfütterung in Theorie und Praxis. *Schriftenreihe Wildbiologie* **4**/33, Wildtier Schweiz, 16 pp.

Putman, R.J. (2008) *A Review of Different Options Available for Collecting and Reporting Cull Data*. Contract report for the Deer Commission for Scotland, Inverness.

Putman, R.J. and Staines, B.W. (2003) *Supplementary Feeding of Deer in Scotland: a review of the extent and geographical patterns of supplementary feeding of wild deer in Scotland; reasons for feeding and an analysis of the balance of advantage/disadvantage*. Report to the Deer Commission for Scotland, Inverness.

Putman, R.J. and Staines, B.W. (2004) Supplementary winter feeding of wild red deer *Cervus elaphus* in Europe and North America: justifications, feeding practice and effectiveness. *Mammal Review* **34**, 285–306.

van Wieren, S.E. and Groot Bruinderink, G.W.T.A. (2010) Ungulates and their management in the Netherlands. In M. Apollonio, R. Andersen and R. Putman (eds.) *European Ungulates and their Management in the 21st Century*. Cambridge, UK: Cambridge University Press, pp. 165–183.

4

Hunting seasons in relation to biological breeding seasons and the implications for the control or regulation of ungulate populations

MARCO APOLLONIO, RORY PUTMAN,
STEFANO GRIGNOLIO AND LUDĚK BARTOŠ

4.1 Introduction

Almost all European countries operate with a principle of restricted hunting periods for some or all species (with the exception of Portugal, where technically the season lasts from 1 June in any year to 31 May of the following year, although most hunting activity is carried out between September and February). There is, however, an enormous diversity in length (and actual time of year) of the permitted season in different countries (even in adjacent countries: e.g. seasons in the three adjacent countries of the Baltics: Latvia, Lithuania and Estonia) and seasons also vary – often quite markedly – between different regions or provinces of one country (e.g. Italy, Austria, Germany).

It is further apparent that such seasons may also show little relation to actual biological breeding seasons (rut, parturition, period of dependency of young) and such mismatch between hunting and biological seasons may have serious consequences. This chapter explores the wide variation in hunting season in different European countries and the implications of the mismatch with biological seasons for welfare, social dynamics – and the ability (or failure) of hunters to regulate ungulate populations.

There are at least three critical times of year in relation to breeding seasons of ungulates:

- the period of the rut (i.e. period between the first and the last copulation in the observed population)
- pre-parturition (i.e. period between late development of embryos and parturition; we take this period as that period between the time when the foetus may reach half of the birth weight and actual birth)

Ungulate Management in Europe: Problems and Practices, eds. Rory Putman, Marco Apollonio and Reidar Andersen. Published by Cambridge University Press.

- the period following parturition when young animals may be nutritionally or socially dependent on the mother.

In this chapter we will first examine the potential social, ecological and evolutionary consequences of hunting within these biologically critical seasons before considering the actual timing of permitted hunting seasons in different European countries and how far these may 'respect' these potential problems. In addition we will consider the potential consequences of a 'mismatch' between hunting seasons and biological seasons (as well as the problem of hunting seasons which are over-conservative) on our ability actually to manage populations of overabundant species, before exploring possible solutions to these various problems.

4.2 Potential problems of hunting during different biological seasons

4.2.1 Potential problems of hunting during the rut – the disruption of reproductive aggregations

While the reproductive systems of European ungulates vary between species and may vary between populations even of the same species, all inevitably involve some degree of aggregation and some mechanism to facilitate male–male competition, female–male meeting and/or female choice.

There is clearly considerable flexibility in the form of reproductive system which may be adopted, between and even within a single species, with precise 'choice' of reproductive strategy dependent on a number of factors, both social (Clutton-Brock *et al.*, 1988a, 1993; Bartoš *et al.*, 1998; Willisch and Neuhaus, 2009) and ecological (Langbein and Thirgood, 1989; Carranza *et al.*, 1995; Thirgood *et al.*, 1999). Thus for example, populations of any given species may be truly territorial (as roe deer; Liberg *et al.*, 1998) or may establish reproductive territories (where males defend a reproductive arena and call to attract females to it, as for example in most populations of fallow deer: Chapman and Chapman, 1975; Langbein *et al.*, 2008), or alternatively simply defend an area of rich resources often visited by females for foraging (red deer: Carranza, 1995; fallow deer: Clutton-Brock *et al.*, 1988b; alpine chamois: von Hardenberg *et al.*, 2000). Males in some species or some populations may defend groups of females, rather than a fixed spatial territory, establishing and defending harems of females (Clutton-Brock *et al.*, 1988a, 1993) or alternatively may simply wander freely in search of oestrous females with which to mate, as in the case of ibex (Willisch and Neuhaus, 2009), mouflon (Türcke and Schmincke, 1965; Briedermann, 1992) and wild boar (Heck and Raschke, 1980; Briedermann, 1986) and as is also reported for some populations of fallow or sika deer (Putman, 1993; Thirgood *et al.*, 1999). In the extreme males may congregate together on a communal display ground or lek to compete directly for access to females (Schaal and Bradbury, 1987; Langbein and Thirgood, 1989; Apollonio *et al.*, 1992; Bartoš *et al.*, 1992, 1998, 2003; reviewed by Thirgood *et al.*, 1999). But whatever the strategy adopted, that strategy delivers the potential for intra-male competition and offers opportunities for females to select mating partners and optimise fitness.

Shooting an individual at such a time may directly cause a temporary, or perhaps more permanent, disruption of the breeding group formed, dispersing animals (which may in consequence join other groups), or in the extreme removing completely the focal male. A female may even abandon mating altogether if her preferred mate is not accessible (Morrison, 1960). A further problem which may arise if the disruptions are frequent is that the area used for the mating aggregation is abandoned by the population and new one/s established. Breeding grounds used by most ungulate species tend to be traditional, with both males and females undertaking deliberate movements to reach them at the beginning of the rut, or sometimes well beforehand. The abandonment of a traditional site may be a gradual and incomplete process causing substantial reduction in the amount of breeding opportunities of single individuals.

Not only spatial but also temporal shifts in the rutting activity can be induced by diurnal hunting: red deer, fallow deer and sika deer males (and therefore females) are active in reproduction all day and night long in protected, undisturbed areas, where their rutting calls are commonly heard in daytime as well as through the night; where they are heavily hunted, reproductive activity is virtually entirely restricted to the night (or at dawn and dusk).

A number of immediate and ultimate consequences for reproduction might arise as the result of various types and degrees of disruption. In species which display a high degree of polygyny and where, therefore, sexual selection can be intense, the amount of time at disposal for mate choice and actual mating can be critical, particularly when females usually have an oestrous period extending only to some 36–48 hours. If hunting disturbance causes mating groups to disband or in some other way reduces this time (for example limiting activity almost exclusively to the night-time) a female can lose the precise fertile time window of her cycle without being mated. This may have different consequences in relation to the species affected and the amount of disturbance.

Failure to conceive

The most serious consequence could even be a failure to conceive during that season. In species which are monoestrous, such as the roe, a proportion of females might theoretically fail to breed at all in any given year simply because of disturbance. However, even amongst polyoestrous species of European ungulates, all have a comparatively short period of repeated oestrus (only wild boar sows show a wide fertile period, being able to conceive throughout the year). If intense hunting reduces the chances of a mating during this time, a proportion of females, even in polyoestrous species, may fail to breed at all.

Disruption of breeding may cause reduction in synchronisation of births

In polyoestrous species, even if all females do eventually breed successfully, continued disturbance may mean the rut is extended/prolonged. This will result in lack of close synchrony of parturition, with calves born over a protracted period. This may increase losses due to increased availability of vulnerable fawns to predators, since these will be available over a longer time period (Linnell *et al.*, 1995; Aanes and Andersen, 1996; Kjellander and Nordstrom, 2003; Jarnemo, 2004; Panzacchi *et al.*, 2009) and/or because late born calves, entering the winter in poor condition because they have not had sufficient time to build up necessary body condition, are thus more susceptible to overwinter mortality (Festa-Bianchet, 1988a; Festa-Bianchet *et al.*, 2000; Côtè and Festa-Bianchet, 2001; Gendreau *et al.*, 2005; Pettorelli *et al.*, 2007).

Disturbance of breeding groups may increase the chance of mating by inferior males

The competition among males that usually takes place at breeding grounds can be important in enabling females to select males with the best genes and to avoid inferior competitors, so any disruption of breeding aggregations is a potential treat to the fitness of the population as a whole.

A possible consequence of any disruption may be an increase of access to females by inferior males, if as a result of disturbance, females are for increased periods out of the control of dominant male/s. During the time that females spend away from dominant male/s they may be mated by inferior males that may transfer genes with lower value and therefore contribute to the production of newborns with a lower fitness. In the extreme, if the dominant male is actually killed, then clearly the probability is much increased that females may be forced to breed with inferior males (in order to breed at all while they remain in oestrus).

In a number of ungulate mating systems one of the major determinants of male mating success is the length of tenure which may be maintained by an individual male of harem or territory or any single defended female. It is therefore not surprising that these highly successful males have the highest chance of being shot during the rut, as they spend more time in rutting activities that not only make them more predictable in terms of use of space, but also more conspicuous (by, for example, roaring, fighting or courting). The consequences of shooting prime males in ungulate populations – with the resultant increase in access to females of less fit males – are widely known (Coltman *et al.*, 2003) and range from an overall reduction of population viability (Mysterud *et al.*, 2005) to a decrease in weight and/or horn or antler size in males (Singer and Zeigenfuss, 2002; Coltman *et al.*, 2003).

It is important to emphasise the strong counter-evolutionary effect of this kind of 'selective' hunting, in that the culling of the fittest is basically the opposite of what might be desirable by sound management. Fever for large trophies often leads to reduction of the proportion of the fully mature males in the population (e.g. in various parts of Central Europe). Absence of prime males may result in lack of fully developed physical traits such as body and antler size, etc., utilised in mate selection. As shown previously, male phenotypic quality affects mate selection (for example, red deer stags with large antlers are preferred for mating; Bartoš and Bahbouh, 2006) and males with large antlers had increased lifetime breeding success in an unhunted population on the Isle of Rum (Kruuk *et al.*, 2002) and also other breeding characteristics, including offspring sex ratio (body size: Røed *et al.*, 2007).

Inappropriate harvesting might thus induce an undesirable evolutionary response when the target characteristic is heritable, while in addition there might well be unexpected effects on genetically correlated traits. Wildlife managers must pay attention in order to plan hunting seasons and establish appropriate hunting practice in order to reduce the genetic effects and the evolutionary implications (Harris *et al.*, 2002; Festa-Bianchet, 2003).

4.2.2 Culling during the period of late pregnancy

We believe that the killing of females in the last stages of pregnancy (i.e. females that have already successfully borne almost all costs implied in the successful development of an embryo) is acceptable in management only if the declared intent of that management is to stabilize populations with a high growth rate or effect a decrease in population size. In other contexts we believe that this practice is undesirable.

Possible damaging effects of harassment

Some types of hunting practices, especially those which employ the use of hunting dogs, but also noisy drive hunts or the use of vehicles, can cause considerable distress in the ungulates that are pursued. Such hunting methods may cause distress not only to the individuals that are actively being hunted but also for other individuals that may be disturbed by the drives or incidentally chased by dogs.

In the extreme such pursuit may lead to abortion of the foetus, but even without that, this stress can induce a number of physiological and behavioural modifications. With respect to pregnant females, high levels of stress (for example long and repeated flights) can cause welfare implications, such as abortion with possible mother's death. While it is very difficult to collect data about effects of stress on reproduction and welfare in free-ranging wildlife, several laboratory studies have produced data linking prenatal stress with disturbances in offspring development and behaviour (see, for example, Paarlberg *et al.*, 1995). Moreover, stress-induced variations, for example in maternal care, can serve as the basis for a non-genomic behavioural transmission of individual differences in stress reactivity across generations (Francis *et al.*, 1999). There are unfortunately no quantitative studies about the consequential impact on overall population reproductive success, but, although it is difficult to quantify the effects, if hunting pressure is intense both in terms of hunting days and in terms of numbers of dogs and beaters this problem may be not irrelevant.

Even if we are unable to quantify such effects, it would seem probable that such effects may well have some impact on overall population dynamics of hunted species. And, as noted, several hunting practices may cause distress not only to the individuals that are actively being hunted but also for other individuals, or even other species that may be disturbed by the drives or incidentally followed by dogs.

Ethical implications

Whatever may be the impact of hunting during late pregnancy on overall population dynamics, there are also arguments against such practice simply from an ethical standpoint. Even in situations where, from a strictly technical point of view, hunting during the period of late pregnancy might be justified (in situations, for example, where there is seen to be a need to effect a significant reduction in a given ungulate population), pursuit or shooting of heavily pregnant females does raise a number of issues of ethics and most importantly is often considered unacceptable by the more general public. Periodic outcry in the newspapers on the 'infanticide' during wild boar hunting in January and on similar occasions bears testimony to a rather general attitude of the public against what is perceived to be 'cruelty' in hunting. Such negative perceptions often become generalised to hunting in general and may help to develop negative views of hunting and management activities overall.

4.2.3 Culling of mature females when young are still dependent on their mothers

Where the hunting seasons which apply in some countries are such that culling of adult females may be permitted during the period of dependency of offspring, there are once again clear implications both from a population dynamics point of view and from purely welfare considerations.

In this context we should recognise a distinction between the period for which the young may be nutritionally dependent on the mother and the period during which the young are socially dependent on the mother (which in social species may be much longer than the period for which they are nutritionally dependent).

Neither of these has, to our knowledge, been adequately defined for any species. Even in terms of nutritional dependency, the recorded period of lactation is not necessarily a particularly good indicator of dependency, in that although females may continue to lactate, and juveniles may continue to take opportunistic advantage of such lactation, for considerable periods, this does not necessarily imply a requirement for that nutritional subsidy. Fallow does, for example, may still be lactating some 7 months after parturition (e.g. Langbein, 1991), but this does not imply that fawns are actually dependent on that milk, or would suffer loss of condition were it not available.

To generalize, we would suggest that nutritional dependence ends when physical growth of the offspring is no longer dependent on mother's energy budget. While it is difficult to identify an actual time period for that nutritional dependency, the effects of enforced early weaning have been widely studied in laboratory and domestic animals. Disruption of mother–infant bonding can induce physical and behavioural problems, including increased neuroendocrine stress responses, augmentation of fear and aggression, and reduced maternal behaviour (e.g. Kikusui *et al.*, 2007, 2008), as well as morphological changes such as myelin formation, dendrite length and spine density in the brain (Ferdman *et al.*, 2007; Kikusui *et al.*, 2007; Nakamura *et al.*, 2008).

It is clear that the significance of lactation in ungulates changes as the young grow and in the later stages of lactation may become more significant in social bonding than in actual nutritional terms. However, even within this context, it is again extremely difficult to determine what may be the actual length of this period of social dependency between offspring and their mother. The time that mother and young stay together cannot be used as any clear measure of such reliance since this may change with environmental conditions and habitat use – and differ from year to year. For example, after hotter summers the mother–young bond of alpine ibex endures significantly longer (Grignolio *et al.*, 2003). Further, the period for which mother and young typically remain together does not necessarily mean that the young are dependent on the mother for that entire period, or that juvenile survival or social integration is actually compromised if the mother is killed before that time.

That said, the implication of killing the mother before the young are fully independent (whether socially or nutritionally) has clear implications from pure welfare considerations, and also has implications for population dynamics.

Culling of the mother could result in death, or loss of fitness of dependent young

Even if there are no objective data clearly establishing the time at which juveniles are no longer nutritionally dependent on the dam's milk (above) we may safely assume that at least three months are necessary for most species of European ungulates if the young are to survive at all. On that basis, any shooting of lactating females before this time carries with it a considerable risk of the death of the dependent young by starvation, unless the juvenile is already accompanying the mother and is shot with it. Best practice would thus dictate that if a hunter is to shoot a female during the period of lactation, he/she must ensure that they also kill any accompanying calf.

Problems arise, however, where the hunter may not be aware that there is a dependent juvenile, because it is not actually accompanying the mother. Immediately after birth, neonates of almost any species may not be accompanying the mother; in addition, in those species (e.g. roe deer, fallow deer) whose anti-predator strategy makes the offspring 'hiders' not 'followers' (Lent, 1974), this period where dependent young do not accompany the mother may be considerably extended. In such situations culling of adult females will commonly lead to orphaning of dependent young, because the hunter is unaware of the existence of those offspring.

Such problems are of course most likely to be most acute early in the calving season simply because the hunter may not even be aware that there is a calf at all. But even later in the season, where offspring are accompanying the mother, a strategy of shooting both mother and calf, however appropriate in theory, may prove hard to achieve in practice. Problems arise with such a strategy simply for technical reasons (because of the need to shoot two animals in quick succession). Because calves usually linger for a few moments in confusion after the death of the mother, many hunters advocate shooting the mother first and then shooting the calf while it is still disoriented. But from a purely welfare point of view (to avoid any risk of leaving an orphaned calf) it is actually more appropriate to shoot the calf first – and risk not being able to shoot the mother too. As an additional problem, in many social species (for example chamois or roe deer) female groups often consist of a number of different females and their young. In this case it is not always possible to determine accurately the mother–young pair. (Winter groups of roe deer, for example, are usually based on a family unit, i.e. a doe and her offspring. Nevertheless, winter groups may fuse or merge and the content of such a group is typically rather unstable. In Kalø, Denmark, Strandgaard (1972) monitored content of winter grouping in a population with a high proportion of marked deer. Group size remained more or less stable with eight members present on average. Nonetheless, 21 different individuals were alternating in these groupings. For example, one doe was seen over the period of a few days two times with her three fawns, five times with only two of them, and three times without them.)

In either case, whether due to lack of awareness that there are dependent offspring, or lack of ability to shoot both mother and accompanying calf, hunting during the period of dependency may lead to premature orphaning of juveniles. If nutritionally deprived, the young animal may die as a consequence, or may be half-starved so that it takes longer to reach good breeding condition (or fails altogether to reach breeding condition). In ungulates, juvenile body mass can be related to maternal care, and body weight is an important factor affecting offspring survival (Clutton-Brock *et al.*, 1985, 1987a, 1987b; Bender *et al.*, 2007, 2008; Carrión *et al.*, 2008; Feder *et al.*, 2008; Stopher *et al.*, 2008). For example, in mountain goats survival to one year of age is greater for heavier female kids than for light ones (Côtè and Festa-Bianchet, 2001). Even if the animal survives, slower growth rate may imply a significant delay in reaching the mature body weight and perhaps the threshold for reaching puberty (see, for example, Hamilton and Blaxter, 1980; Albon *et al.*, 1986); we should note, however, that in bighorn sheep, female orphans and non-orphans had the same weight as yearlings and the same probability of producing their first lamb at two years of age (Festa-Bianchet *et al.*, 1994).

Finally, in polygynous species with highly skewed probability of breeding among males, a reduced adult body weight may result in orphaned males failing to grow to a mature body mass where they are able to secure any mating opportunities at all. Even though Festa-Bianchet *et al.* showed that *female* orphans among bighorn sheep reached the same weight as yearlings as did non-orphans and the same probability of producing their first lamb at two years of age, *male* orphans were smaller as yearlings compared with non-orphans and they were unable to compensate for this early weight difference in later life (Festa-Bianchet *et al.*, 1994). In all cases, early deprivation leads in effect to a reduction of lifetime reproductive success.

Even when the period of nutritional dependence is over, the bond between mother and young in ungulates still has an important social meaning. Often these species are social and females can play a significant role in obtaining access to food (e.g. socially dominant mothers can favour optimal feeding of their young; Veiberg *et al.*, 2004), in teaching population traditions (e.g. migratory route from winter to summer areas: Festa-Bianchet, 1988b; Nicholson *et al.*, 1997; Lamberti *et al.*, 2004) or in proper development of anti-predator behaviour (Childress and Lung, 2003; Li *et al.*, 2009; Pipia *et al.*, 2009). All these various social aspects related to social competence and proper exploitation of environmental resources, as well as the avoidance of potential threats, can be lost with the premature loss of the mother, producing individuals with limited chances of survival and reproductive success with obvious limitations to population recruitment.

In addition, since many animals 'acquire' some of their social status within the group as a consequence of mother's status, a young animal whose mother has been killed before it is socially independent may also suffer from being rather low in the dominance ranking. In red deer calves social rank was related to both body weight and mother rank (Veiberg *et al.*, 2004). Loss of the mother might thus result in an important decrease of social rank. Considering that social dominance is a fundamental aspect of male evolutionary ecology in many polygynous mammals, with lifetime reproductive success strongly related to dominance rank, an artificial modification of social rank is likely to result, at least for males, in a significant alteration in individual life history and breeding success.

4.3 European hunting seasons in relation to biological seasons

These various considerations suggest that there may be significant biological (as well as ethical) issues associated with culling animals during the rut, hunting during late pregnancy, or killing of adult females during the post-parturition period. But in many European countries prescribed hunting seasons do permit hunting at these times. To illustrate we have chosen here to review the open seasons in different countries for red deer, roe deer, moose, chamois and wild boar and explore the possible implications. Tables 4.1–4.5 offer a summary of current seasons operative in different European

Table 4.1 *Open seasons for red deer*

Country	Males	Females
Austria	1.05–31.01	1.05–31.01
Belgium	21.09–31.12 (21.09–30.09 large males only)	1.10–31.12
Croatia	16.08–14.01	1.09–14.01
Czech Republic	1.08–15.01	1.08–15.01
Denmark	1.09–31.01	1.10–31.01
England-Wales-NI	1.08–30.4	1.11–31.03
Estonia	1.09–31.01	1.10–30.11
France	23.08–28.02 all practices; 1.03–31.03 coursing	23.08–28.02 all practices; 1.03–31.03 coursing
Germany (most states)	Adults 1.08–31.01 (or 15.01); subadults 1.06–31.01	Adults 1.08–31.01 (earliest 16.06); subadults 1.06–31.01
Hungary	1.09–31.10 (prime age) or 31.01	1.09–31.01 (old) or 28.02 (young)
Ireland (Republic)	1.09–28.02	1.11–31.01 (28.02 in some counties)
Italy Alps	15.09–31.12 (stopped during rutting period)	15.09–31.12
Italy Apennines	1.08–30.09	1.08–30.09 and 1.02–15.03 or 25.03
Latvia	1.09–31.01	15.08–31.12
Lithuania	15.08–15.10	1.10–31.12
Netherlands	1.08–15.02	1.08–15.02
Norway	10.10–11.11	10.10–11.11
Poland	21.08–28.02	1.10–15.01
Portugal	1.06–31.05	
Romania	10.09–15.12 (prime) or 1.09–15.12	1.09–15.02
Scotland	1.07–20.10	21.10–15.02
Slovakia	1.08–31.12	1.08–31.12
Slovenia	16.08–31.12	1.09–31.12
Spain	September to mid February	
Switzerland	1.08–31.12	1.08–31.12

countries/states; these are illustrative only and should not be taken as definitive, since in some cases seasons vary between provinces within a given country (e.g. Italy, Austria, Germany), or may vary with age-class of animal (for example, distinct seasons in Wallonia, Hungary or Romania for prime age stags and 'poor'/cull stags). Further, seasons are shown here only for adult males and females and calves of the year; in some countries (e.g. Germany, Poland, Slovenia, Estonia), there are distinct (and different) seasons specifically for juveniles/yearlings of both sexes.

Table 4.2 *Open seasons for roe deer*

Country	Males	Females
Austria	1.05–31.12	1.05–31.12
Belgium Wallonia	1.05–15.05 and 1.08–30.11	1.10–30.11
Belgium Flanders	15.05–15.09	15.01–15.03
Croatia	1.05–30.09	1.09–31.01
Czech Republic	16.05–30.09	1.09–31.12
Denmark	16.05–15.07 and 1.10–15.01	1.10–15.01
England-Wales-NI	1.04–31.10	1.11–31.03
Estonia	1.06–30.09	1.09–30.11
Finland	1.09–31.01 and 16.05–5.06	1.09–31.01
France	15.05–31.08 stalking; 1.09–28.02 driving etc.; 1.03–31.03 coursing	1.09–28.02 driving etc.; 1.03–31.03 coursing
Germany (most states)	1.05–15.10	adults 1.09–31.01 subadults 1.05–31.01
Hungary	15.04–30.09	1.10–28.02
Italy Alps	1.09–7.12	1.09–7.12 and 1.02–15.03
Italy Apennines	1.08 or 15.08–30.09	1.08 or 15.08–30.09; 1.01–15.03
Latvia	1.06–30.11	15.08–30.11
Lithuania	1.06–1.11	1.10–31.12
Netherlands	1.05–15.03	1.01–15.03
Norway	10.08–23.12	25.09–23.12
Poland	11.05–30.09	1.10–15.01
Portugal	01.06–31.05	
Romania	15.05–15.10	1.09–15.02
Scotland	1.04–20.10	21.10–31.03
Slovakia	16.05–30.09	1.09–30.11
Slovenia	1.05–31.10	1.09–31.12
Spain	Mid April–31.07	Mid April–31.07
Switzerland	1.05–31.01	1.05–31.01

4.3.1 Culling during the rut

Considering these five more numerous ungulate species, it is clear that legislation does not in general take any real account of possible problems which might arise from hunting during the period of the rut. In fact, the hunting season overlaps the rutting period of moose, chamois and wild boar in all countries. For roe deer and red deer, hunting during the rut is permitted in more than 80% of the countries. In practice, the period of the rut is often actively exploited by hunters in order to increase the ease of gaining access to ungulates, especially males, which will at this time be showing reduced vigilance. In some countries there are even specific

Table 4.3 *Open seasons for wild boar*

Country	Boars	Sows	Subadults
Austria	All year	All year (except if with piglets)	All year
Belgium Flanders	1.10–31.12	1.10–31.12	1.10–31.12
Belgium Wallonia	1.01–31.12/ 01.10–31.12	1.01–31.12/ 01.10–31.12	1.01–31.12/ 01.10–31.12
Croatia	All year	1.07–31.01	All year
Czech Republic	1.08–31.12	1.08–31.12	All year
Denmark	1.10–31.01	1.10–31.01	1.10–31.01
Estonia	All year	All year	All year
Finland	1.06–29.02	1.06–29.02 (except if with piglets)	
France	15.04–14.08 stalking; 15.08–28.02 driving etc.; 1.03–31.03 coursing	15.04–14.08 stalking; 15.08–28.02 driving etc.; 1.03–31.03 coursing	15.04–14.08 stalking; 15.08–28.02 driving etc.; 1.03–31.03 coursing
Germany	15.06–31.01	15.06–31.01	15.06–31.01
Great Britain	Not hunted	Not hunted	Not hunted
Greece	15.09–20.01	15.09–20.01	15.09–20.01
Hungary	All year	1.05–31.12	All year
Italy	Third Sunday Sept.–31.01	Third Sunday Sept.–31.01	Third Sunday Sept.–31.01
Latvia	1.05–31.01	1.05–31.01	1.05–31.01
Lituania	1.05–1.03	1.10–01.02	1.05–1.03
Netherlands	1.07–31.01	1.07–31.01	1.07–31.01
Poland	1.04–28.02	15.08–15.01	1.04–28.02
Portugal	All year	All year	All year
Romania	1.08–15.02	1.08–15.02	1.08–15.02
Slovakia	16.07–31.12	16.07–31.12	16.07–31.01
Slovenia	1.04–31.01	1.08–31.01	All year
Spain	1.10–28.02	1.10–28.02	1.10–28.02
Switzerland	1.07–31.01	1.07–31.01	1.07–31.01

Table 4.4 *Open seasons for moose*

Country	Males	Females
Estonia	15.09–30.11	15.09–30.11
Finland	25.09–31.12	25.09–31.12
Latvia	1.09–15.12	1.09–15.12
Lithuania	1.09–15.11	1.10–15.11
Norway	25.09–31.10	25.09–31.10
Poland*	1.09–30.11	1.10–31.12

* Hunting stopped from 2001.

Table 4.5 *Open seasons for alpine chamois*

Country	Males	Females
Austria	1.06–31.12	16.07–31.12
Croatia	1.09–31.12	1.09–31.12
Czech Republic	1.10–30.11	1.10–30.11
France	23.08–28.02	23.08–28.02
Germany	1.08–15.12	1.08–15.12
Italy	1.08/15.08/15.09–15.11/31.12/ 31.01	1.08/15.08/15.09–15.11/31.12/ 31.01
Romania	15.10–15.12	15.10–15.12
Slovenia	1.08–31.12	1.08–31.12
Switzerland	1.08–31.12	1.08–31.12

traditions of hunting roaring red deer stags (specific calls are used in this context) or rutting alpine chamois.

With very few exceptions (including Norway and the Italian Alps) the legal seasons which apply in almost all European countries allow the shooting of red deer stags during the rut. Scotland has one of the earliest openings of the permitted season for shooting red stags – 1 July; more generally the commencement of the season is in August or September. In some regions of north Italy, in a change of practice in recent years, hunting is stopped for two to three weeks during the actual roaring period. The season for roe bucks also extends through the rut in the majority of countries. In a few instances (e.g. Norway) hunting does not commence until the rut is completed and in Italy too the hunting season for males generally begins from the end of the rut (15 August), but there are regions when it starts earlier. In Denmark, while the buck season starts well before the rut as in many other countries (16 May), the season appears deliberately 'broken' to accommodate an undisturbed rut (between 16 July and 30 September).

Hunting of male moose is allowed during the rut in all European countries where the species is managed. Alpine chamois males (and females) are hunted throughout their range during the rut (November) with no exception. And finally, both male and female wild boar can be hunted during the breeding season (late autumn to early winter, October to December/January) in all European countries; more specifically it is interesting to observe this is the only part of the hunting season that is the same in all countries, which otherwise show very different patterns in their seasons.

4.3.2 Culling during the last weeks of pregnancy

Hunting in the period before parturition is not a problem per se, at least if we do not worry about some subjective ethical reason (see above), but it may become a problem if the method of hunting results in high levels of stress either in the animal culled or others in the area, as in the case of hunting with dogs. In such context we might note that while the season for roe does in France finishes at the end of March, for the final month (1–31 March) the animals may only be hunted by coursing!

Seasons for roe deer more generally are restricted to a period well before parturition (e.g. 1 January to 15 March, or 15 January to 15 March in the Netherlands or Flanders) or do not commence until September or the beginning of October (the majority of countries). Only in Spain, Austria and Switzerland is the season for *mature* females open from April (Spain) or May. It should be noted that females culled at this time may well be near-term or actually have given birth.

In the case of red deer, the closing date for the hind season is in no case later than the end of March and the opening date is late enough to guarantee that pregnant hinds would not be culled in late pregnancy or the period immediately after parturition. In this case, the only exception is Austria with an open season for females starting from 1 May. Note, however, that we have suggested above that any disturbance may have serious consequences for offspring and mother's welfare. In this context pursuit even of male deer during this period may thus involve disturbance, movement and stress to heavily pregnant females. One of the concerns about hunting red stags with dogs in England (now banned but which used to continue until the end of April) was that it might cause disturbance to heavily pregnant hinds.

Female moose and alpine chamois are hunted well outside the period of late pregnancy/parturition time. For moose the close season extends from the end of December to the beginning of September in the more permissive cases; for chamois females are not hunted from end December/January (with the exception of France, 28 February) to beginning of August (16 July in Austria).

By contrast, in the case of wild boar, sow hunting is allowed, in the majority of cases, at a time that virtually guarantees that at least a proportion of mature females are in an advanced stage of pregnancy (31 January or as later as 31 March). In Austria, Estonia and Portugal females can be hunted year-round. In respect of this, however, we do recognise that hunting during the last part of pregnancy may be considered necessary when the aim is to reduce or control ungulate density. For wild boar, culling during this period may be simply inevitable in order to enable hunters to exercise some degree of control over a species which has a very rapid reproductive potential. Despite this we note that there may be implications of extended hunting seasons on other, non-target species. In the same way that hunting red deer stags with hounds in England as late as April potentially caused stress to pregnant females (above), extended hunting of wild boar – particularly by the most usual methods of driving and coursing – might actually result in serious disturbance for other species.

4.3.3 Culling during the period of juvenile dependency

Finally, we note that under the current hunting seasons in many countries, there remains a potential for killing females while young may still be nutritionally or socially dependent on them in at least the three more abundant species: wild boar, roe deer and red deer.

We may broadly assume (over the latitudinal range) that the period of parturition for roe deer is from late April or early May to end of June and that for red deer is from mid May to the end of June. For roe deer it is apparent that there is some considerable variation among geographical areas. Although fawning season is quite 'tight', with 80% of fawns born within 20–30 days of median date of parturition (Irvine, 2004), that median date itself may vary from 11 May to 13 June in different locations (Linnell and Andersen, 1998). There is, however, no simple relationship with latitude, although birth dates do follow some pattern, with southern and Atlantic coast populations giving birth in general before inland, continental ones (Linnell *et al.*, 1998). For red deer, there appears to be far less variation and populations in most areas give birth over the same range of dates (26 May–15 June; Fletcher, 1974). Given these general dates, we may speculate that neonates not accompanying the mother may be orphaned if mothers are shot before say mid June (roe) or mid July (red), while unless culled with the mother, juveniles of either species nutritionally dependent on lactating dams will die if mothers are shot before the end of August (see also Putman, 2008; Apollonio *et al.*, 2010).

On such a basis it is clear that, with seasons for mature red deer females in most countries not opening until September (Italy, Croatia, Slovenia, Hungary, Romania – all central European, with earlier breeding seasons anyway) or October (Denmark, Norway, Sweden, Wallonia (Belgium), Poland, Estonia, Lithuania), cull seasons may be considered outside the period of maximum welfare risk. Some countries delay the commencement of the season even further (November in England and Republic of Ireland). However, we may note that seasons in the Netherlands, Latvia, the Czech Republic, Slovakia and Switzerland open as early as the beginning of August.

In the majority of countries in Europe hunting of adult roe deer females is not permitted before the beginning of September, but in some cases an early open season for mature females is allowed: this is the case in Austria, Latvia,

Switzerland, Spain and the Italian Apennines. In some cases the extension of the season is not great (e.g. from August in the Italian Apennines), but this may nonetheless be in an ecological context that may also lead to early birth dates. In other cases the open season may start from May so there can be little doubt about the potential and actual possibility that female culls may indeed leave orphaned dependent fawns.

Moose as a species does not seem to suffer from any real problems with permitted seasons: open season for adult females ranges from 1 September (Latvia) to 1 October (Lithuania) so we can rule out the possibility that a suckling young could be orphaned as birth dates happen in the last ten days of May to early June (Bowyer *et al.*, 1998). Variation in the start-date for hunting of female alpine chamois is from 16 July (the earliest, in Austria) to 15 October (Romania), with most seasons in the actual alpine range starting in August. In Italy culling opens in September. Most births in the alpine countries are concentrated in the month of May. As a consequence there seems to be some room for problems connected with early shooting of lactating females.

In ten countries (Belgium, Spain, Italy, Lithuania, Poland, Denmark, Czech Republic, Slovenia, Romania, Greece) wild boar sows are not culled before August but in 12 countries sows may be legally taken well before this time: it is surely no coincidence that some of the European countries with the highest number of wild boar (and the consequent highest toll in damage to agriculture) are within the list, as France, Germany and Austria. Finally it is interesting to note that in some countries (e.g. Austria, Finland) killing a sow with dependent piglets is explicitly prohibited. Wild boar have the largest litter size of any European ungulate species with high social interactions. When the mother–young bond is still present, a good practice might be to cull only the piglets. Once again, the problem is even more exaggerated in relation to wild boar, since even the peak period of parturition in this species (when most births will occur) may extend over a period of three months or more (for example in Spain, the birth period occurs from February to April: Fernandez-Llario and Mateos-Quesada, 1998) and there are occasional females giving birth throughout the entire year.

4.4 Conclusions

Not all species are equally susceptible to all the problems explored in our introduction and we must recognise that some of the problems explored are potential rather than necessarily actual.

The particular breeding biology of individual species (both mating systems and timing of reproduction) plays a decisive role in making them more or less vulnerable to the various potential issues we have suggested. It is also clear that breeding systems vary within individual species (so that there is not even a single species-specific system which applies universally) and that the timing of both rutting period and parturition may vary with latitude and geography. It is important to make it clear therefore that there can be no 'hard and fast' general rules.

In addition not all hunting practices have the same kind of impact: single-hunter selective hunting with rifle from fixed blinds or high seats is at the lowest level of impact, whereas large drives with dogs and beaters are the maximum (Swenson, 1982; Kufeld *et al.*, 1988; Kilgo *et al.*, 1998).

Montería is a typical Spanish hunting practice for big game where several packs of dogs are released within at least 500 hectares of forest or scrubland (Carranza, 2010). There are some animal welfare concerns surrounding the use of hounds in hunting, but a positive side of the *montería* system is that each portion of land is hunted only once per year. Moreover, species not forming large breeding associations (such as roe deer) are presumably less subject to hunting disturbance during the rut as any given disturbance potentially affects a few deer per time at maximum. However, the potential for disturbance of non-target species, while rarely taken into account, may represent a relevant source of distress. Thus, for example the potential impact of wild boar hunting with hounds on roe deer during the rut could be important, particularly if it occurs repeatedly in the same areas. These situations may generate further constraints in setting hunting seasons.

Environmental constraints can also be important in the decision of hunting times of some species: in the Alps for instance it is hard to hunt higher than 1500 metres a.s.l. during winter in years of average snow and environmental difficulties suggest that it is inappropriate to stress ungulates further by hunting them in a period of food shortage and harsh climate.

Legislators and wildlife managers must thus consider a number of different factors when setting seasons, which tend to change among different areas. Hunting season, and hunting practice, must reflect management needs and objectives: whether directed for conservation of threatened populations or rare taxa, to support and sustain recreational hunting, or to ensure control of overabundant species, or impacts of ungulates on agriculture and forestry. However, to deliver those without risk of welfare or other problems outlined above, they must also take into account accurate information about (local) seasonality of the rut and of parturition of different species (as well as the risk of hunting seasons of one species causing problems in other non-target species sharing the same habitat), length of harassment and, last but by no means least, social traditions.

Local traditions may play an important, and not always positive, role in the attempt to find more scientifically sound hunting practices for ungulates. For example, hunting red deer stags by imitating the call during the rut is so deeply rooted in the cultural hunting tradition of Central European countries that specific competitions of calling ability are held among hunters; the hunting of alpine chamois during the rut allows hunters who adhere to a more Germanic tradition to obtain with the kill of a mature male his '*Gamsbart*' – the long hairs that are erected along the backbone by the rutting males are exhibited in a dense brush on the traditional mountain hunter cap.

Within such constraints, however, we examine, in conclusion, some possible changes which might be considered in the timing of seasons to try and overcome some of the shortcomings identified earlier in the chapter. A *caveat* is necessary here, however: as we have had occasion to note earlier, not all species in all countries are managed to maintain intact their potential reproductive output or to manage populations for stability or to encourage actual expansion in population size. In many instances ungulate populations are managed instead to try to exercise some control over populations and their impacts and actually to deliver some reduction in population size or distribution. In such contexts it is obvious that the primary focus is not concerned with the conservative management of the population; however, the maintenance of a healthy, balanced population and the avoidance of inhumane hunting practices remains a goal that has a general validity.

4.4.1 Culling during the rut

Because of the various negative implications we have rehearsed at the beginning of this chapter, as a general rule it would be wise not to hunt ungulates at all during their rut. In practice, however, for many reasons this would probably be difficult to establish across the entire range. A partial solution could be the protection of at least some traditional and important breeding ground for any given species, as red deer on the Alps. This kind of approach is obviously limited to situations where there are specific and localised areas where most individuals of a population do reproduce; it would not be applicable in situations where breeding grounds are widely distributed across the whole landscape, as in the case of the red deer in the Scottish Highlands.

An alternative may be to limit hunting just to the second half of the rut, leaving at least some portion of this important biological activity free from disturbance. The choice of the first instead of the last half is advisable because early conception and consequently early birth are linked to better chance of survival of young; in addition it is generally the most competitive, fitter males which are active early in the season. In addition, we believe there should be stringent regulation that males culled should be the poorer individuals (individuals showing signs of age or ill health, or those with poor antlers for their age, poor body conformation) since retention within the population of those males of better quality is crucial to maintaining overall quality and fitness of the population.

More generally, in those countries where hunting does occur during the mating season and where there is a long social tradition associated with this practice, we advise that numbers culled within this period should be at least restricted. This would help to reduce harassment and also to encourage hunters to use other hunting practices and periods. Hunters will be more favourable to suspend culling during the rut if they have successfully hunted during other periods.

4.4.2 Culling during advanced pregnancy

One possible solution to the problems associated with hunting close to the period of parturition is to ban hunting when females are close to giving birth. However, as noted above, actual culling of pregnant females is not a problem per se; rather it is the use of hunting practices which involve pursuit or disturbance of a population of females at this time, whether during culling of females themselves, or during the hunting of males, or even of other ungulate species. Thus consideration should rather be given to banning of particular hunting methods rather than necessarily a complete ban on hunting altogether.

We recognise that there are species like wild boar in which a female is theoretically able to give birth at any time of the year and also at any age (a female of only 8–9 months old is able to reproduce under favourable circumstances). In such cases it is clearly impracticable to ban hunting during periods where at least some females are bound to be heavily pregnant. In the few regions without wild boar damage, managers could minimize the risk of culling sows during late pregnancy by recognising the peak birth months in their particular region and to stop hunting at that time. But such a solution is not widely applicable. In almost every European country where they occur, wild boar are a problem, and attempts to control wild boar populations are failing dismally. For this reason managers and law makers must encourage and facilitate hunting activities aimed at reducing wild boar density and rebuilding a more natural population structure.

Even so, in cases like this when the aim of the management is to effect a major reduction of the population size, welfare and ethical constraints might suggest that culling of mature females should be concentrated where possible in the early stages of pregnancy and only, in extreme cases, during later pregnancy. In addition where culling is required, or permitted later in the period of pregnancy, consideration should again be given to hunting method, and during this time culling should be restricted to only those methods which cause minimum collateral disturbance. In this respect even if a 'surgical cull' (for example, stalking with rifles from high seats) of females during these times could still be considered acceptable, where minimum disturbance is occasioned to other females in the population, any hunting practice causing wider stress (like hunting with hounds) should be avoided.

4.4.3 Culling during the period of juvenile dependency

Logically a priority in this context would be to ensure that no culling of females is allowed during the period when neonatal young may still be concealed and not accompanying the mother. It is difficult to define the actual end of hiding behaviour so a prudent suggestion would be a hunting ban period from the first likely date of parturition in a given area to some three weeks after the latest possible date of parturition. After that period we suggest that any shooting of mature females must be avoided in the time in which juvenile nutrition is primarily dependent on lactation, or, if females must be shot, that the young should always be shot before the mother.

After this first period of strict nutritional dependence, a more ambiguous period begins over which the young are still socially dependent upon the mother. This period is poorly determined and clearly differs between different species. Typically, the social bond between mother and offspring generally continues until at least one year of age in most European ungulate species. We suggest therefore that in a strict conservation perspective, the non-hunting season for adult females should extend at least until the end of strict social dependence. However, this is probably impracticable in most situations, especially for species or populations where a primary aim of management is to control expanding populations and limit damaging impacts. At the very least therefore, we would suggest, as above, if females must be shot, that the young should always be shot before the mother. Moreover it is strongly suggested that some monitoring programme be put in place (as, for example, monitoring of average weight of males and females as yearlings) to assess if any management option that includes the cull of mature females with dependent young has an impact on population quality. Finally we think that during breeding seasons hunting is undesirable. If managers plan to cull during these periods, they must be aware of the potential consequences and they must foresee the best approaches in order to reduce or to avoid negative biological outcomes.

References

Aanes, R. and Andersen, R. (1996) The effects of sex, time of birth, and habitat on the vulnerability of roe deer fawns to red fox predation. *Canadian Journal of Zoology* **74**, 1857–1865.

Albon, S.D., Mitchell, B., Huby, B.J. and Brown, D. (1986) Fertility in female red deer (*Cervus elaphus*): the effects of body composition, age and reproductive status. *Journal of Zoology (London)*, **209**, 447–460.

Apollonio, M., Festa-Bianchet, M., Mari, F., Mattioli, S. and Sarno, B. (1992) To lek or not lek: mating strategies of male fallow deer. *Behavioural Ecology* **3**, 25–31.

Apollonio, M., Andersen, R. and Putman, R. (2010) Present status and future challenges for European ungulate management. In M. Apollonio, R. Andersen and R. Putman (eds.) *European Ungulates and their Management in the 21st Century*. Cambridge, UK: Cambridge University Press, pp.578–604.

Bartoš, L. and Bahbouh, R. (2006) Antler size and fluctuating asymmetry in red deer (*Cervus elaphus*) stags and the probability of becoming a harem holder in the rut. *Biological Journal of the Linnean Society* **87**, 59–68.

Bartoš, L., Zeeb, U. and Mikeš, J. (1992) Lekking behaviour in sika deer. In N. Maruyama, B. Bobek, Y. Ono, *et al.* (eds.) *International Symposium on Wildlife Conservation: Present Trends and Perspectives for the 21st Century*. Tokyo: Japan Wildlife Research Center, pp.205–208.

Bartoš, L., Herrmann, H., Šiler, J., Losos, S. and Mikeš, J. (1998) Variation of mating systems of introduced sika deer. *Revue d'Ecologie – La Terre et la Vie* **53**, 337–345.

Bartoš, L., Šustr, P., Janovskỳ, P. and Bertagnoli, J. (2003) Sika deer (*Cervus nippon*) lekking in a free-ranging population in Northern Austria. *Folia Zoologica* **52**, 1–10.

Bender, L.C., Lomas, L.A. and Browning, J. (2007) Condition, survival, and cause-specific mortality of adult female mule deer in north-central New Mexico. *Journal of Wildlife Management* **71**, 1118–1124.

Bender, L.C., Cook, J.G., Cook, R.C. and Hall, P.B. (2008) Relations between nutritional condition and survival of North American elk, *Cervus elaphus*. *Wildlife Biology* **14**, 70–80.

Bowyer, R.T., van Ballenberghe, V. and Kie, J.G. (1998) Timing and synchrony of parturition in Alaskan moose: long-term versus proximal effects of climate. *Journal of Mammalogy* **79**, 1332–1344.

Briedermann, L. (1986) *Schwarzwild*. Berlin: Neumann-Neudamm.

Briedermann, L. (1992) Reproduction in European mouflon. In F. Spitz, G. Jeneau, G. Gonzalez and S. Aulagnier (eds.) *Ongulées/Ungulates 91*. Toulouse, France: IRGM, pp.557–560.

Carranza, J. (1995) Female attraction by males versus sites in territorial rutting red deer. *Animal Behaviour* **50**, 445–453.

Carranza, J. (2010) Ungulates and their management in Spain. In M. Apollonio, R. Andersen, and R. Putman (eds.) *European Ungulates and their Management in the 21st Century*. Cambridge, UK: Cambridge University Press, pp.419–440.

Carranza, J., Garcia-Muñoz and de Dios Vargas, J. (1995) Experimental shifting from harem defence to territoriality in rutting red deer. *Animal Behaviour* **49**, 551–554.
Carrión, D., García, A.J., Gaspar-López, E., Landete-Castillejos, T. and Gallego, L. (2008) Development of body condition in hinds of Iberian red deer during gestation and its effects on calf birth weight and milk production. *Journal of Experimental Zoology A–Ecological Genetics and Physiology* **309**, 1–10.
Chapman, D.I. and Chapman, N.G. (1975) *Fallow Deer: Their History, Distribution and Biology*. Lavenham, UK: Terence Dalton, 271pp.
Childress, M.J. and Lung, M.A. (2003) Predation risk, gender and the group size effect: does elk vigilance depend upon the behaviour of conspecifics? *Animal Behaviour* **66**, 389–398.
Clutton-Brock, T.H., Albon, S.D. and Guinness, F.E. (1985) Parental investment and sex differences in juvenile mortality in birds and mammals. *Nature* **313**, 131–133.
Clutton-Brock, T.H., Albon, S.D. and Guinness, F.E. (1987a) Interactions between population density and maternal characteristics affecting fecundity and juvenile survival in red deer. *Journal of Animal Ecology* **56**, 857–871.
Clutton-Brock, T.H., Major, M., Albon, S.D. and Guinness, F.E. (1987b) Early development and population dynamics in red deer. I. Density-dependent effects on juvenile survival. *Journal of Animal Ecology* **56**, 53–67.
Clutton-Brock, T.H., Albon, S.D. and Guinness, F.E. (1988a) Reproductive success in male and female red deer. In T.H. Clutton-Brock (ed.) *Reproductive Success: Studies of Individual Variation in Contrasting Breeding Systems*. Chicago, IL: University of Chicago Press, pp.325–343.
Clutton-Brock, T.H., Green, D., Hiraiwa-Hasegawa, M. and Albon, S.D. (1988b) Passing the buck: resource defence, lek breeding and mate choice in fallow deer. *Behavioral Ecology and Sociobiology* **23**, 281–296.
Clutton-Brock, T.H., Deutsch, C.J. and Nefdt, R.J.C. (1993) The evolution of ungulate leks. *Animal Behaviour* **46**, 1121–1138.
Coltman, D.W., O'Donoghue, P., Jorgenson, J.T., *et al.* (2003) Undesirable evolutionary consequence of trophy hunting. *Nature* **426**, 655–658.
Côtè, S.D. and Festa-Bianchet, M. (2001) Birthdate, mass and survival in mountain goat kids: effects of maternal characteristics and forage quality. *Oecologia* **127**, 230–238.
Feder, C., Martin, J.G.A., Festa-Bianchet, M., Bérubé, C. and Jorgenson, J. (2008) Never too late? Consequences of late birthdate for mass and survival of bighorn lambs. *Oecologia* **156**, 773–781.
Ferdman, N., Murmu, R.P., Bock, J., Braun, K. and Leshem, M. (2007) Weaning age, social isolation, and gender, interact to determine adult explorative and social behavior, and dendritic and spine morphology in prefrontal cortex of rats. *Behavioural Brain Research* **180**, 174–182.
Fernandez-Llario, P. and Mateos-Quesada, P. (1998) Body size and reproductive parameters in wild boar, *Sus scrofa. Acta Theriologica* **43**, 439–444.
Festa-Bianchet, M. (1988a) Nursing behaviour of bighorn sheep: correlates of ewe age, parasitism, lamb age, birthdate and sex. *Animal Behaviour* **36**, 1445–1454.
Festa-Bianchet, M. (1988b) Seasonal range selection in bighorn sheep: conflicts between forage quality, forage quantity, and predator avoidance. *Oecologia* **75**, 580–586.
Festa-Bianchet, M. (2003) Exploitative wildlife management as a selective pressure for the life-history evolution of large mammals. In M. Festa-Bianchet and

M. Apollonio (eds.) *Animal Behavior and Wildlife Conservation*. Washington, DC: Island Press, pp.191–207.

Festa-Bianchet, M., Urquhart, M. and Smith, K.G. (1994) Mountain goat recruitment: kid production and survival to breeding age. *Canadian Journal of Zoology* **72**, 22–27.

Festa-Bianchet, M., Jorgenson, J.T. and Réale, D. (2000) Early development, adult mass, and reproductive success in bighorn sheep. *Behavioral Ecology* **11**, 633–639.

Fletcher, T.J. (1974) The timing of reproduction in red deer (*Cervus elaphus*) in relation to latitude. *Journal of Zoology* **172**, 363–367.

Francis, D., Diorio, J., Liu, D. and Meaney, M.J. (1999) Nongenomic transmission across generations of maternal behavior and stress responses in the rat. *Science* **286**, 1155–1158.

Gendreau, Y., Côtè, S.D. and Festa-Bianchet, M. (2005) Maternal effects on post-weaning physical and social development in juvenile mountain goats. *Behavioral Ecology and Sociobiology* **58**, 237–246.

Grignolio, S., Parrini, F., Bassano, B., Luccarini, S. and Apollonio, M. (2003) Habitat selection in adult males of Alpine ibex, *Capra ibex ibex*. *Folia Zoologica* **52**, 113–120.

Hamilton, W.J. and Blaxter, K.L. (1980) Reproduction in farmed deer. I. Hind and stag fertility. *Journal of Agricultural Science, Cambridge*, **95**, 261–273.

Harris, R.B., Wall, W.A. and Allendorf, F.W. (2002) Genetic consequences of hunting: what do we know and what should we do? *Wildlife Society Bulletin* **30**, 634–643.

Heck, L. and Raschke, G. (1980) *Die Wildsauen*. Hamburg, Germany: Paul Parey.

Irvine, J. (2004) *Calving Dates: a Literature Review*. Report to DCS on contract RP28. Summary in *Deer Commission for Scotland Annual Report 2003–04*, pp.52–53.

Jarnemo, A. (2004) Predation processes: behavioural interaction between red fox and roe deer during the fawning season. *Journal of Ethology* **22**, 167–173.

Kikusui, T., Kiyokawa, Y. and Mori, Y. (2007) Deprivation of mother-pup interaction by early weaning alters myelin formation in male, but not female, ICR mice. *Brain Research* **1133**, 115–122.

Kikusui, T., Nakamura, K. and Mori, Y. (2008) A review of the behavioural and neurochemical consequences of early weaning in rodents. *Applied Animal Behaviour Science* **110**, 73–83.

Kilgo, J.C., Labisky, R.F. and Fritzen, D.E. (1998) Influences of hunting on the behaviour of white-tailed deer: implications for conservation of the Florida panther. *Conservation Biology* **12**, 1359–1364.

Kjellander, P. and Nordstrom, J. (2003) Cycles voles, prey switching in red fox, and roe deer dynamics: a test of the alternative prey hypothesis. *Oikos* **101**, 338–344.

Kruuk, L.E.B., Slate, J., Pemberton, J.M., *et al.* (2002) Antler size in red deer: heritability and selection but no evolution. *Evolution* **56**, 1683–1695.

Kufeld, R.C., Bowden, D.C. and Schrupp, D.L. (1988) Influence of hunting on movements of female mule deer. *Journal of Range Management* **41**, 70–72.

Lamberti, P., Mauri, L. and Apollonio, M. (2004) Two distinct patterns of spatial behaviour of female roe deer (*Capreolus capreolus*) in a mountain habitat. *Ethology, Ecology, and Evolution* **16**, 41–53.

Langbein, J. (1991) Effects of density and age on body condition, reproductive performance, behaviour and survival of fallow deer. PhD thesis, University of Southampton, UK.

Langbein, J. and Thirgood, S.J. (1989) Variation in mating systems of fallow deer (*Dama dama*) in relation to ecology. *Ethology* **83**, 195–214.

Langbein, J., Chapman, N.G. and Putman, R.J. (2008) Fallow deer, *Dama dama*. In S. Harris and D.W. Yalden (eds.) *Mammals of the British Isles: Handbook*, 4th edn. London: The Mammal Society, pp.595–604.

Lent, P.C. (1974) Mother-infant relations in ungulates. In V. Geist and F. Walther (eds.) *The Behaviour of Ungulates and its Relation to Management*. Morges, Switzerland: IUCN, pp.14–55.

Li, Z., Jiang, Z. and Beauchamp, G. (2009) Vigilance in Przewalski's gazelle: effects of sex, predation risk and group size. *Journal of Zoology* **277**, 302–308.

Liberg, O., Johansson, A., Andersen, R. and Linnell, J.D.C. (1998) Mating system, mating tactics and function of male territoriality in roe deer. In R. Andersen, P. Duncan and J.D.C. Linnell (eds.) *The European Roe Deer: The Biology of Success*. Oslo: Scandinavian University Press, pp.221–256.

Linnell, J.D.C. and Andersen, R. (1998) Timing and synchrony of birth in a hider species, the roe deer *Capreolus capreolus*. *Journal of Zoology* (*London*) **244**, 497–504.

Linnell, J.D.C., Aanes, R. and Andersen, R. (1995) Who killed Bambi? The role of predation in the neonatal mortality of temperate ungulates. *Wildlife Biology* **1**, 209–223.

Linnell, J.D.C., Wahlstrom, K. and Gaillard, J.-M. (1998) From birth to independence: birth, growth, neonatal mortality, hiding behaviour and dispersal. In R. Andersen, P. Duncan and J.D.C. Linnell (eds.) *The European Roe Deer: The Biology of Success*. Oslo: Scandinavian University Press, pp.257–284.

Morrison, J.A. (1960) Characteristics of estrus in captive elk. *Behaviour* **16**, 84–92.

Mysterud, A., Solberg, E.J. and Yoccoz, N.G. (2005) Ageing and reproductive effort in male moose under variable levels of intrasexual competition. *Journal of Animal Ecology* **74**, 742–754.

Nakamura, K., Kikusui, T., Takeuchi, Y. and Mori, Y. (2008) Changes in social instigation- and food restriction-induced aggressive behaviors and hippocampal 5HT1B mRNA receptor expression in male mice from early weaning. *Behavioural Brain Research* **187**, 442–448.

Nicholson, M.C., Terry Bowyer, R. and Kie, J.G. (1997) Habitat selection and survival of mule deer: tradeoffs associated with migration. *Journal of Mammalogy* **78**, 483–504.

Paarlberg, K.M., Vingerhoets, A.D.J.J.M., Passchier, J., Dekker, G.A. and vanGeijn, H.P. (1995) Psychosocial factors and pregnancy outcome: a review with emphasis on methodological issues. *Journal of Psychosomatic Research* **39**, 563–595.

Panzacchi, M., Linnell, J., Odden, M., Odden, J. and Andersen, R. (2009) Habitat and roe deer fawn vulnerability to red fox predation. *Journal of Animal Ecology* **78**, 1124–1133.

Pettorelli, N., Pelletier, F., von Hardenberg, A., Festa-Bianchet, M. and Côté, S.D. (2007) Early onset of vegetation growth vs. rapid green-up: impacts on juvenile mountain ungulates. *Ecology* **88**, 381–390.

Pipia, A., Ciuti, S., Grignolio, S., Luchetti, S., Madau, R. and Apollonio, M. (2009) Effect of predation risk on grouping pattern and whistling behaviour in a wild mouflon *Ovis aries* population. *Acta Theriologica* **54**, 77–86.

Putman, R.J. (1993) Flexibility of social organisation and reproductive strategy in deer. *Deer* **9**, 23–28.

Putman, R.J. (2008) *A Report on Current Perception, Legal Status and Expectation with Respect to Deer Welfare in Other Countries*. Contract report for the Deer Commission for Scotland.

Røed, K.H., Holand, Ø., Mysterud, A., *et al.* (2007) Male phenotypic quality influences offspring sex ratio in a polygynous ungulate. *Proceedings of the Royal Society B–Biological Sciences* **274**, 727–733.

Schaal, A. and Bradbury, J.W. (1987) Lek breeding in a deer species. *Biology of Behaviour* **12**, 28–32.

Singer, F.J. and Zeigenfuss, L.C. (2002) Influence of trophy hunting and horn size on mating behavior and survivorship on mountain sheep. *Journal of Mammalogy* **83**, 682–698.

Stopher, K.V., Pemberton, J.M., Clutton-Brock, T.H. and Coulson, T. (2008) Individual differences, density dependence and offspring birth traits in a population of red deer. *Proceedings of the Royal Society B–Biological Sciences* **275**, 2137–2145.

Strandgaard, H. (1972) The roe deer (*Capreolus capreolus*) population at Kalø and the factors regulating its size. *Danish Review of Game Biology* **7**, 1–205.

Swenson, J.E. (1982) Effects of hunting on habitat use by mule deer on mixed-grass prairie in Montana. *Wildlife Society Bulletin* **10**, 115–120.

Thirgood, S.J., Langbein, J. and Putman, R.J. (1999) Intraspecific variation in ungulate mating strategies: the case of the flexible fallow deer. *Advances in the Study of Behavior* **28**, 333–361.

Türcke, F. and Schmincke, S. (1965) *Das Muffelwild*. Hamburg, Germany: Paul Parey.

Veiberg, V., Loe, L.E., Mysterud, A., Langvatn, R. and Stenseth, N.C. (2004) Social rank, feeding and winter weight loss in red deer: any evidence of interference competition? *Oecologia* **138**, 135–142.

von Hardenberg, A., Bassano, B., Peracino, A. and Lovari, S. (2000) Male Alpine chamois occupy territory at hotspots before the mating season. *Ethology* **106**, 617–630.

Willisch, C.S. and Neuhaus, P. (2009) Alternative mating tactics and their impact on survival in adult male Alpine ibex (*Capra ibex ibex*). *Journal of Mammalogy* **90**, 1421–1430.

5

The census and management of populations of ungulates in Europe

NICOLAS MORELLET, FRANÇOIS KLEIN, ERLING SOLBERG AND REIDAR ANDERSEN

5.1 Introduction

Over recent decades managers have had to cope with populations of ungulates which have substantially increased both their range and density in Europe (see Apollonio *et al.*, 2010, and references therein). In fact, of the 20 different species of free-ranging ungulates in Europe, only fallow deer *Dama dama* populations have remained comparatively stable in the last decades. This situation has been perceived by some people as a positive change, leading to greater opportunities for observation of native ungulates by the general public, as well as increased opportunities for those interested in hunting. Unfortunately this marked change in status also has some drawbacks where there is perceived to be an increase in ungulate–human conflicts. Ungulates may cause significant damage to farming and forestry (see Chapter 6 this volume), collisions with vehicles (see Chapter 8), and the spread of disease (Simpson, 2002; Chapter 7 this volume). In addition, there is a growing concern that large populations of ungulates may have a substantial effect on ecosystem function (e.g. Danell *et al.*, 2006). Consequently, in many countries society considers that managers should control ungulate populations through hunting in order to meet specific management objectives (Sinclair, 1997). In practice, the management of most populations of ungulates consists of setting harvest quotas for hunting. To be able to determine the harvest plan, most managers would like to require a reliable assessment of the demographic situation, and generally they consider estimates of population density or indices of density as the minimal information required for this (Williams *et al.*, 2002). Hence, many approaches have been used to assess population size with differences in methodology often dictated by habitat, cost and practical constraints.

Ungulate Management in Europe: Problems and Practices, eds. Rory Putman, Marco Apollonio and Reidar Andersen. Published by Cambridge University Press.

In this chapter, we will first give an overview of the various management objectives found in Europe, before we look at the variety of census methods used to reach the objectives; then we review the advantages and drawbacks of the different techniques. We will argue that the classical monitoring approach which focuses solely on the estimation of animal abundance needs to be expanded to also include changes in quality and quantity of the habitat and the interaction between the two; population density per se provides no information on the relationships between the population and its habitat.

5.2 One Europe – many management objectives

Because there are marked differences in the species present, in their distribution and local densities, and because there are marked cultural differences in attitudes (Chapter 1), management aims and practices vary widely between different countries. Consequently, the census methods used in the various countries will inevitably reflect these various differences in objective and management practice.

A summary of the management objectives in nearly 30 different European countries shows that there is a huge variation between countries, as well as, in some cases, lack of consistency in management objectives even between different parts of the same country (see Apollonio *et al.*, 2010, and references therein); it is equally clear that there may be differences in objectives declared for management of different species within any one particular country, and even within a particular species within that country.

Variations between countries reflect, in part, differences in patterns of land ownership, and differences in the relative importance attached to other land-use interests. Thus, in most European countries management of ungulates includes some sort of reference to other sectors of society, like agriculture, forestry and transportation (i.e. deer–vehicle collisions). In Germany and a number of other Central European countries, for example, control of population densities in order to prevent unacceptable levels of damage to crops and forests is the main management objective (e.g. Wotschikowsky, 2010). That is, crops and forests are the main target for ungulate management, and the tolerance level of damage to forestry is often low.

Several other countries, like Belgium (Wallonia), Finland and Italy, also state that their management objectives are to control population density in order to maintain acceptable damage levels to forestry and agriculture, and reduce deer–vehicle collisions, but in practice their main focus is the management of sporting game populations (Apollonio *et al.*, 2010). In the same way, while damage control is mentioned as a major management objective also in Hungary (Csányi and Lehoczki, 2010), the main focus for management of all species (including wild boar) is in practice the selective shooting to increase quality of trophy animals, and densities of ungulates within forested and other areas are tolerated at levels far above those which would be accepted if damage limitation was indeed the primary aim.

In many countries (like France, Norway, Romania, Sweden, Switzerland) the management objectives are primarily directed towards promoting sustainability, while taking into account forestry, agricultural priorities and other human interests. Other countries, like Denmark, state that their objectives are related to the principles of sustainable use of resources, implying that hunting should not decrease population sizes, but make no reference to damage levels. Both the Czech Republic and Slovakia have a clear ecosystem-oriented objective and see hunting as a maintenance of their cultural heritage, as stated in the Czech Game Management Act from 2003 (see also Bartoš *et al.*, 2010):

> . . . the game management should be taken to mean those activities carried out on wild land, directed towards managing wild game as a part of the ecosystem, and the activities of those associations whose objectives are the maintenance and development of hunting traditions and customs as a part of the Czech national cultural heritage.

As already noted, we also find that management objectives may vary between species. In Croatia and Portugal management of chamois and Iberian wild goat, respectively, is entirely directed towards conservation, while management of wild boar and mouflon are equally oriented towards sport and recreation activities. In some countries we also find spatial variation in management objectives even within a species. In the northern part of Portugal red deer are managed for conservation purposes, that is, as prey for the threatened Iberian wolf population, whereas in areas with no wolves red deer are managed for venison production.

Clearly, seeing the great span in management objectives across Europe, and the great variation in objectives even between species and within species in a particular country, it is not surprising that ungulate managers have used a wide range of census methods. It is, however, surprising that most countries have used methods aiming to elucidate absolute numbers or actual densities, while achieving their actual management objectives may not in practice be dependent on such data. Let us take a brief overview of the most familiar methods, and see for which species they are used in the different European countries. References to specific countries are in most cases drawn from Apollonio *et al.* (2010).

5.3 One Europe – many census methods

In Europe, a surprisingly high number of countries lack a formal coordination of their ungulate management, and consequently lack any formal agreement of methods to be used for population census of each species.

In some countries, like Austria, Denmark, Germany, Greece, Portugal and the UK outside Scotland, there is no scientific or systematic monitoring of ungulate populations initiated by the state (although some census may be undertaken by managers within individual game management districts/ hunting areas).

In many countries (e.g. Austria, Germany) even these 'estimates' are based on hunting bags, assuming a fixed sex ratio and population growth rate. Quotas are then defined based on recent harvest records and on the amount of damage to forest regrowth and forest stands.

It is also extremely common to find that different census methods may be being used in different hunting areas or game management districts, with no real coordination or consistency of methods. In Italy, for example, seven different census methods are used in red deer censuses, and six methods for roe deer and fallow deer, with the choice of census methods depending on local conditions and local preferences.

This inconsistency of method is particularly clearly demonstrated in Poland. Here the popular snow tracking counts that were used in the nineteenth and twentieth centuries have proven to be unreliable in estimating animal numbers. Drive counts revealed 1.1 to 3.5 times as many animals as did snow tracking surveys (Jędrzejewska *et al.*, 1994), while even drive counts themselves are known to seriously underestimate animal numbers (Maillard *et al.*, 2010). Managers then tested various methods including counts of moose from aircraft, modified snow tracking on transects, counts of the roaring red deer stags during the rut, capture–mark–recapture methods for wild boar, and estimation of roe deer numbers by faecal pellet group counts (Wawrzyniak *et al.*, 2010). At present there is no 'single' census method employed consistently in all areas and the official data on ungulate numbers in Poland derive from a combination of various census methods and common-sense guesses done in forest districts and national parks. Unfortunately, the situation in Poland seems to be the norm in Europe, rather than the exception.

Interestingly, nearly all European countries report that they are aware of the shortcomings in the census methods they are using (for a more detailed description of use see Apollonio *et al.*, 2010 and individual country chapters therein), as clearly stated in Hungary where they expect spring numbers to be educated guesses or guesstimates. Still, both in Hungary and Croatia it is compulsory for each game management unit to report the spring population size of all ungulate species occurring on their area.

Wildlife managers are faced with two main options when trying to decide the status of their populations: either (i) try to estimate absolute numbers or density, or (ii) use methods that could give some reliable relative index of annual variation in animal numbers or density. Perhaps surprisingly, many of the census methods used in Europe belong to the first category, that is they are expected to give absolute numbers or absolute densities (Table 5.1).

Ungulate census methods may also be broadly separated into two main groups: direct methods that involve observing, counting or classifying animals to provide estimates of population size, and indirect methods which relate the presence of signs of animals (e.g. tracks in snow/soft ground, faeces) to animal density. Direct census methods include methods like hunter observations, aerial counts, capture–mark–recapture methods, open hill counts, counts at artificial feeding places, salt licks or mating grounds, vantage point counts and drive counts. Indirect methods rely upon estimates from faecal pellet group counts, snow tracking, cohort analyses, etc.

Many of these same methods, in addition to others, can also be used to attain relative indices of animal abundance, like faecal pellet group counts, wildlife triangle transects (Finland), observation of marks, large-scale deer observation monitoring, variable techniques of transect counts, counts at restricted areas where animals are gathered, vantage point counts, spotlighting, and animal vital rates and assessment of changes in habitat quality and quantity expected to reflect animal density. Overall, European wildlife managers are using a total of 18 different census methods (many of them are variants of a certain type): 9 indirect methods and 9 direct methods (Table 5.1).

Irrespective whether the goal is to reveal absolute or relative densities, vital elements in judging the applicability of any census method is their estimator bias, precision and accuracy. Using the target analogy (Figure 5.1), where the centre is the true value, 5.1(a) denotes an imprecise and unbiased situation, that is, it will be hard to estimate the true value (i.e. population size), whereas 5.1(b) represents a situation which is precise and biased, again leaving the estimation of the true value difficult. An optimal census method with high accuracy should be both precise and unbiased. As the real number of animals within an area is seldom obtained (Nichols, 1992; Yoccoz *et al.*, 2001), the true accuracy of any method is hard to resolve.

We will look at the various census methods used by European wildlife managers to obtain absolute and relative numbers of animals. We focus deliberately on simple, robust and repeatable methods which may readily be understood by, and used by, managers; we deliberately exclude methods

Table 5.1 *Overview of the indirect (I) and direct (D) census methods used in Europe, the expected output of their use, and countries using the methods*

Census techniques	Type (I/D)	Absolute numbers	Relative abundance	Used by (European countries)
Faecal pellet group counts	I	✓	✓	Lithuania, Estonia, Scotland, Portugal, Sweden
Track counts on bare ground	I		✓[a]	Portugal, Romania
Snow tracking	I	✓		Romania, Estonia
Wildlife triangle transect	I		✓	Finland
Observations of marks	I		✓	Croatia, Romania
Cohort analyses	I	✓	✓	Spain
Vocalisation of animals	I	✓		Italy, Spain
Animal vital rates	I		✓	France, Norway, Slovenia
Habitat quality	I		✓	France, Norway, Slovenia
Deer observation monitoring	D		✓	Norway, Sweden, Finland
Capture–mark–recapture from ground or air	D	✓		Poland, Switzerland
Strip transects/line transects/total counts from aerial surveys	D	✓	✓	Lithuania, Finland, Norway, Sweden, Scotland
Open hills counts	D	✓		Scotland, Greece, Switzerland, Spain
Count at feeding places/salt licks/mating grounds	D	✓	✓	Finland, Hungary, the Netherlands, Romania, Slovakia
Transect counts	D		✓	Italy, Spain, the Netherlands, Sweden
Vantage point counts	D	✓	✓	Belgium, Scotland, Hungary, Slovakia, Italy, the Netherlands, Portugal
Spotlighting	D		✓	Portugal, Switzerland, Italy
Drive counts	D	✓		Scotland, Italy, Poland, Switzerland, the Baltic countries

[a] Presence and absence.

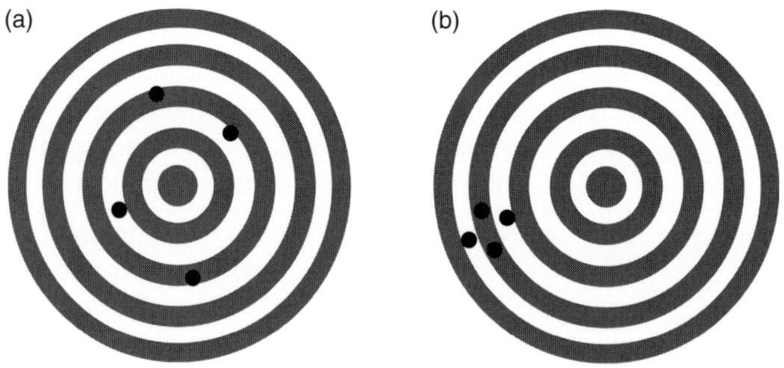

Figure 5.1 (a) Imprecise and unbiased versus (b) precise and biased.

which may be used by various scientific projects, but which are too complex or too time consuming to be used in routine management. Thus, we exclude the indirect estimation of animal density using the genotyping of animals from faeces or hair samples (Taberlet *et al.*, 1999; Piggott and Taylor, 2003). The latter method provides the possibility of estimating population size using capture–mark–recapture models. This technique has been successfully used in estimations of brown bear populations in Scandinavia (Solberg *et al.*, 2006b). Although this method may be relevant for estimating population levels of less abundant species of large ungulates, and may be interesting in the scientific arena, it is unlikely to be appropriate for management purposes due to the high cost of this approach, especially in the context of an abundant species.

Over the last 30 years, evidence has accumulated for problems of bias and imprecision in censusing ungulates by all the widely used methods (e.g. Mayle and Staines, 1998; Redfern *et al.*, 2002 for aerial counts; Gaillard *et al.*, 2003 for ground counts). Here we will address the census methods most commonly used by wildlife managers in Europe to estimate population size, and briefly describe their advantages and drawbacks (Tables 5.2 and 5.3). Depending on their use, some methods are used to derive both absolute numbers and relative densities; however, we allocate the described methods to one of these classes of techniques and divide each class into direct methods and indirect methods.

5.4 Census techniques used for estimating absolute numbers of ungulates

5.4.1 Direct census methods

Drive counts

Drive counts were used extensively in France in the 1970s and 1980s, but the use of this method has dropped at least to derive absolute densities following recognition that its use led to an increasing rate of underestimation with increasing population density (Maillard *et al.*, 2010).

Table 5.2 *An overview of some census methods/techniques used by European wildlife managers to attain absolute and relative densities of ungulates, and advantages and drawbacks of the various census methods*

Census method/ techniques	Advantages	Drawbacks
Faecal accumulation rate	– low disturbance on population – adapted in all habitat types	– no measure of accuracy and variable precision – no measure of detection probability – need the estimation of the defecation rate – risk of confusion between species – low cost-effectiveness – the method assumes a stable relationship between the amount of dung present and the number of animals
Faecal standing crop	– low disturbance on population – adapted in all habitat types	– no measure of accuracy and variable precision – no measure of detection probability – need the estimation of the defecation rate – risk of confusion between species – low cost-effectiveness – the method assumes a stable relationship between the amount of dung present and the number of animals – need the estimation of the decay rate (variable between habitat and meteorological condition)
Snow track counts	– low disturbance on population – adapted in all habitat types during snowing period – easily applied method	– no measure of accuracy and variable precision – no measure of detection probability – requires sufficient snow conditions or soft ground – risk of confusion between species – depending on method, requires an estimation of animal daily travel distance – low cost-effectiveness

Table 5.2 (*cont.*)

Census method/ techniques	Advantages	Drawbacks
Cohort analysis		– to calculate the population size in any past year, data need to cover a period extending beyond the lifespan of animals born in that year; this requirement is problematic in large ungulates because of their long lifespan – variations in age structure from harvested animals are notoriously difficult to interpret, and can provide very misleading results without independent data on birth and mortality rates
Vocalisation of animals		– no measure of accuracy and variable precision – risk of double counting but also: – risk of underestimation with increasing density
Hunter/gamekeeper observations	– cost-effective	– no measure of accuracy and precision – shown to severely underestimate number of animals
Strip transect/line transect from airplane or helicopter	– large areas observed – weak disturbance on population – cost effective	– no measure of accuracy and variable precision – no measure of detection probability for strip transect – risk of underestimation in closed habitat
Capture–mark–recapture from air	– large areas observed – improvement of accuracy – a measure of detection probability	– no measure of accuracy and variable precision – risk of large underestimation – requires marked animals well distributed at the scale of the study of interest – low cost-effectiveness – high disturbance on population during marking – not suitable in closed habitat

Table 5.2 (*cont.*)

Census method/ techniques	Advantages	Drawbacks
Capture–mark–recapture from ground	– best method to estimate population size in terms of accuracy and precision – a measure of detection probability – possibility to estimate some parameters of the population dynamic – well adapted in all habitats	– low cost-effectiveness – high disturbance on population during marking (and eventually during remarking if it is not re-sighting)
Counts in open areas/artificial feeding places	– often cost effective	– no measure of accuracy and precision
Vantage point counts		– sampling error may be large – greatly affected by vegetation type
Drive counts		– increasing rate of underestimation with increasing population density – sampling error affected by beater skill and numbers

The method is still widely used, however, in Italy, Poland, Portugal, Switzerland, and in the Baltics.

In Italy, the accuracy of the method has been tested using capture–mark–recapture techniques (radio-collared deer), and the average underestimate is estimated to be 20–25% of the actual population (M. Apollonio, pers. comm.). However, most likely the accuracy will be affected by the skill and number of beaters, and if the population size is estimated using subsamples of the population, as is often the case with drive counts, sampling error may be large.

Ground counts

In many countries attempts to estimate animal numbers are made from ground counts (see, for example, Daniels, 2006) whereby rows of observers traverse an area attempting to record all animals present. Such counts may be of reasonable (although largely untested) accuracy in completely open range

Table 5.3 *An overview of census methods used by European wildlife managers exclusively to attain relative densities of ungulates, and advantages and drawbacks of the various census methods*

Census methods	Advantages	Drawbacks
Track counts on bare ground		– can only be used for estimating presence or absence of animals
Observation of marks		– can only be used for estimating presence or absence of animals
Wildlife triangle scheme		– unknown accuracy and precision during winter
Spotlighting		– the assumption of full detectability seems unrealistic
Line transect counts	– can catch trends of animal variation over time on a specific study area – low cost-effectiveness	– the irrelevance of the absolute value (preventing comparison between different sites) – the absolute necessity to respect standardised protocols
Deer observation monitoring	– well suited for solitary animals – active involvement of hunters	– not well suited for group-forming species

(such as in the Highlands of Scotland), where observers familiar with the land can traverse the area on high ground offering good visibility into all areas. However, the method is inappropriate for census of animals in concealing habitats such as woodland, where visibility is poor. Ratcliffe (1984), for example, estimated that counts of red deer in coniferous forests in Scotland may underestimate true density by a factor which at best is 2 and at worst 16 times the number actually counted.

Vantage point counts

Some attempt may be made to overcome the problems of detectability of animals in such situations by undertaking counts of animals attracted to open spaces within forests, or to artificial feeding sites.

Thus, so-called vantage point counts are used in many European countries, in habitats which are too dense to count from the ground. In Belgium (Flanders) this method is used to estimate roe deer numbers during the hunting season and in Belgium (Wallonia) the same technique is used for estimating red deer numbers. In Scotland, Hungary and Slovakia vantage points are used for estimating red deer in concealing habitats. Vantage point counts are the most popular census method in Italy, and used for both total counts and counts in sample areas for all ungulates except ibex. In the Netherlands the technique is used for spring counts of roe deer, and in Portugal it is the most popular method for counting red deer and Iberian goat, but is used for counting roe deer, wild boar and mouflon as well.

Using repeated counts in blocks of woodland habitat, Ratcliffe (1984) was able to estimate red deer numbers before and after a cull, by this vantage point method. When the difference in population estimates were compared with the actual number removed during the cull, the two figures showed close agreement. Trials in lowland broadleaved forests, however, proved unsuccessful due to poor sighting ability (Mayle and Staines, 1998).

A variant of such methodology is to assess numbers of animals drawn to salt licks, or artificial feed. In Finland fallow deer are counted when using artificial feeding stations during winter. This is believed to give absolute numbers. In Hungary, the Netherlands and Romania wild boar are counted at artificial feeding places. In Slovakia counts of animals around artificial feeding places and salt licks are used for most ungulate species.

In all cases the assumption is that all animals are attracted to these sites, and that all animals occur at the sites at the same time. More generally the main drawback of all such methods, like other census methods that use subsamples to estimate total population size, is that the sampling error can be very large.

Spotlighting and thermal imaging

Other methods may be used to try and increase detectability of animals in direct ground counts (or transect counts, below), for example by using thermal imagers (Gill *et al.*, 1997) or spotlights. An evaluation of this last technique, however, showed that population size was seriously underestimated (McCullough, 1982; Fafarman and DeYoung, 1986; Collier *et al.*, 2007). Using a closed mark–recapture design, Collier *et al.* (2007) explored the efficiency of spotlights for detecting deer by operating thermal imagers and spotlights simultaneously. Spotlights detected only 50.6% of the deer detected by thermal imagers. Relative to the thermal imager, spotlights failed to detect 44.2% of deer groups, and detection probabilities for spotlight observers varied considerably.

Estimation of absolute population size – mostly an unrealistic aspiration?

In general, we might argue that if the partial detectability of individuals in the population is constant over space and time, we can use direct counting methods to monitor the minimum number of individuals present in the population. However, the assumption of a partial but stable detectability seems unrealistic. The probability of detecting an animal is likely to vary in space and time, depending on the structure of the habitat, the topography, the behaviour of animals and so on (Caughley, 1974; Samuel *et al.*, 1987; Schwarz and Seber, 1999; Redfern *et al.*, 2002). Thus, in most cases it is unrealistic to try to estimate absolute population size with a method that does not explicitly integrate a consideration of sampling error nor actually attempt to measure the detectability of individuals. Incorporating the distance sampling method (Buckland *et al.*, 1993) may overcome these problems, but possibly makes the analysis of such data too complex for widespread use.

Aerial surveys

Aerial surveys include a number of different methods. Indeed, by air it is possible to perform, for example, strip transects, line transects, total counts or capture–mark–recapture. Generally counts are made from helicopter or fixed-wing aircraft, with a pilot and some observers flying at a relatively low altitude (10–100 m for the helicopter and 75–150 m for fixed-wing aircraft, above ground level) and at various speeds (50–110 km/h for the helicopter and 80–240 km/h for fixed-wing aircraft). The helicopter has the advantage of being able to fly at lower speed and altitude than fixed-wing aircraft.

Aerial counts using fixed-wing aircraft are used occasionally in Lithuania and Finland for estimating moose numbers on winter grounds (Andersone-Lilley *et al.*, 2010, Ruusila and Kojola, 2010), and regularly in Norway for estimating musk ox and wild reindeer numbers on open land.

For a long time only fixed-wing aircraft were used, but gradually use of fixed-wing aircraft has been replaced by counts from helicopter. Daniels (2006), comparing faecal pellet counts, ground counts and helicopter counts, concluded that the last was most likely to minimize errors while maximizing cost efficiency. Managers in Finland use helicopter surveys for counting forest reindeer, and in Scotland, red deer populations on the open hill are now regularly counted from helicopter. In Sweden absolute numbers of moose are counted in sampling units of 400 ha from helicopters during winter. Each sampling area is censused twice by two different crews immediately after each other, the position of each sighted moose is determined exactly by GPS technique, and then proportion of moose seen by both crews can be used to calculate visibility. In recent years trials with counting red deer from the air have also been performed, and the results seem promising (Liberg *et al.*, 2010).

In the case of aerial surveys, many authors recognise that aerial estimates are imprecise (Gasaway *et al.*, 1985; Messier, 1991) and inaccurate, because the observer misses a significant and unknown number of animals on the transect (Caughley, 1974; LeResche and Rausch, 1974; Caughley *et al.*, 1976; Bartmann *et al.*, 1986; Beasom *et al.*, 1986; Pollock and Kendall, 1987; Samuel *et al.*, 1987; Cogan and Diefenbach, 1998; Redfern *et al.*, 2002; Potvin and Breton, 2005).

These authors have shown that aerial transect survey data yield biased estimates due to problems of differential visibility, observer experience, group size (red deer in Scotland may mass into groups of 500 or more when disturbed by a helicopter), vegetation cover and so on. For example, during aerial surveys, Beasom *et al.* (1986) obtained low accuracy, ranging between 26% and 40%, by estimating the mean percentage of marked deer seen. In fact, to our knowledge, the true accuracy of these approaches has never been estimated due to the difficulty of obtaining the true number of ungulates in a given population. However, Daniels (2006) suggested that the accuracy can be improved by the use of digital photography for counting animals in open terrain, a method widely used when counting wild reindeer in Norway.

Hunter/gamekeeper observations

In Latvia and Croatia estimates of ungulates are based on all year round observations by hunters, despite both countries mentioning danger of double counting and unknown estimation error as drawbacks. In the Czech Republic and in parts of Slovakia (where gamekeepers are present) annual census is generally carried out by visual survey during a fixed period in spring. The same approach is used in many parts of England where any census is attempted at all (e.g. Putman and Langbein, 1999).

This figure is then used to plan the following year's harvest without any control for the accuracy of the census figure produced, nor any real consideration given as to whether the proposed harvest may be effective in maintaining steady population size or achieving an agreed increase or decrease in population.

In a particularly spectacular example, hunter and gamekeeper observations were employed to estimate the population size of roe deer in an area of about 1000 ha comprising two separate forests (176 ha and 164 ha each) of mixed wood in Denmark. Subsequently, the population was hunted to extinction over a period of 7 months between 15 May and 31 December, with more than 90% of the roe deer shot between 1 October and 31 December, to obtain a definitive count. Using this approach, the estimated population size of about 70 animals was found to be underestimated by a factor of 3 as a total of 213 animals was shot in the area (Andersen, 1953). Clearly, estimating density of ungulate populations with this sort of approach is not a realistic and professional goal.

5.4.2 Methods based on sampling a proportion of the population

Line transect counts

Transect methods in general include different techniques, some of which deal with detectability and some which do not. In the case of strip transects it is assumed that all animals are detected out to a fixed distance on either side of the centreline. By contrast in line transects there is no assumption that all animals are detected and the method actually calculates its own 'detection function' (Buckland *et al.*, 1993). The method requires that the observer travelling along a transect line records the perpendicular distances (or the sighting distance and sighting angle) of all animals visible from the line. The distribution of observation distances enables calculations of the probability of detecting the animal as a function of perpendicular distance from the line (detection function) which may then be used to estimate density of populations.

Two key assumptions implicit in the line transect approach are that the detection probability is the same for all animals and that all animals present along the transect line are detected (Schwarz and Seber, 1999). However, few studies have assessed the performance of line transect methods on a population of known size (but see Vincent *et al.*, 1996), hence precision has still to be satisfactorily evaluated. Indeed, Southwell (1994) concluded for macropods that a 'density-dependent bias likely resulted from failure to count all animals on or close to the line at high density due to counting saturation'.

Capture–mark–recapture (CMR) techniques

In practical management CMR techniques are only used occasionally in Poland and Switzerland for estimating wild boar densities. However, CMR methods embrace a large family of techniques (Buckland *et al.*, 2000, for a review), and are among the most reliable methods available to estimate population size of free-ranging ungulates. Among CMR methods, the Lincoln–Petersen method is the most popular. This technique requires catching and marking some animals (M animals are marked). These animals are then released back into the population. The second step consists of recapturing n animals directly or using a different method such as resighting or harvest among the unknown population size (N). The random sample of size n from the population yields *m* marked and *u* unmarked animals. Finally, assuming marked and unmarked are equally represented in the sample *n*, we can estimate population size by the Petersen estimator nM/m. Hence, the population size is obtained by equating the proportion of marked animals for the complete population to the corresponding proportion in the sample (Buckland *et al.*, 2000).

This principle of equating proportions caught from similar subgroups underlies many of the more complex capture–recapture models. All CMR-type methods are, however, dependent on a number of different assumptions: the population size is stable over the course of the census leading to a closed population (i.e. with no birth, death and immigration–emigration), marks are not lost and all marks are reported, and all animals have the same probability of capture in each sample. Some studies have attempted to estimate the performance of these approaches (e.g. McCullough and Hirth, 1988; Vincent *et al.*, 1996, for the Lincoln–Petersen method), generating contrasting conclusions with regards to precision and accuracy. In fact, when locating or counting individuals on plots, observer bias may occur as the ability to detect animals varies considerably between observers and some animals may not be detectable at all (Schwarz and Seber, 1999). Moreover, for the Lincoln–Petersen method, it may be necessary to capture and mark two thirds of the population in order to obtain reliable estimates of population size (Strandgaard, 1967; Gaillard *et al.*, 1986).

Animal vocalisations

Vocalisation of animals is used to estimate number of rutting red deer stags on their breeding grounds in Italy and Spain (e.g. Mazzarone *et al.*, 1991), based on the method of Bobek *et al.* (1986). This method requires the estimation of the number of stags with harems that roar, the number of stags with harems that do not roar, the number of solitary stags that roar, the number of solitary stags that do not roar and any other social unit. Unfortunately, it is doubtful whether this method can generate precise and accurate estimates of the population size. Indeed, animals are not marked and it is impossible to compensate for missed animals. Moreover, the probability of observing stags that roar is higher than for those that do not, so we certainly underestimate the number of silent animals. Finally, as the proportion of stags roaring and the sex ratio are unknown, it is extremely doubtful whether we can extrapolate the information to estimate population size, although it could allow managers to monitor the number of roaring stags (see Malgras and Maillard, 1996).

In France, Reby *et al.* (1998) proposed a variant of this method, using vocalisations of roe deer to estimate population size. This approach involves walking along transects to observe and disturb animals present on a given study plot. In response to the disturbance, a certain proportion of animals bark. The survey generates different categories of animals: those observed or not and those animals that barked or did not. The total number of individuals that barked is considered as the 'marked' sample of the population. The proportion of marked individuals in the population is estimated from the proportion of animals barking in the subsample of individuals observed. Population size is then estimated by the classical Lincoln–Petersen approach, dividing the number of marked individuals by the proportion of marked animals, with the possibility of correcting for bias. This technique, based on capture–mark–recapture logic, requires validation of different assumptions, notably that the observed animals should have the same probability of vocalising as those disturbed but not seen. Moreover, observers walking along a transect may hear the same individual several times, so there is a risk of double-counting. This difficulty is particularly acute in the case of individuals living in groups, as different individuals may vocalise, but not at the same time. Furthermore, we can envisage a risk of underestimation with increasing population size: that is, a problem of saturation or an increase in error if too many deer are present at the same place.

5.4.3 Indirect methods

Faecal pellet group counts

Faecal pellet group counts are popular in the Baltics, where they are used both in spring (Lithuania) and winter (Estonia) in combination with double snow track counting to estimate roe deer numbers (Andersone-Lilley *et al.*, 2010). In Scotland the technique is occasionally used for estimating red deer and especially roe deer in concealing habitats. Both in Portugal and Sweden faecal count methods are used for estimating roe deer numbers; in Sweden the method is currently under evaluation. The method is more widely used to estimate relative abundance (below) rather than to estimate actual (absolute) abundance. There are in fact a number of different approaches towards the estimation of ungulate abundance based on the counting of faecal pellet groups along transects or within plots. We can distinguish two main techniques: the faecal standing crop (FSC) or the faecal accumulation rate (FAR) (Putman, 1984). The FSC is obtained by counting the number of accumulated pellet groups within randomly distributed sample quadrats, or along fixed transect lines. If rates of defecation and faecal decay are known, the number of pellet groups counted per unit area allows estimation of the population size.

In contrast to FSC, the FAR does not require the estimation of the decay rate leading to a simplification of the method and removing a source of error. The FAR is measured by initially clearing the sample areas of all ungulate faeces and then re-examining the same areas after a fixed time to determine the number of pellet groups accumulated during this interval. It is assumed that the interval is sufficiently short that no faeces could have decayed during that time, hence the problem of the decay rate is circumvented. However, the interval should be sufficiently long to ensure that faecal accumulation is likely. In that sense, it is recommended to first estimate the mean decay rate and defecation rate.

While the precision of these different techniques has been estimated several times, the conclusions are variable. For Mandujano and Gallina (1995) the FAR precision (CV) ranged from 13.5% to 24.7%, for Campbell *et al.* (2004) precision ranged from 9% to 23% for FSC and from 11% to 29% for FAR, for Smart *et al.* (2004) precision was about 40% for FSC and about 46% for FAR. Daniels (2006) found that precision ranged from 5% to 16% for FSC.

Recently, it has been suggested that combining faecal counts with line transect techniques could improve these methods (Marques *et al.*, 2001). Here, the number of pellet groups located within the area surveyed is modelled as a function of the perpendicular distance of detected pellet groups from the transect line. This overcomes the need to detect all pellet groups within the area, as the technique estimates a detection function (Buckland *et al.*, 1993): that is, the probability of detecting a pellet group at a certain distance from the line. When using line transect surveys of faecal pellet groups, precision was higher and was estimated to be around 10% in Marques *et al.* (2001).

Campbell *et al.* (2004) have shown that the precision of both FAR and FSC declines with declining pellet group density. This is for two reasons:

(i) Ungulate dung is never uniformly distributed but shows significant aggregation. Thus calculation of 'average dung density' is strongly affected by the number of zero plots (plots in which no pellet groups are counted at all) in any given sample. This problem of zero plots clearly declines as pellet density overall increases.
(ii) Most analyses assume a statistically normal distribution. Faecal pellet group counts are not normally distributed and more properly should be analysed with non-normal statistics (White and Eberhardt, 1980).

Some papers have tried to compare population estimates generated by different methods. For example, Mandujano and Gallina (1995) compared track counts, distance sampling (on the ground) and faecal pellet group methods. Smart *et al.* (2004) compared distance sampling using thermal imaging from a vehicle and faecal pellet group methods, while Daniels (2006) used ground, helicopter, infra-red helicopter and faecal pellet group methods. While these papers are interesting for comparing the precision of different methods under the same field conditions, these authors had no means of assessing method accuracy. Smart *et al.* (2004) tried to estimate accuracy with simulated data, but for a real valid evaluation of potential bias for faecal pellet group counts, we still need a comparison with real numbers or good population estimates (see Putman and Bailey, 1981).

Potentially, sampling design, environmental conditions, and variable detection efficiency of pellet groups in different habitats and by different observers affect the accuracy of faecal pellet group counts (Neff, 1968). Further, it is clear that defecation rates are not uniform in space, differing in relation to habitat use and activity rhythm and while sampling design must attempt to take this source of variation into account, this is not always explicitly considered. Subsequent 'translation' of those pellet group counts into population estimates is also dependent on a number of assumptions, not all of which are necessarily valid (or tested).

Firstly, defecation rates may vary over the year, reflecting seasonal variation in diet (see Mayle *et al.*, 2000). Moreover, if defecation rates are estimated on captive animals (as is often the case, as they are more easily observed), significant differences may exist between captive and wild individuals (Rogers, 1987). Decay rate (needed to estimate period of dung accumulation in FSC methods) may vary both spatially (Harestad and Bunnell, 1987; Marques *et al.*, 2001) and seasonally (MacCracken and Van Ballenberghe, 1987; Rogers, 1987). In the light of these differing sources of variation, it seems doubtful that counting faecal pellet groups is an appropriate method to accurately estimate the absolute density of ungulates.

Snow track counts

Snow track counts were the most commonly used method in Poland throughout the nineteenth and twentieth centuries to estimate relative abundance of ungulates. However, more recent analyses have shown that such a method leads to serious underestimation of animal numbers (see references in Maillard *et al.*, 2010). The method is now largely abandoned, although in Romania the method has been obligatory since 2007 (Micu *et al.*, 2010). In Estonia a double snow-tracking method is used: removing all tracks in day one and counting new tracks in day two, which is in line with the recommendations described in Silveira *et al.* (2003) and Mandujano (2005).

In general, in order to use track counts to offer an estimate of actual number, the method requires an estimate of the daily travel distance of the surveyed species in order to estimate the probability of an animal crossing a transect within the 24-hour period prior to the survey (Stephens *et al.*, 2006). Hence this method also requires that only fresh tracks be recorded; that is, within the last 24 hours. The precision of this method seems to be poor, especially when estimating populations at very low densities (Stephens *et al.*, 2006). The precision of density estimates, evaluated by simulations, is predicted to be affected by population density, total survey effort and distribution of survey effort (Stephens *et al.*, 2006).

However, Mandujano (2005) has proposed that the snow track count method may be improved by carrying out a calibration with estimates of density obtained from a more accurate and precise approach. Indeed, in his study, Mandujano (2005) obtained a good correlation between direct deer counts using distance sampling on the ground and track counts.

Snow track counts may overestimate abundance, since the same animal can be counted more than once. Second, the use of roads for track counts can bias results due to the associated human disturbance which encourages animals to avoid roads. Finally, for groups of animals, it is extremely difficult to evaluate the number of individuals from tracks and there is potential for error arising from misidentification of tracks when different species co-occur in the same study area. Hence, this method is probably of limited interest in terms of generating estimates of absolute density.

Cohort analyses and ageing of hunted animals

Cohort analysis (also known as virtual population analysis, VPA) is a method for population reconstruction applied to age-specific harvest data (catch-at-age data) that was originally developed in fisheries sciences (Ricker, 1940; Fry, 1949; Gulland, 1965; Pope, 1972), but since has also been applied to terrestrial species including ungulates (e.g. red deer, Lowe, 1969; white-tailed deer, McCullough, 1979; moose, Fryxell *et al.*, 1988; Ferguson, 1993; Solberg *et al.*, 1999; roe deer, Lowe and Thompson-Schwab, 2003; reindeer, Eberhardt and Pitcher, 1992; and sika deer *Cervus nippon*, Ueno, 2008). The method is based on the principle that individuals born in a particular year (a cohort) will die at a later age, and given that most of these animals can be retrieved (e.g. through harvesting, or collecting jaws from predator kills; e.g. Vucetich and Peterson, 2004) and age determined, the minimum number of individuals in the cohort can be calculated. By combining the information from several cohorts with independent estimates of natural mortality rate managers can reconstruct the population size in a given year. The method, however, depends on the assumption that most animals which die will be recovered, and also that the population unit is sufficiently large that neither immigration nor emigration have a significant effect.

In Norway this method is used on a regular basis to reconstruct the population development in a number of moose populations, from which most harvested animals are age determined each year (e.g. Solberg *et al.*, 2006a). The precision of the population estimates has been validated by comparison with independent population indices and aerial surveys (e.g. Fryxell *et al.*, 1988, Solberg *et al.*, 1999). However, to calculate the population size in any past year, data need to cover a period extending beyond the lifespan of animals born in that year. This requirement is problematic in large ungulates because of their long lifespan. Although there exist several methods to estimate the size of incomplete cohorts (e.g. Fryxell *et al.*, 1988; Ueno *et al.*, 2009), this means that population estimates get less reliable for the most recent years. Accordingly, the cohort analysis is more useful for retrospective analyses of the population development than as a management tool to determine harvest quotas etc. on an annual basis (e.g. Ferguson, 1993).

However, the age structure of harvested animals may independently offer some indication of the development of game populations given independent estimates of recruitment rates (e.g. Caughley, 1977). In Spain, population trends and even absolute population size of wild boar and red deer in non-fenced areas are calculated simply by having some estimation of the mean age of the individuals shot along consecutive years, and a reduction in the mean age of the animals shot over the years is regarded as a useful tool to detect overhunting (Carranza, 2010). Similarly, statistics on wild boar culled or found dead for other causes are used to check the accuracy of population estimates (cohort analysis) in Switzerland. However, variations in age structure from harvested animals are notoriously difficult to interpret, and can provide very misleading results without independent data on birth and mortality rates (e.g. Caughley, 1977). We therefore advise against the use of harvest age structure as the only source of information to guide management decisions.

5.5 Census techniques used to derive relative abundance

5.5.1 Direct methods

Many of the methods suggested above in estimation of absolute numbers may be used effectively in estimation of relative abundance (or in detection of trends in overall abundance). Here, however, we focus specifically on methods which have been developed specifically towards generating indices or estimates of relative abundance.

Line transect counts

To measure the relative abundance of roe deer the kilometric index was developed in France (Vincent *et al.*, 1991). The kilometric index is derived as the number of deer observed per kilometre of transect sampled on foot, from standardised transects, beginning at dawn or before dusk. This index has been validated against CMR estimates of an intensively studied population (Vincent *et al.*, 1991).

It is critical that indices of abundance which are used routinely for management purposes are validated and employed rigorously in relation to the defined protocol. In France, for example, some managers have frequently failed to adopt such a rigorous approach in their deer monitoring. Some managers have carried out their animal observations from a vehicle rather than on foot. Others have tried to combine different protocols designed for the monitoring of hares, generally at night with the aid of a spotlight, and of roe deer, even though the latter is designed to be performed during the three hours following sunrise or preceding sunset. Clearly, if we want to obtain reliable information on temporal trends in population abundance change it is essential to implement a rigorous protocol and in a standardised way which does not alter over the years. If this is not done, apparent changes in populations over time may be more a function of differences in recording methodology than actual changes in population size.

Hunter/gamekeeper observations

In Norway, Sweden and Finland a 'moose observation monitoring' system was established in the 1970s and 1980s (Lavsund *et al.*, 2003). In Norway, the system was later introduced in red deer management as well (Mysterud *et al.*, 2007), and in Finland, white-tailed deer and roe deer are also recorded. At present, data from the observation monitoring are used in more than 80% of the areas where moose are hunted in Norway, and a database of more than 4 million observations is currently available. The observation monitoring is a systematic recording and collecting of sex and age (calf or adult) of animals observed by the hunters during the hunting season, from which several indices of population structure and density are calculated. Most important are the 'animal seen per hunter-day' as an index of population density, and 'calves per female' and 'female per male' as indices of recruitment rate and adult sex ratio, respectively.

Indices based on hunter moose observations have been validated against independent estimates from cohort analyses (Solberg and Sæther, 1999; Solberg *et al.*, 2002) and aerial surveys (Ericsson and Wallin, 1999; Sylvèn, 2000). The moose is probably well suited for such census methods as it is solitary living and mainly utilises the forest habitat, reducing the variance caused by chance observations of larger groups. In red deer, which commonly form temporary groups and utilise open fields, variation in animals seen per hunter effort has performed less well as an index of variation in relative abundance (Mysterud *et al.*, 2007), possibly because of the large variation in deer seen caused by chance events and climate-related variation in the proportion of deer using fields. Moreover, Bonenfant *et al.* (2005) have shown temporal variations in the probability of observing the mother–calf pair in the field for red deer, and they question the validity of such an approach due to strong biases in count ratio-based methodologies.

It is suggested that further improvements of moose and red deer recordings as a management tool can be obtained by controlling for weather condition, hunting methods and quota structure, and if the data are analysed separately for smaller periods (e.g. week, day) of the hunting season, because this will reduce the impact of annual variation in harvesting rates and harvest conditions on the estimates. Making observation data available at a lower spatio-temporal level (days and hunting unit) may also open other estimation techniques, such as change-in-ratio and catch-effort methods (Williams *et al.*, 2002; Solberg *et al.*, 2005). In Norway, this census method is used in combination with a set of indicators for ecological change, giving a reliable set of methods for managing the large populations of moose and red deer.

5.5.2 *Indirect methods*

Faecal pellet group counts

Counts of the relative abundance of faecal pellet groups may offer a more robust index of relative population size than when they are used to attempt to estimate absolute population size. In a recent paper, Forsyth *et al.* (2007) found that three indices (total pellets, pellet groups and pellet frequency) change positively and approximately linearly with increasing deer density.

Despite this, these authors encourage managers wishing to use pellet counts as an index of abundance to first evaluate the relationship between the index and population density (Forsyth *et al.*, 2007).

Counting tracks and marks

Track counts on bare ground is not widely used by wildlife managers in Europe to estimate absolute or relative numbers, but may be used in Portugal and Romania for estimating simple presence or absence of wild boar. *Observations of marks* are used in Croatia, where hunting management programmes have approved an approach for game census which includes year-round observations of marks on trunks and around feeding places. Managers in Romania occasionally utilise the same method.

Wildlife triangle transect line system

While estimation of absolute or relative abundance of animals from (individual) counts of slots recorded in mud or snow may not be considered a good proxy for estimation of absolute or relative animal number, counts of regularly used trackways have been shown to be closely related to relative animal abundance.

The wildlife triangle scheme was developed in Finland in 1988, and is expected to provide annual information on abundance levels and changes in 30 wildlife species in about 1600 localities throughout Finland (Lindén *et al.*, 1996). In each locality, a triangle with 4-km sides is visited by three persons (mainly hunters) twice a year (summer: grouse species; winter: mainly ungulates, carnivores, mustelids). Track densities for each species are converted to abundance values for each 50 × 50 km grid, and differences in relative densities of animals between different grids are calculated using the abundance ratio, assuming there is no difference between the ratio of track densities and the ratio of animal densities (Pellikka *et al.*, 2005). Although the overall detectability in summer is high (Brittas and Karlbom, 1990), less is known about the factors affecting the statistical properties (accuracy, precision) of the abundance estimates regarding the ungulates monitored by counting tracks in winter.

A similar approach is adopted by the index of Mayle *et al.* (2000). The method involves walking a minimum distance of 1 km round each of a number of sample woodlands in the area to be surveyed, recording the number of obvious deer pathways crossing the woodland edge (tracks left where deer regularly leave the woodland cover to feed beyond the woodland edge). Wherever perimeter fencing constitutes an effective barrier to deer this length should not be included in the assessment. While this methodology may not be of utility to discriminate between species where more than one deer species (or other ungulate) may be present within an area, it has proven very effective in providing a rapid survey of relative ungulate abundance overall.

Indicators of ecological change

The concept of indicators of ecological change is based on the observation that increasing populations of large mammals are often faced with decreasing per-capita resources, and as a result, density-dependent responses may be discovered in the animal (phenotypic responses) and their resources (habitat responses) (e.g. Eberhardt, 2002; Morellet *et al.*, 2007). As defined by Waller and Alverson (1997) indicators of ecological changes should be 'efficient and reliable indicators capable of serving as early warning signs of impending ecological change'.

Animal vital rates In France, Norway and Slovenia several indicators are used to track ecological changes that occur either with the animal or the habitat. The French roe deer group has developed and validated various indicators to monitor roe deer populations in relation to their habitat (Maillard *et al.*, 1999). Different indicators such as female reproductive success (Boutin *et al.*, 1987; Vincent *et al.*, 1995), body mass of fawns (Maillard *et al.*, 1989; Gaillard *et al.*, 1996), cohort jaw length (Hewison *et al.*, 1996), and hind foot length of fawns (Toïgo *et al.*, 2006) enable managers to monitor changes over the year in animal performance.

Norway has used a similar system since 1991 to monitor the variation in age-specific body mass and fecundity (by ovary sectioning) in moose and red deer populations as part of the monitoring programme for cervids. All primary data are collected by the hunters, and the programme has contributed to several management improvements. The census system has revealed spatial variation in physical condition among areas, affecting both age of first reproduction and fecundity in older age groups, and temporal variation in physical condition caused by variation in climate and density-dependent food limitation (e.g. Solberg and Sæther, 1999; Mysterud *et al.*, 2001a, 2001b; Solberg *et al.*, 2002; Langvatn *et al.*, 2004; Herfindal *et al.*, 2006). The programme also monitors annual variation in abundance and reproduction of wild reindeer, using ground counts and aerial surveys (e.g. Solberg *et al.*, 2001; Aanes *et al.*, 2003). In addition, a national health surveillance programme for wild cervids is established in Norway, which systematically registers diseases and causes of death in 65 selected municipalities, and samples serum and tissue samples for screening purposes.

In Slovenia adaptive management of red deer populations was launched in 1976 over an area of about 140 000 ha of the Dinaric Karstland in west-central Slovenia. Browsing intensity of young trees, body weights of calves and yearlings of both sexes, reproduction rates of adult and yearling hinds, as well as the general health conditions of the population have been registered as model parameters to inform future management strategy.

Habitat quality and browsing pressure In France, a browsing index has been developed to monitor changes over years in habitat quality and the interaction between the population and its habitat (Morellet *et al.*, 2001, 2003). The browsing index closely tracked a roe deer population of known size, and although the species-specific browsing rates differed widely, bramble (*Rubus* sp.) could be reliably used to assess total browsing pressure of this species.

To monitor changes in habitat quality over years, a method for moose browsing assessments (MBA) was developed in Norway in the mid 1980s (Solbraa, 1987, 1998). This method has been used extensively in the moose areas for estimating variation in browse production (density of trees <3 m) and winter browsing pressure (on five deciduous and two coniferous tree species) by irregular intervals. An average browsing pressure above 40–50% (i.e. removal of >40% of all annual twigs) is considered to be overbrowsing in the sense that this will seriously reduce the future production of browse from the species in question. However, although the MBA is intended to predict serious food limitation and subsequent declines in moose condition and fecundity, this ability has not yet been validated.

A recent evaluation of a similar, though simpler, system introduced in Sweden in 2000 (ÄBIN) concluded that annual variation exists in browsing pressure despite no similar variation in population density (Kjellander, 2007). A possible explanation is that variation in snow cover affects the access to more preferred plants in the field layer, indicating that adjusting for variation in climate or smoothing over years may be necessary to reveal changes in density-related browsing pressure under boreal conditions (Kjellander, 2007).

5.6 One Europe – a united management approach?

In a few cases in Europe, the objective of census for ungulates is simply to reveal presence or absence of a particular ungulate species. In such cases all kinds of census methods may be used. However, in most other cases the fulfilment of wider management objectives relies on knowledge of relative spatial and temporal number of animals. Despite the fact that in the majority of cases management only requires some idea of relative abundance, and an ability to monitor trends, a surprisingly large number of wildlife managers have strived to collect absolute numbers or absolute densities of their populations. Commonly management must satisfy a number of separate objectives (sometimes including sporting management) and in such cases it is necessary to have some initial idea of actual abundance in order to develop sustainable harvesting strategies for the population in question and balance this with other management interests.

Such data are hard to obtain, and at present only CMR techniques and distance sampling techniques have provided satisfactory results (Buckland *et al.*, 1993; Schwarz and Seber, 1999). Both these techniques have, however, serious limitations in practical local or regional management situations, as they are costly and time consuming (CMR) or rely on sophisticated data analysis and two-dimensional habitats. In a few cases in Europe where we find restricted populations in open terrain, total counts may be used. For wild reindeer and musk ox inhabiting treeless terrain in restricted areas in Norway, absolute numbers are normally easily acquired by aerial surveys. For ungulates in mountainous or open habitats like ibex, chamois, mouflon, and red deer on the hills, line transect counts may be also used given the right conditions. However, in general we urge managers to drop searching for absolute numbers or densities, and instead using their efforts to obtain reliable relative indices.

Given the variety of habitats used by European ungulates, and the great span in population sizes and distribution, as well as differences in management objective, it is not possible to select a single uniform census technique which would be appropriate, or even optimal, in all circumstances. However, the use of indices of abundance has been widely debated (Anderson, 2001, 2003; Engeman, 2003). The index of abundance (c) is the product of the abundance (N) and a detection or encounter probability (p): such that $c = pN$. Implicitly, when using an index, we assume a constant detection probability across habitat types, observers and many other factors. So, it is strongly recommended that managers use validated indicators in a standardised way, paying particular attention to the repeatability of a protocol and to observer bias. Indeed, to ensure representative observations, managers should avoid collecting data based on convenience criteria, as for example along roads for the kilometric index, or indicators based on restricting sampled area on criteria of access, visibility or level of density.

Exploring the relationship between the sampling effort and the repeatability of the results using resampling methods, Loison *et al.* (2006) compared population estimates of chamois from CMR and indices of population size as the mean number of animals observed on a line transect surveyed repeatedly. Both in an exponentially increasing population and in a stable population CMR estimates and indices from line track counts revealed the same trend or lack of trend, respectively. It was concluded that the line track counts may be used for monitoring trends in population size (Loison *et al.*, 2006). However, the index is only reliable if calculated over many surveys the same year (10 in exponentially increasing populations and 3 in stable populations), and over several years. Furthermore, using such methods it is vital to use observers with similar and adequate qualifications in population counts, as variation in observer experience may strongly bias population estimates (Garel *et al.*, 2005). Thus, for ungulates in open habitats the use of line transects following standardised protocols may give reliable relative indices of population size.

Included in the wildlife objectives in most European countries are some statements about the state of the habitat used by the ungulates. In some countries we find phrases like 'the sustainability of the habitat should be maintained'; in other countries 'acceptable damage levels' are included in the management objectives. Definition of what damage levels are actually acceptable varies widely. Some wildlife managers consider all browsing to be damage to a tree stand, while in other cases when income from selling hunting licences is higher than selling tree products, landowners and managers tolerate higher levels of browsing. Nevertheless, some sort of biological indicators to assess the interaction between ungulates and their habitats seem to be a natural expansion of the toolkits available for ungulate managers in the future. However, identification of suitable indicators is not straightforward. In Chapter 6 (this volume) Reimoser and Putman show that in the majority of situations, impacts are affected by a wide variety of environmental factors and only loosely related to ungulate density – and indeed suggest that control of ungulate populations and control of impacts often should be considered quite distinct objectives; thus deciding which species or species groups to use as indicators of ecological change should be done carefully.

Referring to the management objectives in the different European countries, managers require information on trends in both population abundance and habitat quality, allowing them to interpret annual fluctuations of the interaction between these two components. Population dynamics of ungulates is strongly influenced by a combination of stochastic variation in the environments, and population density inducing changes in life history traits (Sæther, 1997). Indeed, populations of ungulates are well known to be subject to density dependence (e.g. Putman *et al.*, 1996; Sæther, 1997; Gaillard *et al.*, 2000).

Density dependence is defined as the functional dependence of a demographic rate on changes in population size (Williams *et al.*, 2002) and provides a way to measure population–environment relationships. Based on the concept of density dependence, it is possible to build several indicators to monitor populations of ungulates (Morellet *et al.*, 2007). All parameters which respond to changes in relative density (i.e. changes in population size for a given habitat quality) can be viewed as candidate ecological indicators. The philosophy of this approach consists of assessing the state of the relationship between a population and its habitat along the continuum from colonisation to saturation by the monitoring of a set of indicators of ecological change (Morellet *et al.*, 2007). This set of indicators should include, at least, indicators of relative animal abundance, of individual performance (reproduction, mortality, phenotypic quality), of habitat quality and of the impact of the population on that habitat. This approach was first developed to manage roe deer populations in France (Cederlund *et al.*, 1998).

However, when using vital rates in a population–environment assessment, it is important to know the form of the density regulation, and the strength of the density regulation when the population reaches saturation. The commonly used assumption in population ecology of loglinear density regulation is not always valid, as in some species density regulation happens close to a saturated level, as for ibex (Sæther *et al.*, 2002), which may result in long-term fluctuations in population size. In other ungulates (e.g. Soay sheep: Coulson *et al.*, 2001) we may find strong over-compensatory winter mortality leading to large annual fluctuations in population size.

In France, Norway and Slovenia we find an approach based on monitoring the annual fluctuations in a set of indicators, which allow a quantification of the changes that occur in the population–environment system over time. The assumption is that, all things being equal, the spatial and temporal trends observed in the indicators will allow managers to identify the effect of their management actions on the demographic status of the population, thus establishing an active adaptive management strategy. It should, however, be noted that this approach first of all applies to the relatively dense ungulate populations currently found in most of Europe. At the lower range of densities, indicators of the population–environment relationship are unlikely to be very sensitive to variation in density, and will therefore be less able to indicate whether the population is low or in fact critically close to extinction.

5.7 *Towards adaptive management*

In practice, managers need to set out some expectations or goals to monitor and manage ungulate populations. These expectations should be based, we believe, on animal performance, population productivity, and/or habitat quality. Other indicators dealing with sociological and ecological problems such as car collisions, ecosystem biodiversity, overgrazing and predator abundance in relation to the status of ungulate populations, could be considered once the appropriate tools are developed and validated.

Then, the approach consists of monitoring change over years in individual performance, population productivity, and habitat quality and/or herbivore impact on the habitat. In this way managers monitor the population–environment system with a view to achieving predefined goals. Managers have to measure this set of indicators for a given period to be able to detect temporal variation in the different compartments of the population–environment system, and to reveal possible time delays in the response of individual performance to variation in habitat quality. The temporal variation of this set of ecological indicators can be quantified and compared with predefined goals to assess if the hunting pressure is high enough or not. Then, a new management decision is taken in terms of new hunting quotas for the next year in order to converge towards the predefined goals. This approach is more or less equivalent to a trial and error process during the first years of monitoring, but the understanding of the population–environment system increases with the accumulation of information over the years.

This process bears some resemblance to adaptive management. Indeed, in adaptive management, the information on the system response to management is gathered continuously so that this information is used to improve biological understanding and to inform future decision making (Nichols *et al.*, 1995; Shea *et al.*, 1998; Williams *et al.*, 2002). The term adaptive refers to managers learning about systems as they attempt to manage them (Lancia *et al.*, 1996). Adaptive management deals with scientific uncertainty by incorporating a set of models representing competing hypotheses about system responses to management (Runge and Johnson, 2002). We believe that the management of ungulates should take advantage of this sort of approach by improving the monitoring of the population–environment system, and in order to collect a larger and more reliable amount of data, hunters should be stimulated to take part in the data collection.

References

Aanes, R., Sæther, B.E., Solberg, E.J., *et al.* (2003) Synchrony in Svalbard reindeer population dynamics. *Canadian Journal of Zoology–Revue Canadienne de Zoologie* **81**, 103–110.

Andersen, J. (1953) Analysis of a Danish roe-deer population (*Capreolus capreolus* (L)) based upon the extermination of the total stock. *Danish Review of Game Biology* **2**, 127–155.

Anderson, D.R. (2001) The need to get the basics right in wildlife field studies. *Wildlife Society Bulletin* **29**, 1294–1297.

Anderson, D.R. (2003) Response to Engeman: index values rarely constitute reliable information. *Wildlife Society Bulletin* **31**, 288–291.

Andersone-Lilley, Z., Balciauskas, L., Ozolins, J, Randveer, T. and Tonisson, J. (2010) Ungulates and their management in the Baltics. In M. Apollonio, R. Andersen, and R. Putman (eds.) *European Ungulates and their Management in the 21st Century*. Cambridge, UK: Cambridge University Press, pp. 103–128.

Apollonio, M., Andersen, R. and Putman, R. (2010) *European Ungulates and their Management in the 21st Century*. Cambridge, UK: Cambridge University Press, 604 pp.

Bartmann, R.M., Carpenter, L.H., Garrott, R.A. and Bowden, D.C. (1986) Accuracy of helicopter counts of mule deer in pinyon-juniper woodland. *Wildlife Society Bulletin* **14**, 356–363.

Bartoš, L., Kotrba, R. and Pintíř, J. (2010) Ungulates and their management in the Czech Republic. In M. Apollonio, R. Andersen and R. Putman (eds.) *European Ungulates and their Management in the 21st Century*. Cambridge, UK: Cambridge University Press, pp. 243–261.

Beasom, S.L., Leon III, F.G. and Synatzske, D.R. (1986) Accuracy and precision of counting white-tailed deer with helicopters at different sampling intensities. *Wildlife Society Bulletin* **14**, 364–368.

Bobek, B., Perzanowski, K. and Zielinski, J. (1986) Red deer population census in mountains: testing of an alternative method. *Acta Theriologica* **31**, 423–431.

Bonenfant, C., Gaillard, J.-M., Klein, F. and Hamann, J.L. (2005) Can we use the young:female ratio to infer ungulate population dynamics? An empirical test using red deer *Cervus elaphus* as a model. *Journal of Applied Ecology* **42**, 361–370.

Boutin, J.M., Gaillard, J.-M., Delorme, D. and Van Laere, G. (1987) Suivi de l'évolution de la fécondité chez le chevreuil (*Capreolus capreolus*) par l'observation des groupes familiaux. *Gibier Faune Sauvage* **4**, 255–265.

Brittas, R. and Karlbom, M. (1990) A field evaluation of the Finnish 3-man chain: a method for estimating forest grouse numbers and habitat use. *Ornis Fennica* **67**,18–23.

Buckland, S.T., Anderson, D.R., Burnham, K.P. and Laake, J.L. (1993) *Distance Sampling: Estimating Abundance of Biological Populations*. London: Chapman and Hall.

Buckland, S.T., Goudie, I.B.J. and Borchers, D.L. (2000) Wildlife population assessment: past developments and future directions. *Biometrics* **56**, 1–12.

Campbell, D., Swanson, G.M. and Sales, J. (2004) Comparing the precision and cost-effectiveness of faecal pellet group count methods. *Journal of Applied Ecology* **41**, 1185–1196.

Carranza, J. (2010) Ungulates and their management in Spain. In M. Apollonio, R. Andersen and R. Putman (eds.) *European Ungulates and their Management in the 21st Century*. Cambridge, UK: Cambridge University Press, pp. 419–440.

Caughley, G. (1974) Bias in aerial survey. *Journal of Wildlife Management* **38**, 921–933.

Caughley, G. (1977) *Analyses of Vertebrate Populations*. New York: John Wiley and Sons.

Caughley, G., Sinclair, R. and Scott-Kemmis, D. (1976) Experiments in aerial survey. *Journal of Wildlife Management* **40**, 290–300.

Cederlund, G., Bergqvist, J., Kjellander, P., *et al.* (1998) Managing roe deer and their impact on the environment: maximising the net benefits to society. In R. Andersen, P. Duncan and J.D.C. Linnell (eds.) *The European Roe Deer: The Biology of Success*. Oslo: Scandinavian University, pp. 337–372.

Cogan, R.D. and Diefenbach, D.R. (1998) Effect of undercounting and model selection on a sightability-adjustment estimator for elk. *Journal of Wildlife Management* **62**, 269–279.

Collier, B.A., Ditchkoff, S.S., Raglin, J.B. and Smith, J.M. (2007) Detection probability and sources of variation in white-tailed deer spotlight surveys. *Journal of Wildlife Management* **71**, 277–281.

Coulson, T., Catchpole, E.A., Albon, S.D., *et al.* (2001) Age, sex, density, winter weather and population crashes in Soay sheep. *Science* **292**, 1528–1531.

Csányi, S. and Lehoczki, R. (2010) Ungulates and their management in Hungary. In M. Apollonio, R. Andersen and R. Putman (eds.) *European Ungulates and their Management in the 21st Century*. Cambridge, UK: Cambridge University Press, pp. 291–318.

Danell, K., Bergström, R., Duncan, P. and Pastor, J. (2006) *Large Herbivore Ecology, Ecosystem Dynamics and Conservation*. Cambridge, UK: Cambridge University Press.

Daniels, M.J. (2006) Estimating red deer *Cervus elaphus* populations: an analysis of variation and cost-effectiveness of counting methods. *Mammal Review* **36**, 235–247.

Eberhardt, L.L. (2002) A paradigm for population analysis of long-lived vertebrates. *Ecology* **83**, 2841–2854.

Eberhardt, L.L. and Pitcher, K.W. (1992) A further analysis of the Nelchina caribou and wolf data. *Wildlife Society Bulletin* **20**, 385–395.

Engeman, R.M. (2003) More on the need to get the basics right: population indices. *Wildlife Society Bulletin* **31**, 286–288.

Ericsson, G. and Wallin, K. (1999) Hunter observations as an index of moose *Alces alces* population parameters. *Wildlife Biology* **5**, 177–185.

Fafarman, K.R. and DeYoung, C.A. (1986) Evaluation of spotlight counts of deer in south Texas. *Wildlife Society Bulletin* **14**, 180–185.

Ferguson, S.H. (1993) Use of cohort analysis to estimate abundance, recruitment and survivorship for Newfoundland moose. *Alces* **29**, 99–113.

Forsyth, D.M., Barker, R.J., Morriss, G. and Scroggie, M.P. (2007) Modeling the relationship between fecal pellet indices and deer density. *Journal of Wildlife Management* **71**, 964–970.

Fry, F.E.J. (1949) Statistics of a lake trout fishery. *Biometrics* **5**, 27–67.

Fryxell, J.M., Mercer, W.E. and Gellately, R.B. (1988) Population dynamics of Newfoundland moose using cohort analysis. *Journal of Wildlife Management* **52**, 14–21.

Gaillard, J.-M., Boisaubert, B., Boutin, J.M. and Clobert, J. (1986) L'estimation d'effectifs à partir de capture-marquage-recapture: application au chevreuil (*Capreolus capreolus*). *Gibier Faune Sauvage* **3**, 143–158.

Gaillard, J.-M., Delorme, D., Boutin, J.M., Van Laere, G. and Boisaubert, B. (1996) Body mass of roe deer fawns during winter in 2 contrasting populations. *Journal of Wildlife Management* **60**, 29–36.

Gaillard, J.-M., Festa-Bianchet, M., Yoccoz, N.G., Loison, A. and Toïgo, C. (2000) Temporal variation in fitness components and population dynamics of large herbivores. *Annual Review of Ecology and Systematics* **31**, 367–393.

Gaillard, J.-M., Loison, A. and Toïgo, C. (2003) Variation in life history traits and realistic population models for wildlife management. In M. Festa-Bianchet and M. Apollonio (eds.) *Animal Behavior and Wildlife Conservation*. Washington, DC: Island Press, pp. 115–132.

Garel, M., Cugnasse, J.-M., Gaillard, J.-M., *et al.* (2005) Effect of observer experience on the monitoring of mouflon population. *Acta Theriologica* **50**, 100–114.

Gasaway, W.C., Dubois, S.D. and Harbo, S.J. (1985) Biases in aerial transect surveys for moose during May and June. *Journal of Wildlife Management* **49**, 777–784.

Gill, R.M.A., Thomas, M.L. and Stocker, D.R. (1997) The use of portable thermal imaging for estimating deer population density in forest habitats. *Journal of Applied Ecology* **34**, 1273–1286.

Gulland, J.A. (1965) Estimation of mortality rates. In *Annex to Arctic Fisheries Working Group Report*. ICES C.M. Doc. 3. Copenhagen: International Council for the Exploration of the Sea.

Harestad, A.S. and Bunnell, F.L. (1987) Persistence of black-tailed deer fecal pellets in coastal habitats. *Journal of Wildlife Management* **51**, 33–37.

Herfindal, I., Sæther, B.-E., Solberg, E.J., Andersen, R. and Høgda, K.-A. (2006) Population characteristics predict responses in moose body mass to temporal variation in the environment. *Journal of Animal Ecology* **75**, 1110–1118.

Hewison, A.J.M., Vincent, J.P., Bideau, E., Angibault, J.M. and Putman, R.J. (1996) Variation in cohort mandible size as an index of roe deer (*Capreolus capreolus*) densities and population trends. *Journal of Zoology* **239**, 573–581.

Jędrzejewska, B., Okarma, H., Jędrzejewski, W. and Milkowski, L. (1994) Effects of exploitation and protection on forest structure, ungulate density and wolf predation in Białowieża Primeval Forest, Poland. *Journal of Applied Ecology*, **31**, 664–676.

Kjellander, P. (2007) *Utvärdering av ÄBIN*. (In Swedish) Report 1. Jönköping, Sweden: Skogstyrelsen, 90 pp.

Lancia, R.A., Braun, C.E., Collopy, M.W., *et al.* (1996) ARM! For the future: adaptive resource management in the wildlife profession. *Wildlife Society Bulletin* **24**, 436–442.

Langvatn, R., Mysterud, A., Stenseth, N.C. and Yoccoz, N.G. (2004) Timing and synchrony of ovulation in red deer constrained by short northern summers. *American Naturalist* **163**, 763–772.

Lavsund, S., Nygren, T. and Solberg, E.J. (2003) Status of moose populations and challenges to moose management in Fennoscandia. *Alces* **39**, 109–130.

LeResche, R.E. and Rausch, R.A. (1974) Accuracy and precision of aerial moose censusing. *Journal of Wildlife Management* **38**, 175–182.

Liberg, O., Bergström, R., Kindberg, J. and von Essen, H. (2010) Ungulates and their management in Sweden. In M. Apollonio, R. Andersen and R. Putman (eds.) *European Ungulates and their Management in the 21st Century*. Cambridge, UK: Cambridge University Press, pp. 37–70.

Lindén, H., Helle, E., Helle, P. and Wikman, M. (1996) Wildlife triangle scheme in Finland: methods and aims for monitoring wildlife populations. *Finnish Game Research* **49**, 4–11.

Loison, A., Appololinaire, J., Jullien, J.-M. and Dubray, D. (2006) How reliable are total counts to detect trends in population size of chamois *Rupicapra rupicapra* and *R. pyrenaica? Wildlife Biology* **12**, 77–88.

Lowe, V.P.W. (1969) Population dynamics of the red deer (*Cervus elaphus* L.) on Rhum. *Journal of Animal Ecology* **38**, 425–457.

Lowe, V.P.W. and Thompson-Schwab, J.D.D. (2003) Using cohort analysis to reconstruct the size and structure of deer populations in forestry with special reference to roe deer (*Capreolus capreolus* L.). *Forestry* **76**, 437–447.

MacCracken, J.G. and Van Ballenberghe, V. (1987) Age- and sex-related differences in fecal pellet dimensions of moose. *Journal of Wildlife Management* **51**, 360–364.
Maillard, D., Boisaubert, B. and Gaillard, J.-M. (1989) La masse corporelle: un bioindicateur possible pour le suivi des populations de chevreuils. *Gibier Faune Sauvage* **6**, 57–68.
Maillard, D., Gaultier, P. and Boisaubert, B. (1999) Revue de l'utilisation des différentes méthodes de suivi des populations de chevreuils en France. *Bulletin Mensuel de l'Office National de la Chasse* **244**, 30–37.
Maillard, D., Gaillard, J-M., Hewison, M., *et al.* (2010) Ungulates and their management in France. In M. Apollonio, R. Andersen and R. Putman (eds.) *European Ungulates and their Management in the 21st Century*. Cambridge, UK: Cambridge University Press, pp. 441–474.
Malgras, J. and Maillard, D. (1996) Analyse spectrale et biologie des populations: analyse de l'activité de brame chez le cerf élaphe (*Cervus elaphus* L.). *Comptes Rendus de l'Académie des Sciences III: Science de la Vie* **319**, 921–929.
Mandujano, S. (2005) Track count calibration to estimate density of white-tailed deer (*Odocoileus virginianus*) in Mexican dry tropical forest. *Southwestern Naturalist* **50**, 223–229.
Mandujano, S. and Gallina, S. (1995) Comparison of deer censusing methods in tropical dry forest. *Wildlife Society Bulletin* **23**, 180–186.
Marques, F.F.C., Buckland, S.T., Goffin, D., *et al.* (2001) Estimating deer abundance from line transect surveys of dung: sika deer in southern Scotland. *Journal of Applied Ecology* **38**, 349–363.
Mayle, B.A. and Staines, B.W. (1998) An overview of methods used for estimating the size of deer populations in Great Britain. In C.R. Goldspink, S. King and R. J. Putman (eds.) *Population Ecology, Management and Welfare of Deer*. Manchester, UK: British Deer Society/Universities' Federation for Animal Welfare, pp. 19–31.
Mayle, B.A., Peace, A.J. and Gill, R.M.A. (1999) *How Many Deer? A Field Guide to Estimating Deer Population Size*. Forestry Commission Field Book 18. London: Her Majesty's Stationery Office.
Mayle, B.A., Putman, R.J. and Wyllie, I. (2000) The use of trackway counts to establish an index of deer presence. *Mammal Review* **30**, 233–237.
Mazzarone, V., Siemoni, N., Pedone, P., Lovari, C. and Mattioli, L. (1991) A method of red deer (*Cervus elaphus* L.1758) census during the roaring period in a forested area of the northern Apennines (central Italy). In S. Csányi and J. Ernhaft (eds.) *XX Congress International Union of Game Biologists*, Transaction Part 1. Gödöllö, Hungary: University of Agricultural Sciences, pp. 140–145.
McCullough, D.R. (1979) *The George Reserve Deer Herd: Population Ecology of a K-selected Species*. Ann Arbor, MI: University of Michigan Press.
McCullough, D.R. (1982) Evaluation of night spotlighting as a deer study technique. *Journal of Wildlife Management* **46**, 963–973.
McCullough, D.R. and Hirth, D.H. (1988) Evaluation of the Petersen-Lincoln estimator for a white-tailed deer population. *Journal of Wildlife Management* **52**, 534–544.
Messier, F. (1991) The signifiance of limiting and regulating factors on the demography of moose and white-tailed deer. *Journal of Animal Ecology* **60**, 377–393.
Micu, I., Náhlik, A., Neguş, S., Mihalache, I. and Szabó, I. (2010) Ungulates and their management in Romania. In M. Apollonio, R. Andersen and

R. Putman (eds.) *European Ungulates and their Management in the 21st Century.* Cambridge, UK: Cambridge University Press, pp. 319–337.

Morellet, N., Champely, S., Gaillard, J.-M., Ballon, P. and Boscardin, Y. (2001) The browsing index: new tool uses browsing pressure to monitor deer populations. *Wildlife Society Bulletin* **29**, 1243–1252.

Morellet, N., Ballon, P., Boscardin, Y. and Champely, S. (2003) A new index to measure roe deer (*Capreolus capreolus*) browsing pressure on woody flora. *Game and Wildlife Science* **20**, 155–173.

Morellet, N., Gaillard, J.M., Hewison, A.J.M., *et al.* (2007) Indicators of ecological change: new tools for managing populations of large herbivores. *Journal of Applied Ecology* **44**, 634–643.

Mysterud, A., Langvatn, R., Yoccoz, N.G. and Stenseth, N.C. (2001a) Plant phenology, migration and geographical variation in body weight of a large herbivore: the effect of a variable topography. *Journal of Animal Ecology* **70**, 915–923.

Mysterud, A., Yoccoz, N.G., Stenseth, N.C. and Langvatn, R. (2001b) Effects of age, sex and density on body weight of Norwegian red deer: evidence of density-dependent senescence. *Proceedings of the Royal Society B–Biological Sciences* **268**, 911–919.

Mysterud, A., Meisingset, E.L., Veiberg, V., *et al.* (2007) Monitoring population size of red deer *Cervus elaphus*: an evaluation of two types of census data from Norway. *Wildlife Biology* **13**, 285–298.

Neff, D.J. (1968) The pellet-group count technique for big game trend, census, and distribution: a review. *Journal of Wildlife Management* **32**, 597–614.

Nichols, J.D. (1992) Capture-recapture models: using marked animals to study population dynamics. *BioScience* **42**, 94–102.

Nichols, J.D., Johnson, F.A. and Williams, B.K. (1995) Managing North American waterfowl in the face of uncertainty. *Annual Review of Ecology and Systematics* **26**, 177–199.

Pellikka, J., Hannu, R. and Linden, H. (2005) Monitoring wildlife richness: Finnish applications based on wildlife triangle censuses. *Annales Zoologica Fennica* **42**, 123–134.

Piggott, M.P. and Taylor, A.C. (2003) Remote collection of animal DNA and its applications in conservation management and understanding the population biology of rare and cryptic species. *Wildlife Research* **30**, 1–13.

Pollock, K.H. and Kendall, W.L. (1987) Visibility bias in aerial surveys: a review of estimation procedures. *Journal of Wildlife Management* **51**, 502–510.

Pope, J.G. (1972) An investigation of the accuracy of Virtual Population Analysis using cohort analysis. *Research Bulletin of the International Commission for the Northwest Atlantic Fisheries* **9**, 65–74.

Potvin, F. and Breton, L. (2005) Testing 2 aerial survey techniques on deer in fenced enclosures: visual double-counts and thermal infrared sensing. *Wildlife Society Bulletin* **33**, 317–325.

Putman, R.J. (1984) Facts from faeces. *Mammal Review* **14**, 79–97.

Putman, R.J. and Bailey, R.E. (1981) Estimation of large ungulate populations from faecal accumulation; an evaluation of the various methods using fallow deer. *Journal of Applied Ecology* **18**, 697–702.

Putman, R.J. and Langbein, J. (1999) *Deer and their Management in the New Forest.* Technical Review and Management Plan developed for the Deputy Surveyor, New Forest. London: Forestry Commission, 152 pp.

Putman, R.J., Langbein, J., Hewison, A.J.M. and Sharma, S.K. (1996) Relative roles of density-dependent and density-independent factors in the population dynamics of British deer. *Mammal Review* **26**, 81–101.

Ratcliffe, P.R. (1984) Population density and reproduction of red deer in Scottish commercial forests. *Acta Zoologica Fennica* **172**,191–193.

Reby, D., Hewison, A.J.M., Cargnelutti, B., Angibault, J.M. and Vincent, J.P. (1998) Use of vocalizations to estimate population size of roe deer. *Journal of Wildlife Management* **62**, 1342–1348.

Redfern, J.V., Viljoen, P.C., Kruger, J.M. and Getz, W.M. (2002) Biases in estimating population size from an aerial census: a case study in the Kruger National Park, South Africa. *South African Journal of Science* **98**, 455–461.

Ricker, W.E. (1940) Relation of 'catch per unit effort' to abundance and rate of exploitation. *Journal of Fishery Resource Board, Canada* **5**, 43–70.

Rogers, L.L. (1987) Seasonal changes in defecation rates of free-ranging white-tailed deer. *Journal of Wildlife Management* **51**, 330–333.

Runge, M.C. and Johnson, F.A. (2002) The importance of functional form in optimal control solutions of problems in population dynamics. *Ecology* **83**, 1357–1371.

Ruusila, V. and Kojola, I. (2010) Ungulates and their management in Finland. In M. Apollonio, R. Andersen and R. Putman (eds.) *European Ungulates and their Management in the 21st Century*. Cambridge, UK: Cambridge University Press, pp. 86–102.

Samuel, M.D., Garton, E.O., Schlegel, M.W. and Carson, R.G. (1987) Visibility bias during aerial surveys of elk in northcentral Idaho. *Journal of Wildlife Management* **51**, 622–630.

Schwarz, C.J. and Seber, G.A.F. (1999) Estimating animal abundance: review, III. *Statistical Science* **14**, 427–456.

Shea, K. and the NCEAS Working Group on Population Management (1998) Management of populations in conservations, harvesting and control. *Trends in Ecology and Evolution* **13**, 371–375.

Silveira, L., Jácomo, A.T.A. and Diniz-Filho, J.A.F. (2003) Camera trap, line transect census and track surveys: a comparative evaluation. *Biological Conservation* **114**, 351–355.

Simpson, V.R. (2002) Wild animals as reservoirs of infectious diseases in the UK. *Veterinary Journal* **163**, 128–146.

Sinclair, A.R.E. (1997) Carrying capacity and the overabundance of deer. In W.J. McShea, H.B. Underwood and J.H. Rappole (eds.) *The Science of Overabundance: Deer Ecology and Population Management*. Washington, DC: Smithsonian Institution Press, pp. 380–394.

Smart, J.C.R., Ward, A.I. and White, P.C.L. (2004) Monitoring woodland deer populations in the UK: an imprecise science. *Mammal Review* **34**, 99–114.

Solberg, E.J. and Sæther, B.E. (1999) Hunter observations of moose *Alces alces* as a management tool. *Wildlife Biology* **5**, 107–117.

Solberg, E.J., Sæther, B.-E., Strand O. and Losion A. (1999) Dynamics of a harvested moose population in a variable environment. *Journal of Animal Ecology* **68**, 186–204.

Solberg, E.J., Jordhoy, P., Strand, O., *et al.* (2001) Effects of density-dependence and climate on the dynamics of a Svalbard reindeer population. *Ecography* **24**, 441–451.

Solberg, E.J., Loison, A., Ringsby, T.H., Sæther, B.E. and Heim, M. (2002) Biased adult sex ratio can affect fecundity in primiparous moose *Alces alces*. *Wildlife Biology* **8**, 117–128.

Solberg, E.J., Grotan, V., Rolandsen, C.M., Broseth, H. and Brainerd, S. (2005) Change-in-sex ratio as an estimator of population size for Norwegian moose *Alces alces*. *Wildlife Biology* **11**, 163–172.

Solberg, E.J., Rolandsen, C.M., Heim, M., *et al.* (2006a) Elgen i Norge sett med jegerøyne – En analyse av jaktmaterialet fra overvåkingsprogrammet for elg og det samlede sett elg-materialet for perioden 1966–2004. (In Norwegian with an English summary) *Norwegian Institute for Nature Research, Rapport* **125**, 197 pp.

Solberg, K.H., Bellemain, E., Drageset, O.-M., Taberlet, P. and Swenson, J.E. (2006b) An evaluation of field and non-invasive genetic methods to estimate brown bear (*Ursus arctos*) population size. *Biological Conservation* **128**, 158–168.

Solbraa, K. (1987) Elgbeite på busker og trær i nordre Østfold og Åsnes. In S. Myrberget (ed.) *Elgen og skogbruket*. (In Norwegian). Oslo: Norsk Skogbruk.

Solbraa, K. (1998) *Elg og skogsbruk: biologi, økonomi, beite, taksering, forvaltning*. (In Norwegian). Biri, Norway: Skogsbrukets Kursinstitutt.

Southwell, C. (1994) Evaluation of walked line transect counts for estimating macropod density. *Journal of Wildlife Management* **58**, 348–356.

Stephens, P.A., Zaumyslova, O.Y., Miquelle, D.G., Myslenkov, A.I. and Hayward, G.D. (2006) Estimating population density from indirect sign: track counts and the Formozov-Malyshev-Pereleshin formula. *Animal Conservation* **9**, 339–348.

Strandgaard, H. (1967) Reliability of the Petersen method tested on a roe deer population. *Journal of Wildlife Management* **31**, 643–651.

Sylvèn, S. (2000) Effects of scale on hunter moose *Alces alces* observation rate. *Wildlife Biology* **6**, 157–165.

Sæther, B.-E. (1997) Environmental stochasticity and population dynamics of large herbivores: a search for mechanisms. *Trends in Ecology and Evolution* **12**, 143–149.

Sæther, B.-E., Engen, S., Filli, F., *et al.* (2002) Stochastic population dynamics of an introduced Swiss population of the ibex. *Ecology* **83**, 3457–3465.

Taberlet, P., Waits, L.P. and Luikart, G. (1999) Noninvasive genetic sampling: look before you leap. *Trends in Ecology and Evolution* **14**, 323–327.

Toïgo, C., Gaillard, J.-M., Van Laere, G., Hewison, A.J.M. and Morellet, N. (2006) How does environmental variation influence body mass, body size and body condition? Roe deer as a case study. *Ecography* **29**, 301–308.

Ueno, M. (2008) Analyses of population dynamics and establishment of harvest-based population estimation methods in the sika deer (*Cervus nippon*) in Hokkaido, Japan. PhD thesis, Division of Environmental Resources, Graduate School of Agriculture, Hokkaido University, Japan.

Ueno, M., Matsuishi, T., Solberg, E.J. and Saitoh, T. (2009) Application of cohort analysis to harvest data of large terrestrial mammals. *Mammal Study* **34**, 65–76.

Vincent, J.P., Gaillard, J.-M. and Bideau, E. (1991) Kilometric index as biological indicator for monitoring forest roe deer populations. *Acta Theriologica* **36**, 315–328.

Vincent, J.P., Bideau, E., Hewison, A.J.M. and Angibault, J.M. (1995) The influence of increasing density on body weight, kid production, home range and winter grouping in roe deer (*Capreolus capreolus*). *Journal of Zoology* **236**, 371–382.

Vincent, J.P., Hewison, A.J.M., Angibault, J.M. and Cargnelutti, B. (1996) Testing density estimators on a fallow deer population of known size. *Journal of Wildlife Management* **60**, 18–28.

Vucetich, J.A. and Peterson, R.O. (2004) The influence of top-down, bottom-up and abiotic factors on the moose (*Alces alces*) population. *Proceedings of the Royal Society B–Biological Sciences* **271**, 183–189.

Waller, D.M. and Alverson, W.S. (1997) The white-tailed deer: a keystone herbivore. *Wildlife Society Bulletin* **25**, 217–226.

Wawrzyniak, P., Jędrzejewski, W., Jędrzejewska, B. and Boro, T. (2010) Ungulates and their management in Poland. In M. Apollonio, R. Andersen and R. Putman (eds.) *European Ungulates and their Management in the 21st Century*. Cambridge, UK: Cambridge University Press, pp. 223–242.

White, G.C. and Eberhardt, L.E. (1980) Statistical analysis of deer and elk pellet-group data. *Journal of Wildlife Management* **44**, 121–131.

Williams, B.K., Nichols, J.D. and Conroy, M.J. (2002) *Analysis and Management of Animal Populations*. San Diego, CA: Academic Press.

Wotschikowsky, U. (2010) Ungulates and their management in Germany. In M. Apollonio, R. Andersen and R. Putman (eds.) *European Ungulates and their Management in the 21st Century*. Cambridge, UK: Cambridge University Press, pp. 201–222.

Yoccoz, N.G., Nichols, J.D. and Boulinier, T. (2001) Monitoring of biological diversity in space and time. *Trends in Ecology and Evolution* **16**, 446–453.

6

Impacts of wild ungulates on vegetation: costs and benefits

FRIEDRICH REIMOSER AND RORY PUTMAN

6.1 Introduction

It is recognised that wild ungulate species can have a profound effect on their environment and that this may often cause conflict with human land-use objectives (e.g. Eiberle and Nigg, 1983, 1987; Putman and Moore, 1998; Fuller and Gill, 2001; Putman, 2004). Whereas in the past much of the focus on damage by ungulates was in relation to damage to agriculture (Stahl, 1979; Gossow, 1983; Putman and Kjellander, 2002), damage to forestry through browsing and bark stripping is clearly also a major and increasing problem in many European countries (e.g. Mitchell *et al.*, 1977; Mayer and Ott, 1991; Gill, 1992a, 1992b; Donaubauer, 1994; Kuiters *et al.*, 1996). In agriculture impacts from wild boar especially are described once again as an ever-growing problem (e.g. Schley and Roper, 2003; Arnold, 2005; Wildauer, 2006; Wildauer and Reimoser, 2007a, 2007b; Apollonio *et al.*, 2010a).

In some countries, there is also increasing concern being expressed about damage to conservation habitats (Reimoser, 1993, 2002; Putman and Moore, 1998; Reimoser *et al.*, 1999; SNH/DCS, 2002; Putman, 2004; Casaer and Licoppe, 2010; van Wieren and Groot Bruinderink, 2010).

Collisions of ungulates with motor vehicles (accidents with cars, trains, etc.) are also increasing (Groot Bruinderink and Hazebroek, 1996; Putman *et al.*, 2004; Chapter 8, this volume). Sickness transfer by wild ungulates to domestic animals and humans is a severe problem in some regions. Both these issues are, however, treated in other chapters within this volume (Chapters 7 and 8) and therefore the present chapter will focus specifically upon impacts on agriculture, forestry and conservation habitats.

Ungulate Management in Europe: Problems and Practices, eds. Rory Putman, Marco Apollonio and Reidar Andersen. Published by Cambridge University Press.

In an earlier volume (Apollonio *et al.*, 2010b) an attempt was made to draw together information on ungulate populations, their impacts, and current management practice in each of 28 different European countries; this compilation of information permits for the first time an overview of the impacts caused by ungulates across Europe as a whole, and differences in management approach adopted to counteract actual damage. In this chapter therefore we present an overview of impacts and damage caused by ungulates in different European countries, the different concepts and types of assessment and management measures adopted to mitigate problems (and the extent to which different approaches to management may successfully address the problems – or may, in the event, act to exacerbate damage!).

The chapter ends with recommendations for future ungulate and habitat management to prevent damage more efficiently – urging that approaches to the management of populations and the management of impacts are viewed as different (if linked) objectives (and that different methods may be more effective in pursuit of those different ends), as well as emphasising a need for a more holistic approach to management which integrates management of ungulate populations and their impacts within the fuller context of overall land-use strategy and planning.

6.2 Impacts of ungulates on human and natural ecosystems; when are impacts considered damaging?

6.2.1 Ungulates and the way they may affect ecosystems

Grazing and browsing from wild ungulates have always played a role in determining the structure and dynamics of natural ecological systems, both in terms of their immediate, present-day influence on the ecological functioning of those communities and as a powerful selection pressure in the original development of such systems.

In most natural temperate systems, the actual density of large herbivores is relatively low (even without the historical intervention of humans in elimination of many of the larger species or attempted regulation of the population size of others). Density-dependent mechanisms and social factors restrict the density to levels at which, while their impact as selective forces on individual plants may still remain, their impact on the immediate dynamics, species composition and species dominance of whole communities is less obvious. Herbivores in general remove < 10% of the above-ground primary production from *any* natural community (more commonly nearer 5%) and large herbivores on their own – as distinct from smaller rodent or invertebrate herbivores – generally consume far less than this. Where populations do reach sufficient density, however, they may indeed have a marked impact on their vegetational environment. This increased impact of grazing and browsing, seed predation and rooting may have serious implications for management, which will be considered below.

Before embarking on this assessment of negative impact, however, it is perhaps worth noting by way of introduction that we should be cautious of over-reaction. Damage caused by large ungulates, whether to commercial or conservational interests, merely represents the extreme expression of a whole suite of general changes, subtle deflections in the structure and dynamics of natural communities in response to herbivory. We consider it 'damage' when the consequences are extreme and/or conflict with human interests or management objectives, but in fact from the very outset the presence of herbivores has a number of profound effects on the whole structure and ecological functioning of any ecosystem. Grazing and browsing, as well as trampling, or rooting by wild boar, may all have many positive, facilitative effects within natural or managed communities, as well as a *potential* for causing damage – and reduction of population levels in response to perceived damage may itself result in even greater disturbance to the system in other ways.

In order to set this into context we review briefly (after Putman, 1986a, 1996a, 2004; Reimoser, 1986; Reimoser and Gossow, 1996) the range of different (purely neutral, *ecological*) impacts that herbivores may have on the vegetational component of those ecosystems of which they are a part (see also introduction to Chapter 9). These impacts are, as above, neither intrinsically damaging nor beneficial (which 'judgements' depend entirely on human-contrived opinions or objectives); they are simply effects.

By feeding in one place and dunging in another, large herbivores create discontinuities in nutrient flows through the system – and the fact that many of the system's nutrients are taken out of circulation for a period (retained in the body tissues of the herbivore itself until it dies) imposes further heterogeneity in nutrient availability.

Herbivores may independently affect the productivity of the vegetation browsed: while heavy levels of grazing or browsing may suppress growth rates (by simply leaving the plant insufficient leaf area of photosynthetic tissue to operate at maximum efficiency), lighter levels of off-take commonly result in an actual increase in productivity, stimulating production of side shoots, unfurling of new leaves, etc.

All these changes in turn will result in pronounced changes in plant species composition and relative abundance at all levels.

Grazing may also have a *direct* effect on species composition of a given system. It may lead to actual changes in species composition, with elimination from the community of species particularly sensitive to damage, or others particularly palatable to the grazers which thus incur a particularly high level of 'attack'. At the same time we commonly see an expansion in range and abundance of species which are very tolerant to defoliation, or have specific defences against attack: spines, thorns, or chemical defences rendering them less palatable. And, as species composition of the community changes in response to the pressure of herbivory, grazing may also affect overall diversity within the system. By reducing the dominance of particularly vigorous or aggressive species, herbivores may reduce the effects of competition on other weaker competitors and thus enhance the overall species richness; through feeding, dunging or trampling they may also create gaps within closed swards for the establishment of ephemerals.

By elimination of graze-sensitive species, heavy grazing can also act to reduce diversity: driving the community towards a species-poor assemblage of a few hardy and resistant species. In fact we may see that (actually as a general rule in ecological systems) maximum diversity is associated with intermediate levels of disturbance – neither so high as to cause destruction of the system or elimination of many of its species, yet not so low that its effects are unregistered.

In effect, and depending on circumstances, the impacts of ungulates on plant species may result in: (i) a decrease in diversity and/or abundance; (ii) an increase in diversity and/or abundance; (iii) changes in structure without change in diversity or abundance; or (iv) no ascertainable influence (Reimoser *et al.*, 1999). Which of these different responses is recorded in any particular instance depends on (i) the type of 'disturbance' (e.g. the nature, intensity and duration of the impact of ungulates on soil and plants), and (ii) the 'reaction' of the respective system (e.g. soil and plants).

However, the type of reaction is also found to depend on the initial situation at the time of the disturbance (soil status, germination conditions, vegetation density, species composition, browsing attraction and browsing tolerance of plants, competition between plant species, existing seed trees, light and growth conditions, direction of development, etc.). In turn, the initial situation and thus the predisposition of plant communities to ungulate impact can be markedly affected by the type of land management and land management practice: human impacts (e.g. silvicultural or agricultural measures, pollution) often provide the impetus that changes the effect of ungulates on both natural and man-made ecosystems (Reimoser, 1986; Ellenberg, 1988; Reimoser and Ellenberg, 1999).

Finally, grazing and browsing by ungulate populations over a prolonged period may also have a profound effect on the physical three-dimensional architecture of ecological communities. One of the most striking features of any woodland that has suffered heavy grazing over a protracted period – besides the lack of any significant regeneration – is the notable absence of a middle storey. As a consequence of years and years of heavy browsing, shrubby species of the understorey become stunted and 'hedged' – in the end are completely eliminated, leaving little vegetation between the ground layer and a distinct browseline on the underside of the canopy trees themselves, where any foliage below the canopy within reach of a browsing herbivore will be removed (see, for example, Putman, 1986a).

These effects on the physical structure of plant communities, as well as effects on species composition and productivity, may in their turn have significant implications on the abundance and diversity of other animal and plant species dependent on these same (altered) habitats.

Thus, directly or indirectly, the impacts of large herbivore populations may affect diversity and abundance of populations of butterflies (Pollard and Cooke, 1994; Petley-Jones, 1995; Feber *et al.*, 2001) or other invertebrates (Putman *et al.*, 1989; Stewart, 2001; Wallis de Vries *et al.*, 2007); smaller mammals (Hill, 1985; Flowerdew and Ellwood, 2001); birds (e.g. Fuller, 2001; Gill and Fuller, 2007), and their predators (Hirons, 1984; Tubbs and Tubbs, 1985; Petty and Avery, 1990).

In considering the impact of ungulates on vegetation, soils and the wider community, therefore, it is appropriate to note that such effects are not all entirely negative. It was after all the grazing first of the huge sheep flocks of the medieval graziers, subsequently by rabbits, that permitted and maintained the superb diversity and richness of the chalk downlands of southern England. Lowland and upland heath, dominated by dwarf shrubs such as *Calluna* or *Erica* or *Vaccinium*, are equally 'artificial', created by deforestation and maintained by heavy grazing pressure of domestic livestock or native ungulates. Much of the structure and character of some of the ancient woodlands we may value today has also developed as a direct consequence of a long history of livestock grazing, which by altering patterns and rates of regeneration has profoundly altered the age structure of the woodland trees, and has also significantly modified the species composition of the ground flora. And excessive reduction of that grazing pressure can lead to loss of diversity and scrub encroachment.

Conservation bodies themselves recognise this and in many instances maintain a regime of controlled grazing specifically to sustain the management system under which the current diversity of many systems has developed and on which its continuance is entirely dependent. Seasonal or permanent grazing is indeed becoming an increasingly popular management tool for maintenance of natural habitats in conservation areas in many European countries (e.g. Bakker, 1989; Mitchell and Kirby, 1990; van Wieren, 1991; Wallis de Vries, 1998; Bokdam, 2003; see also Chapter 9 in this volume). In the same way, domestic pigs or wild boar are increasingly commonly being introduced to areas of woodland in order to provide some degree of soil disturbance, increasing the number of niches for seedling regeneration and enhancing overall regeneration success (again see Chapter 9).

6.2.2 When is an impact damage?

When do ungulate impacts on vegetation become damage? Not every twig or leaf browsed is damage to a plant; not every plant damaged is damage to a vegetation stand. As we have suggested above, the concept of 'damage' and 'benefit' depends on resource targets set by different interest groups. Thus, the core of the problem is not 'conflict' between vegetation and wildlife, but between differing human interests (Gossow and Reimoser, 1985; Putman, 2004). Conflicts are most evident on areas with forestry or agricultural production defined as the primary land-use objective, although, as already noted, in some countries there is also a high profile given to instances where ungulates are believed to be causing 'damage' to conservation habitats. In general, 'damage' – in the sense of 'a problem caused by an unwanted condition' – is a subjective human value judgement where impacts recorded are assessed against some defined goal or management objective. The same is true for 'benefit' as a value judgement of the effect of a given level of impact assessed against some desired or preferred state.

The obvious corollary of such recognition is that if we are meaningfully to attribute damage in any ecological system, we must assess impacts against a clearly defined set of aims or objectives – some 'desired condition'. This requires in turn that operational limits and critical loads are specified. Only too often, land managers immediately equate impact with damage, without proper assessment of whether or not recorded impacts truly compromise their economic or other objectives, so that 'damage' is immediately presumed when observers simply record heavy impact.

For example, in forestry, the browsing of 1000 trees/ha may be considered significant damage when we have only 2000 trees/ha. It is no damage when we have 10 000 trees/ha and we need for the future only 1500/ha. Further, such analysis makes the presumption that the initial browsing has a damaging effect on the individual tree in the first place; in many cases it is apparent that light levels of browsing have no effect on growth or survival of such trees, as long as the leading shoot is not browsed.

From a species-neutral perspective, the term damage has no meaning. A problem needs an owner – it is the problem owner who defines 'damage'. For example, while heavy browsing will be a problem both for the forester and for the trees, other plants, as well as animals and people that seek open spaces, will welcome the demise of the trees. Attitudes towards browsing damage depend primarily on the forest operator's view of economic priorities. Landowners are more tolerant of browsing damage if income from hunting is the priority, for example when timber prices are low or timber extraction costs are high. Similar factors influence conclusions as to whether any particular level of recorded impact may be considered damage or non-damaging in relation to grazing damage in grasslands or arable lands (Putman and Kjellander, 2002).

To define 'damage' in relation to nature conservation aims and targets requires definition of nature conservation values. Hence it is necessary to define what constitutes a favourable conservation condition. A favourable condition of a habitat has been defined as occurring when 'the specific structure and functions which are necessary for its long-term maintenance exist and are likely to continue for the foreseeable future and the conservation status of its typical species is also favourable' (UK Monitoring Network and NATURA 2000 Coordinators Group 1997, unpublished). Damage would then be any change which induces a condition that can no longer be described as favourable.

6.2.3 Problems of assessment and wrongful attribution of damage

As well as emphasising here that damage is only 'damage' if some recorded impact is in conflict with some clearly defined objective of management, it is also important to stress that the ecological and economic implications of that recorded impact must also be more formally assessed. We must be careful of equating apparent damage with actual long-term economic loss. One of the reasons it is so hard to assess the significance of damage caused is that much of the immediate damage may be repaired.

In forestry, because of the long delay from the cause (influence of ungulates) to the effect (damage or benefit) of often some decades, the assessment of damage is much more difficult than in agriculture (Schwarzenbach, 1982; Reimoser, 1986; Gill, 1992a, 1992b; Reimoser and Reimoser, 1997). Hasty and false inferences about damage frequently result in conflicts between foresters, landowners, hunters, nature conservationists, federal authorities, and even tourists. In many cases there is a large divide between perception and reality.

Pursuing our forestry example above, we may note that, unless they have actually been killed outright, damaged trees will in many cases recover completely – and while growth may be checked in some instances, in others growth rates may be unaffected or even faster than those of unbrowsed peers! Browsing by red or roe deer of lateral shoots of Norway spruce (*Picea abies*) and Sitka spruce (*Picea sitchensis*) in Scotland has been shown in a number of studies to result in an actual increase in growth rates by comparison with unbrowsed controls, and even removal of terminal buds does not necessarily result in permanent damage (Staines and Welch, 1984; Cousins, 1987) and similar effects are reported also for a number of native broadleaved species (e.g. Cousins, 1987).

In agriculture, too, apparent damage may not necessarily equate with actual long-term economic loss (see Putman, 1986b; Doney and Packer, 1998; Putman and Moore, 1998; Putman and Kjellander, 2002). Winter-sown cereal fields in southern England might have 30% of the area of crop field grazed back by roe deer during the vegetative phase; physical damage may also cause sizeable flattened areas within the crop. However, although quite high levels of grazing were recorded to vegetative parts of the crop early in the season, with up to 30% of the total crop area affected, this may in practice prove of no economic significance by harvest. In both Putman's and Doney and Packer's analyses, damaged areas of the crop showed evidence of a compensatory increase in rate of growth to catch up with ungrazed treatments by the time of harvest. Early grazing at low intensity also encouraged tillering within the crop, with an actual increase in the number of grain-bearing stems by harvest. Ears reached maturity at the same height and by the same date as those of undamaged plants; individual ears were somewhat smaller in size, but overall grain yields per square metre were not significantly reduced and were in some cases increased. Thus actual economic significance of damage at harvest may be far less than would appear from assessment of the extent of actual damage caused at the outset. Similar results are presented by Kamler *et al.* (2005) and Cerkal *et al.* (2006).

In order to claim some impact of ungulates as damaging, therefore, we need to be able to demonstrate

- that the perception of impact is real, not simply 'presumed'
- that the recorded impact is in fact due to ungulates and not some other agency (it is not uncommon for ungulates to be accused of damage caused

by other herbivore species, or even damage due to physical factors such as frost or desiccation)

- that the recorded 'instantaneous' impact has long-term ecological or economic consequences on the condition of the vegetation type affected
- that these consequences conflict with clearly defined objectives/aims for the desired condition of this vegetation type.

With such preamble, we would not wish to suggest that impacts of ungulates are never damaging. We recognise that where populations of large herbivores do reach sufficient density, they may indeed have a marked impact on their vegetational environment and in certain situations there may be a case for controlling the level of grazing where impact has risen to such a level that it conflicts with other management objectives determined for a particular site. Simply, here we try to emphasise that we should be more cautious, and more objective, in our assessments, stressing that impacts do not always and inevitably equate to damage – and (below) may even be seen in some instances as beneficial.

6.2.4 Beneficial impacts

As we have noted, the impacts of ungulates within natural or managed systems may be beneficial as well as potentially damaging (although here again we must reiterate that the judgement that an impact is 'beneficial' is as subjective as any conclusion that it is 'damaging', and can only be assessed with reference to defined objectives of management). Possible beneficial impacts of ungulates range from the treading in of seeds into the ground and their dispersal, through selective browsing of unwanted competing species (for instance blackberry competing against tree species; Reimoser *et al.*, 1997), to improving regeneration conditions as a result of their droppings and redistribution of nutrients or by providing regeneration niches by making holes in established swards, or breaking up accumulated litter through trampling and rooting. However, only scant research data concerning the positive effects of ungulates in the ecosystem exist (e.g. Putman, 1986a; Wolf, 1988; van Wieren, 1991; Reimoser *et al.*, 1999) and, in contrast to the situation in respect of the negative impacts, positive ones have rarely been sought. Benefits to the forest from ungulates have also hardly been recognised in forestry practice – indeed this has been considered to be impossible.

As we have already noted (and see again Chapter 9), this appreciation of the many positive effects which may result from grazing and trampling impacts has increasingly led to active use of seasonal or permanent grazing regimes in the management of many conservation habitats, deliberately exploiting ungulate herbivores as 'environmental engineers' to deliver some desired habitat condition or landscape condition (e.g. Bakker *et al.*, 1983; Bakker, 1989; van Wieren, 1991; Wallis de Vries *et al.*, 1998).

6.3 Actual impacts reported in different European countries

In general, the wildlife–ungulate problem in cultivated landscapes comprises two main aspects: (i) damage to humans and human activities by wild ungulates (impacts on forestry and agricultural products) and (ii) harm to ungulate populations and their habitats by human activities (fragmentation and losses of habitat, disturbance impacts) that may cause problems in conservation, animal welfare, and sustainable hunting. In the reports from different European countries collated in Apollonio *et al.* (2010b) we note that impacts reported were largely focused upon negative consequences of ungulates (different kinds of damage by the animals to agriculture, forestry or conservation habitats); few contributors discussed benefits arising from ungulate populations and their impacts (but see Reimoser and Reimoser, 2010; Putman, 2010).

In some countries impact of human activity on ungulate populations is also of increasing concern, where populations of taxa with high conservation importance may be threatened by direct hunting or land-use changes, or where age and sex structure of more abundant species may be profoundly altered by selective harvesting of trophy males (Apollonio *et al.*, 2010a).

6.3.1 Data available

Although impacts of ungulates are described as a problem in many European countries (Apollonio *et al.*, 2010b) much of this appeared to us subjective, and in relatively few cases were countrywide monitoring systems in place to record these impacts systematically and provide objective data. Even where these do exist, different monitoring methods are employed in different situations.

The situation in the United Kingdom can function as an example to show how fragmentary are the data available in most cases:

(i) There is no countrywide system for monitoring damage to agriculture. Such a system was maintained in England and Wales for a brief period between 1987 and 1989 (the COSTER project, see Putman and Moore, 1998), but did not extend to Scotland or Northern Ireland and was subsequently discontinued. In other countries local surveys on impacts on agriculture on managed areas may exist but no federal authorities are in charge of collating all data (see for example Apollonio *et al.*, 2010c, for the situation in Italy). In consequence, most information available (e.g. Putman, 1986b; Putman and Moore, 1998; Packer *et al.*, 1999; Putman and Kjellander, 2002) derives from single individual studies, not an ongoing system of monitoring.

(ii) Similarly there is a regular monitoring system for damage *only in state-owned forests* (Forestry Commission forests). No formal monitoring of ungulate impacts is carried out in any systematic way in privately owned forests.

(iii) Once again, there is no countrywide system for monitoring impacts on conservation systems *except where these are in designated sites* (SAC, SSSI, etc), when routine impact assessments are carried out by the competent authorities (Natural England; Scottish Natural Heritage) approximately every 5 years.

Finally there is no countrywide system for monitoring road traffic accidents or disease status of deer and other ungulates (see Chapters 7 and 8). While individual surveys are carried out at intervals (e.g. Putman *et al.*, 2004; Langbein and Putman, 2006; Langbein, 2007 for deer–vehicle collisions), these are not considered part of an ongoing system of monitoring.

The situation for other European countries is shown in Table 6.1, based on data presented in Apollonio *et al.* (2010b). In the 28 countries for which we have collated information, national (countrywide) monitoring systems or statistics on impact by wild ungulates (rooting, grazing, browsing, fraying, bark stripping, etc.) or damage (economic costs, compensations, etc), only exist in six countries in relation to agriculture, and also in six countries related to forestry (predominantly Scandinavian and Eastern European countries; Table 6.1). One country (Greece) lacked any qualitative information on a regional scale, in both agriculture and forestry, probably due to low ungulate densities without significant problems (see Table 6.2). In relation to impacts on conservation habitats no country undertakes any systematic monitoring at national level; for the most part monitoring, if any occurs at all, is limited to specific conservation areas (i.e. in Austria, Baltics, Belgium, Great Britain, Netherlands, Slovakia, Spain, Sweden, Switzerland; see Table 6.2).

The greatest number of national monitoring systems exist for vehicle collisions (minimum numbers) or traffic accidents by wild ungulates (in 10 of 25 countries, Apollonio *et al.*, 2010b). Very little monitoring exists for disease impacts. The countries surveyed hardly have systematic national

Table 6.1 *Existing national schemes of countrywide monitoring systems for recording impacts (or monetary damage) by wild ungulates (at least for one species)*

Country	Agriculture	Forest	Vehicle collisions	Source
Austria		X	X	Reimoser and Reimoser (2010)
Baltic countries			X	Andersone-Lilley *et al.* (2010)
Belgium				Casaer and Licoppe (2010)
Croatia				Kusak and Krapinec (2010)
Czech Republic				Bartoš *et al.* (2010)
Denmark				Andersen and Holthe (2010)
Finland	X	X	X	Ruusila and Kojola (2010)
France	X		X	Maillard *et al.* (2010)
Germany			X	Wotschikowsky (2010)
Great Britain				Putman (2010)
Greece				Papaioannou (2010)
Hungary	X	X		Csányi and Lehoczki (2010)
Italy				Apollonio *et al.* (2010c)
Netherlands			X	van Wieren and Groot Bruinderink (2010)
Norway			X	Andersen *et al.* (2010)
Poland				Wawrzyniak *et al.* (2010)
Portugal				Vingada *et al.* (2010)
Romania				Micu *et al.* (2010)
Slovakia	X	X		Findo and Skuban (2010)
Slovenia	X	X	X	Adamic and Jerina (2010)
Spain				Carranza (2010)
Sweden		X	X	Liberg *et al.* (2010)
Switzerland	X		X	Imesch-Bebié *et al.* (2010)

disease surveillance for wild ungulates except for epidemic diseases such as swine fever or foot and mouth disease. In the Netherlands this monitoring is undertaken and organised on a national level by the landowners and samples are collected by the hunters during the regular cull or, when necessary, by veterinarians (van Wieren and Groot Bruinderink, 2010).

6.3.2 Significance of impacts recorded

While comprehensive, countrywide monitoring systems often do not exist, 'snapshot' figures of impact or damage are often available from 'one-off' surveys, published at least for certain specific years, which may make it

Based on the information presented in Apollonio *et al.* (2010b), Table 6.2 contains figures of impact and damage caused by ungulates for European countries. The table gives an overall impression of what impacts are addressed in different countries, the very different methods employed in monitoring those impacts (resulting in considerable lack of homogeneity in information), and emphasises the clear lack of certainty about impact and damage caused by ungulates in most cases. In many countries data collection systems (where these exist) differ even between neighbouring provinces, making it still more difficult to collate any meaningful information at a national level. Therefore in many countries there are no national statistics with quantitative data available on the ecological impact and economic damage of wild ungulates.

Table 6.2 *Figures of impact and damage caused by ungulates for European countries*

Country	Agriculture	Forest	Conservation habitats
Austria Reimoser and Reimoser (2010)	No national statistics. Most problems and increasing damage by wild boar. Only locally is damage caused by other ungulate species (esp. red deer). Compensation is the responsibility of the hunters of the hunting district in which the damage arises.	Every 5 years national statistics on impact are available (national forest inventory). 280 million trees (8% of trees in Austria with more than 5 cm breast height diameter) are recorded with impact by bark stripping (mainly red deer). 36% of forest regeneration area damaged by browsing or fraying (mainly roe deer, red deer, and chamois). Calculated damage costs on average €218m/year (1990–1999, browsing, fraying, bark stripping; all present ungulate species). Again compensation is the responsibility of the hunters of the hunting district in which the damage arises. Damage costs are only in part claimed by the forest owners.	Conservation areas have different goals and methods for monitoring ungulate impact on vegetation. But most national parks agree on a system whereby maximum 50% area of each plant-community the vegetation structure may be shaped by ungulates (by browsing, fraying, bark stripping, trampling etc.), and an area of at least 50% of these 'natural' plant communities (particularly forest communities) shall be able to regenerate and grow up without significant change by ungulate impact.

Baltic countries (3) Andersone-Lilley *et al.* (2010)	No national statistics. No data available on the amount of compensation paid. Wild boar can cause significant damage to agriculture. In Lithuania, red deer causes most damage to agriculture after wild boar. In Estonia, the red deer density is very low and no significant damage is done to forestry or agriculture. In Latvia, red deer cause less damage than moose and roe deer. Damage caused to agriculture or forestry by game species should be compensated by the users of the hunting rights in that specific area. In practice, instead of direct compensation, hunters often choose to eliminate troublesome individuals or reduce the density of a particular species that causes damage.	No national statistics. Damage (mainly by moose and red deer in forest plantations) not registered centrally. In mixed spruce–deciduous stands moose may destroy aspen, oak and ash saplings, thus changing the future composition of forest stands. Biggest economic impact of moose is the damage to pine plantations and spruce forests. In Estonia (1991) moose damage was found in 2424 ha of young pine plantations, 13 541 ha of pine plantations, and 12 778 of middle-aged spruce forest. Roe deer usually do not cause significant damage. In deciduous stands, their selective browsing can have an impact. The forested area is constantly increasing due to reforestation of former agricultural lands; therefore, the overall damage of ungulates becomes negligible.	In protected areas wild boar can destroy orchids and their habitats, and can cause damage to ground-nesting birds. In Lithuania wild boar root up *c.* 0.4% of pure pine stands, 2.9% mixed stands, 0.9% mixed spruce–deciduous stands and 2.4% deciduous stands every year. At the same time, their digging activity is favourable for diversifying forest stands and increasing small-scale biodiversity.

Table 6.2 (*cont.*)

Country	Agriculture	Forest	Conservation habitats
Belgium Casaer and Licoppe (2010)	No national statistics. Main damage is caused by wild boar (corn fields and pastures). Costs of damage must be paid by the hunters or by a group of hunters adjoining the damaged fields.	No national statistics. Fencing is required whenever foresters want to convert previous monocultures of pine forest into mixed or deciduous forests. In Wallonia forestry administration developed a monitoring network of bark stripping damage, covering public areas on the distribution area of red deer. The annual bark stripping rate is higher than 2% of trees. Compensation is payable by those having the hunting rights.	No current systems are in place for monitoring of the impact of ungulates on natural biodiversity, although it is suggested that this willl be an important aim of the next decade. In nature reserves compensation would be paid by the Flemish authorities, but no compensation has ever been paid.
Croatia Kusak and Krapinec (2010)	No national statistics. Total yearly damage by game animals estimated €685 000 (incl. damage to agricultural and forest crops, vehicles). True amount of damage is difficult to measure because hunters compensate amounts of damage by venison or crops rather than by cash. Main damage by wild boar (95%), red deer. Regional data show damage of €0.5–3/ha of hunting area.	No national statistics. Damage from ungulates is generally considered to be negligible. Reason: fencing of old oak stands (stands which are in regeneration). In beech stands ungulates do not cause significant damage.	No information available.

Czech Republic Bartoš *et al.* (2010)	No national statistics. It is the responsibility of the hunting ground user to compensate for the damage that is caused in the hunting ground to standing field crops or forest stands. If the hunting right is exercised by an association, its members are liable for the damage compensation jointly and severally. In reality, financial compensation for the damage by the hunting ground users is low.	No national monitoring of wildlife impact. Financial compensation for damage is low (2002: > €2m; 2003: €1m – Report of the Ministry of Agriculture). The total damage to forest stands (under control of the state forest company) caused by ungulates has been estimated to reach up to €1.5 billion every year.	No information available.
Denmark Andersen and Holthe (2010)	No national statistics. Red deer cause most damage. There is no compensation, since the hunting rights for each landowner are treated as a form of compensation. Red deer, fallow deer and sika deer may be shot outside hunting season if they are found inside properly fenced agricultural areas and fruit orchards.	No national statistics. Red deer cause most damage by bark stripping young spruce and pine forests. No compensation (hunting rights for each landowner used as a form of compensation). Foresters are concerned that deer numbers will jeopardise their goal of restoring more stable, resilient and natural forest ecosystems by causing browsing damage on broadleaved tree species.	No information available.
Finland Ruusila and Kojola (2010)	Amount of ungulate compensation in 2006 was €0.26m. Main damage caused by moose and white-tailed deer. Government compensation is	Amount of ungulate compensation in 2006 was €3.2m. Main damage caused by moose and white-tailed deer. Government compensation is	No information available.

Table 6.2 (*cont.*)

Country	Agriculture	Forest	Conservation habitats
	paid to private landowners, but not to state-owned forestry (money comes from licence fees charged to hunters).	paid to private landowners, but not to state-owned forestry (money comes from licence fees charged to hunters).	
France Maillard *et al.* (2010)	Fédération Départementale des Chasseurs (FDC) is in charge of raising funds from hunters for compensation to farmers. The damages are declared by the farmers and the compensation is estimated by experts employed by the FDC. Hunters had to pay €21 6340 000 in 2004/2005 for total damage. Wild boar is responsible for 87% of the total amount paid for big game damage. Red deer may cause high levels of crop damage (10% of the total amount paid).	Up to now no national statistics. Forest owners are now in a position to demand compensation for forest damage, or funds to protect their trees. However, these compensations for the private foresters depend on some specific conditions: damage to the trees by deer must be proved, and the local hunters must have shot the minimum quota of deer for the year. This system is being set up; currently compensation for damage to forestry is provided only very rarely.	No information available.
Germany Wotschikowsky (2010)	No national statistics. Main damage by wild boar (esp. to cornfields and meadows). Damage has to be compensated for by the community of the landowners of a hunting district, but usually the compensation is regulated in the lease contract, with the result that the leaser	No national statistics (different monitoring methods depending on state). Over the country as a whole, browsing pressure is considered generally high according to most foresters (mainly roe deer).	No information available.

	(hunter) will have to pay. No compensation is paid for landowners who execute the hunting themselves.	Bark stripping caused by red deer is a local problem in winter and increasing also in early summer.	
Great Britain Putman (2010)	No national statistics. Over the country as a whole, economic losses due to deer are assessed as small (only local significance). For England only, total cost of damage are estimated €6.56m/year (range 1.66–8.34). No established system for paying compensation.	No national statistics. Perhaps a larger impact of ungulates than in agriculture but damage also tends to be localised and concentrated only in certain areas. Damage due to deer in (coniferous) plantations in England and Wales rarely exceeded 5–10% (see Wray, 1994; Putman, 2004). There is no established system for paying compensation.	Problems seen as generally local. In one survey of impacts in (English) National Nature Reserves 45% of site managers recorded an impact, only 18% reported difficulty in meeting management goals (Putman, 1996b). However, more recently deer implicated in failure of a significant proportion of designated sites (SSSI or SAC) to achieve favourable condition. All designated conservation sites now subject to routine Habitat Condition Monitoring every 5 years.
Greece Papaioannou (2010)	No official records of any kind of damage related to wild ungulates. The national and local media sporadically report limited damage to corn crops, caused by wild boars. There is no compensation system regarding damage, except the damage to crops caused by wild	No official records. In general, apart from sporadic cases of low scale damage, wild ungulate populations in Greece do not cause damage to agricultural production, habitats or productive forests. Low population densities of wild ungulates are considered to be	Probably no significant impact (low ungulate densities).

Table 6.2 (*cont.*)

Country	Agriculture	Forest	Conservation habitats
	boar within the controlled hunting reserves.	the main reasons for this phenomenon.	
Hungary Csányi and Lehoczki (2010)	Compensations paid for agricultural game damages in 2005: €4 564 000 (National Game Management Database), but the proportion of this caused by ungulates (esp. by wild boar, red deer) is unknown. According to the Game Act the party exercising hunting right is responsible for damages caused by game. Responsibilities can involve the compensation of damages and also the contribution to prevention measures (e.g. payments for fencing).	Forest area totally damaged by ungulates (replanting necessary) 430 ha; partly damaged 6500 ha (2003). Red deer and roe deer cause most damage. National statistics show compensations paid for forest damages €585 000 (2005). The party exercising hunting right is responsible for damages.	No information available.
Italy Apollonio *et al.* (2010c)	No national statistics. Reliable estimates of the amount of damage by ungulates to agriculture or forestry are not available, being affected by the pattern of provincial administrations which collect data with different methods or do not have data at all. About 90% of damages attributed to	No national statistics. In the Alps (esp. Eastern Alps), red deer may cause considerable damage (esp. browsing). Damage to trees and commercial forestry are not normally compensated.	No information available.

	ungulates are caused by wild boar with a total economic value of probably more than €10m/year. Damage to crops and orchards dominates overall. In most areas, farmers receive compensation for damage to crops (paid by the provinces).		
Netherlands van Wieren and Groot Bruinderink (2010)	Compensation for ungulate damage (esp. by wild boar, red deer) is small (€25 000–53 000/year for the period 2001–2004). All ungulate species fall within the list of protected species (compensation comes from the Fauna Fund of the Ministry). The total amount paid for damage compensation by the Fauna Fund for all species was €6 177 000 (2004) and €4 239 000 in 2003 (major share was for damage by geese).	Plays only a little role. No further information available.	Policies are being developed to allow red deer more and more as joint-users on certain lands. The Fauna Fund made agreements with 50 landowners to provide opportunities for red deer to use their land. As compensation they receive in total about €35 000/year.
Norway Andersen *et al.* (2010)	No national statistics. Most serious losses seem to be connected to spring grazing by red deer on grazing fields for livestock. There are also reports on roe deer damage to strawberry fields caused by browsing on plants and trampling on the plastic ground cover. There is no state	No national statistics. Locally significant, but regionally moderate, impacts caused by red deer and roe deer. Heavy browsing and bark stripping on forest (esp. Scots pine) in wintering areas for moose. In the last two decades (wood prices low) landowners often have been willing to trade the costs of	No information available.

Table 6.2 (*cont.*)

Country	Agriculture	Forest	Conservation habitats
	compensation system for damage to forest and agriculture. However, landowners can apply to the municipality for economic support to prevent damage through specific programmes of action in local areas with high impact.	forest damage for a high density of moose.	
Poland Wawrzyniak *et al.* (2010)	No recent data on damage caused by ungulates. Hunting clubs and managers are obliged to pay remuneration for damage caused by game species. In hunting season 2002/2003, hunting clubs paid compensations to the owners of 13 200 ha of crops destroyed mainly by red deer and wild boar. Damage caused by protected species (European bison) is compensated by state budget.	National statistics only for state forests. In hunting season 2002/2003 damage by ungulates (mainly red and roe deer) was noted on 24% of all young forests (1–20 years) managed by state forests. Most of the recorded damage (17%) was graded as slight. No such data are available for the private forests (17% of Polish forest cover). In 2003, protection of young stands against damage was applied on 109 400 ha (chemical and mechanical repellents and 14 600 ha fenced). The costs of protecting against ungulates are covered by state forests. In the whole of Poland, those costs exceeded €11m in 2002 and €15m in 2003.	No information available.

Portugal Vingada *et al.* (2010)	No national statistics. Extent and type of damage differs according to region. Main problems are caused by wild boar, increasing problems by red deer.	No national statistics. No further information available.	No information available.
Romania Micu *et al.* (2010)	No national statistics and no quantitative data. Damage mainly caused by wild boar (esp. in cornfields or potato patches close to forests). Because of high costs of prevention, many of the agricultural property owners do not defend against game damages.	No national statistics and no quantitative data. The level of damage in forests is probably small owing to the low density of ungulates. Compensation is rarely given by the responsible authorities and only if all preventive measures have been taken. Most of the game damage is done by red deer and wild boar and the least by roe deer and chamois.	No information available. Game damage in plantations as well as in naturally regenerating areas cannot be estimated properly because damage caused by pasturing cannot be separated from game damage.
Slovakia Findo and Skuban (2010)	Average annual damage reported over the period of 2001–2003 to farm crops was €130 000, while in 2005 this increased to €320 000 (Hunting Statistics 1968–2005). Damage is caused mainly by wild boar and red deer, and evaluated based on expert opinion. Owners and users of hunting ground have to compensate for damage.	Since 1960 damage by ungulates is annually assessed. Since 2000 increasing browsing damage reaches €250 000, also increasing peeling damage €100 000 (2005). Most damage is caused by red deer, further by roe deer, mouflon, and fallow deer. There is a legal duty for forest owners to protect trees against game damage. On average 20 000 ha of forest are	Within protected areas some rare plants are damaged or even locally exterminated e.g. English yew (*Taxus baccata*).

Table 6.2 (*cont.*)

Country	Agriculture	Forest	Conservation habitats
	If protective measures have not been taken by the owners the damage is not fully compensated.	annually protected while the costs exceed €1 429 000. Hunters have to compensate for damage to forest trees (government compensates for protected species such as bison and moose).	
Slovenia Adamic and Jerina (2010)	Extent of damage and compensation claims increasing during the last 30 years (damage by wild boar and red deer increased whereas impacts of roe deer decreased). In the period 1998–2000 the compensation for damage by wild boar has reached €460 000 (i.e. about 60% of all agricultural damage claims in Slovenia). In 2005, 52% of reimbursed damage were related to cereal crops, 43% to pastures, and 5% to others (orchards, vineyards). Compensations must be paid by local hunters clubs.	Browsing damage by ungulates is regularly checked on state level. Impacts of ungulates might seriously reduce the process of natural regeneration and affect strategic issues of forest management. Until now the claims for ungulate damages in forests have not been reimbursed, except in few cases of extreme browsing of young spruce plantations, and in the cases of repeated winter peeling of bark in spruce pole stands.	No information available.

Spain Carranza (2010)	No national statistics. Damage to agricultural and horticultural crops mostly caused by wild boar. In most cases, the compensation/solution by the administration is in the form of special permits to cull animals in the area where damages take place.	No national statistics. Serious damage on deciduous or coniferous species (mostly caused by roe deer) is confined to the north half of Spain because forest plantations are scarce in the south.	In Mediterranean forests and open woodlands (dehesas), deer overabundance (mainly red deer) can jeopardise natural shrub species, and limit the regeneration of native tree species.
Sweden Liberg *et al*. (2010)	No recent estimates of costs due to game damage on agricultural crops are available. Ungulates caused an annual loss of about €1.0m between the years of 1980–1987. The overall area of crops reported as suffering damage (as a proportion of the area grown in any region) rarely exceeds 5% and is usually lower than 1%. Since 1995, it is no longer possible to obtain compensation for damage by ungulates either to agricultural crops or to forest crops.	National monitoring only of moose impact (not of red deer and roe deer). Main problem is moose damage on economically important forest trees, mainly Scots pine. Top shoot browsing, stem breaking, or bark stripping on Scots pine on 12% of main stems/year (trees 1–4 m tall); (goal 2%). Accumulated damage level (all damage irrespective of time of the damage) is 40–50%, i.e. 40–50% of the pine stems have a damage caused by moose (2006). Economic costs related to the above reported impact of ungulates are mainly unknown. Moose impact on pine wood quality is estimated at least €50m/year (2005). No compensation is paid for forest damage caused by ungulates.	In parallel to the growing ungulate populations and stronger focus on biodiversity, the interest in the impacts of ungulates on ecosystem dynamics and ecosystem characteristics increased (esp. impact of moose, roe deer, wild boar). Existing knowledge gap on variety of impacts (lack of good historical data).

Table 6.2 (*cont.*)

Country	Agriculture	Forest	Conservation habitats
Switzerland Imesch-Bebié *et al.* (2010)	National statistics only for wild boar: 2003, €1.6m; 2004 (after extension of hunting season), €1.0m. Wild boar damages to agriculture are assessed either by the cantonal game wardens or special damage experts. Hunters have to pay at least a part of the compensation (20–25%). Damages are only compensated if the farmer has taken some minimal prevention measures.	No statistics on economic damage costs. Problems with red deer, roe deer, chamois, and in some areas also with ibex. Forestry goal: on 75% of forest area (each canton) natural regeneration has to be ensured using locally adapted tree species without protective measures. No compensation of damage (costs must be paid by the forest owners themselves)	Investigations in the Swiss National Park showed positive effects of ungulate impacts on plant biodiversity (Filli and Suter, 2006).

possible to extrapolate some estimate of average ecological or economic cost of ungulate impacts. In other cases data are available for specific *regions* within a country (see, for example, Putman and Kjellander, 2002, where data are available for agricultural damage in some individual Swedish counties, but not all), or 'smaller' specific sites which have been the subject of more detailed research studies (see, for example, Fonseca *et al.*, 2007; Vingada *et al.*, 2010).

But even from these more restricted surveys (restricted in area of coverage, or restricted to a limited timeframe) assessment of the actual significance of ungulate damage in different contexts is extremely difficult, since assessment methods are different within and between the countries. It is equally hard to put any figure on the economic value of damage to agricultural or forest crops since different systems exist in different countries for payment of compensation – and in many countries no direct compensation is offered (Table 6.2). We are therefore unable at present to offer any overall estimates of the economic significance of damage across Europe as a whole. However, in the paragraphs that follow we will attempt to present some broader synthesis.

Agriculture

In most countries, there is no system of regular or stratified monitoring of agricultural damage, and monetary compensation figures are the only available indicators of actual damage levels. Even then, some countries do not have any information about game damage in agriculture; either no compensation is paid to the farmers or such payments are not registered on a regional or countrywide level (e.g. Austria).

In general, however, it would appear that damage due to *cervids* is rarely of significance at a national level. This is not to suggest that these animals do not cause significant damage; rather that such damage tends to be extremely patchy and localised – significant on a farm-by-farm or even field-by-field basis rather than on a larger, regional or national scale (see, for example, Doney and Packer, 1998; Putman and Kjellander, 2002; Putman, 2004). Even where instantaneous grazing impacts appear high, loss of yield is often negligible due to compensatory growth within the crop (see above).

Thus, such published data as are available suggest that *overall*, loss of yield in most arable crops due to grazing of vegetative parts of the plants by roe deer was likely to be insignificant. In the Czech Republic, loss of vegetative parts of maize crops during the summer months resulted in a decrease in fresh weight of ears at harvest of only 2.6%. Since, in addition, grazing by deer affected less than 0.7% of the crop, the effective loss of yield for the crop as a whole was less than that, at 0.15% (Obrtel and Holisova, 1983; Obrtel *et al.*, 1984). Kaluzinski (1982) calculated that despite high densities of roe deer in agricultural areas in western Poland, consumption of vegetative parts of cereal crops outside the growing season was <1% and would not significantly influence yield – and that, although later damage, involving direct removal of ripening ears, caused measurable and irrecoverable loss, it was nonetheless an insignificant proportion of the crop as a whole. Overall grain yields per square metre in fields of wheat and barley grazed by roe or fallow deer in the south of England were not significantly reduced and were in some cases increased (Putman, 1986b; Doney, 1999).

Recent data recorded for damage to cereals in different counties in Sweden confirm that here, too, the overall area of crops reported as suffering damage (as a proportion of the area grown in any region) never exceeds 5% and is usually lower than 1% (Putman and Kjellander, 2002). Again, however, damage at a local or farm level can be significant, with up to a 26% loss of yield in unprotected oat crops against fenced controls.

Timing of damage in relation to the growth stage and growth characteristics of the crop will, however, clearly affect the economic significance of any damage caused, since it will markedly influence the degree of crop recovery possible after grazing ceases. Damage caused by trampling or rolling of the larger species (fallow or red deer) in visiting cereal fields late in the season may also be of real significance, when the opportunity for compensatory growth is past (Putman, 1989; Doney, 1999).

In addition, in most of the studies summarised in these paragraphs, we are dealing with damage largely due to red, fallow and roe deer. In Sweden, as elsewhere in Scandinavia, much of the damage recorded is due to the significantly larger moose. (Of damage reported here for Sweden, over 98% is attributable to moose – and, as elsewhere in Europe, recorded damage from other species is low). Finally, it is clear from studies reported in Apollonio *et al.* (2010b) that while damage levels reported from deer species (even moose) are comparatively low, far more significant levels of damage may be experienced – and on a wider, regional scale – where there are established populations of wild boar (e.g. Schley and Roper, 2003; Arnold, 2005; Wildauer, 2006; Wildauer and Reimoser, 2007a, 2007b; Apollonio *et al.*, 2010a).

The more significant impact of boar seems in part to relate to their comparatively high local abundance, but also to the fact that the digestive physiology of non-ruminant pigs compels them to depend primarily on high energy foods. In consequence, even if this species is in fact able to survive in mountain range relying on wild vegetation only (Herrero *et al.*, 2005), if they can choose between natural vegetation and agricultural crops they almost invariably go for the latter; agricultural crops represent an important component of the diet throughout Europe (Schley and Roper, 2003; Fonseca *et al.*, 2007; see also chapters in Apollonio *et al.*, 2010b).

This is obviously even more apparent in predominantly agricultural landscapes where the native vegetation has all but been eliminated; here the subsistence of wild boar is completely dependent upon the cultivated crops (Herrero *et al.*, 2006). Additional damage may be caused to permanent grassland by rooting activities while searching for food (Wilson, 2004; Schley *et al.*, 2008). Rooting on meadows indeed often results in more significant, long-term problems for the farmers than rooting on arable land (since it may create an uneven surface and thus cause problems for mowing) and thus damage assessment is much more difficult on grasslands (Wildauer, 2006).

Even if no precise data at a national level are available for most countries, it has been estimated that the damage to agriculture caused by wild boar in Europe is in excess of 80 million euros per year (F. Morimando, pers. comm.)

Forestry

Much as in the case of agriculture, where national data are available at all (Table 6.2), data for forest impacts are variously recorded in different countries in relation to area damaged, number of trees damaged, proportion of damaged trees or/and monetary compensation figures. Some countries apply tolerance limits that should not be exceeded, e.g. Sweden (2% of main stems per year – top shoot (leader) browsing, stem breaking, and bark stripping), and Switzerland (on 75% of forest area, natural regeneration has to be ensured using locally adapted tree species without protective measures).

When considering economic cost as reflected by damage compensation in the different countries it is important to consider that the compensation awarded is not an accurate estimate of actual economic cost of damage because (in so many cases) compensation is nowhere near 100%. For example in Austria about 20% of the calculated damages are really paid by the hunters to the forest owners.

Even if we try and base assessments of the economic significance of ungulate damage on more direct assessments of browsing impact on trees themselves, we must recognise that in very few countries are national inventories of forest impacts undertaken on any regular basis and, further, that the methods used to calculate the damage often are not able to show the real damage to a forest, particularly browsing damage to natural forest regeneration (see Sections 6.2.2 and 6.2.3).

The next paragraphs can thus offer some examples only. Based on official 'assessment tables' (Pollanschütz, 1995) for Austria, the monetary damage to forests by browsing, fraying and barking stripping was calculated at on average 218 euros/ha yearly, with at least 10 000 km^2 (25% of total forest area) damaged per year (Reimoser, 2000). These tables in many cases do not show the real damage; they are more or less an assessment convention between hunters (responsible for damage compensation) and stakeholders of forest owners. The tables can give realistic damage figures only for spruce afforestations on clear-cuts. But they are also used for other tree species, natural regeneration, and shelterwood cutting. Not all these calculated damages are compensated to the forest owners. The main cause of damage (about 70%) is heavy browsing of top twigs (leader shoots) of young trees by roe deer, red deer and chamois that are living in forests. The level of annual ungulate damage in forests is about 50% of the economic value of hunting in Austria (Reimoser and Reimoser, 2010).

Whereas at the countrywide level about 25% of the Austrian forest area is more or less damaged per year, at a local or landowner level damage can be much more significant, with up to 100% damaged area in small private forests. The *National Forest Inventory* of Austria registered bark stripping on 8% of forest trees that have more than 5 cm breast height diameter (that is in total 280 million trees) (Büchsenmeister and Gugganig, 2004). Of sample areas with regeneration, 36% were classified as damaged by wild ungulates, based on comparisons of site-dependent regeneration targets and thresholds (minimum density of undamaged trees required, tree species, maximum browsing intensity) with the current status (Schodterer, 2004).

In Sweden (Liberg *et al.*, 2010), forest damage caused by moose is surveyed locally or regionally through a method called ÄBIN ('Moose Browsing Survey'; www.skogsstyrelsen.se). This method has also recently been included in the Swedish National Forest Inventory. ÄBIN is a package of methods for estimating forest damage, browse abundance and recent (last winter's) and accumulated browsing. However, the main aim is to quantify the impact on main stems of Scots pine (*Pinus sylvestris*). Main stems are stems which will form the future stand and damage is defined as a negative impact on stem quality, induced by top shoot browsing, stem breaking and bark stripping. The method does not include estimation of damage in terms of actual consequent loss of timber production.

Damage level recorded varies between areas, with a range of 1–25% between counties or smaller areas. Data from the Swedish National Forest Inventory 2003–2004 indicate a mean level of 12% for the country, with a range of 9–25% between large regions. This figure means that on average 12% of the main stems of Scots pine are damaged each winter in young forests (1–4 m tall) and with at least 10% Scots pine. Some parts of the forestry sector, especially forest companies, have at present a goal of not more than 2%. The accumulated damage level (all damage irrespective of time of the damage) is 40–50%; that is, 40–50% of the pine stems have damage caused by moose (National Board of Forestry data of 2006). The damage level is not only related to moose density, but also to the characteristics of the forest (Bergström *et al.*, 1995). This is especially true on large (e.g. national) scale, on which there only is a weak correlation between damage level and moose densities alone.

In Hungary another approach is employed (Csányi and Lehoczki, 2010). The Hungarian terminology distinguishes two forms of forest damage: (i) quantitative damage (afforestation is fully destroyed and it should be replanted) and (ii) qualitative damage (some proportion of saplings or young trees is damaged but they can recover). The area of a given forest falling into either category is determined by Forest Service personnel when controlling the status of forest plantations. According to Forest Service data, between 1989 and 2004 both categories of forest damage were declining and were rather stable in the last years. Except for the declining period of the early 1990s no association of ungulate numbers and forest damage could be found. About 3.36% (430 ha) of quantitative forest damage and 66.24% (6500 ha) of qualitative damage could be attributed to game in 2003.

In his review paper in 1992, Gill (1992a, 1992b) reviewed some attempts that have been made to assess the cost of browsing and bark stripping to timber production. After adjusting for inflation and exchange rates (1992), estimates of the costs of browsing range from £0.73–0.98 per ha per year for browsing by moose in Sweden (Jantz, 1982) to £85.23 per ha per year for red and roe deer browsing in Germany. Both of these estimates were intended to represent serious, but not catastrophic, damage, but the variability in the results serves to underlie the difficulties in accurately assessing the cost.

Nothing much seems to have changed and the situation today is more or less the same. The cost of bark stripping is perhaps easier to estimate because it occurs later in the rotation and is readily quantifiable (Speidel, 1980). Even this, however, requires making some assumptions about likely extent of economic loss due to staining of internal timber from fungal infection – and this can only partially be predicted from characteristics such as wound size and tree growth rate.

Conservation

While a number of contributors to Apollonio *et al.* (2010b) noted that in their countries objectives for ungulate management sought a balance between, on the one hand, maintenance of viable populations of ungulates for hunting, and, on the other hand, restricting impact on agriculture, forestry and conservation habitats (e.g. Carranza, 2010; Casaer and Licoppe, 2010; van Wieren and Groot Bruinderink, 2010), only in the United Kingdom and in Austria was any formal assessment reported of ungulate impacts in conservation areas (Table 6.2). Especially from the viewpoint of nature conservation, determination of target or threshold values for objective damage assessment (e.g. density and species targeted for forest regeneration) is particularly difficult (Reimoser *et al.*, 1999). However, in some countries there is recognition that grazing by deer and other wild ungulates may be very important in helping to increase plant diversity or improved expansion of forest on grasslands (Schütz *et al.*, 2000; Filli and Suter, 2006).

By contrast there are also a number of instances where deer are believed to be causing 'damage' to conservation habitats (see, for example, Callander and Mackenzie, 1991; Hunt, 2000; SNH/DCS, 2002). Concerns are primarily expressed about impacts on broadleaved woodland in England and some parts of Scotland (where browsing pressure may in some instances be sufficient to suppress altogether all unfenced regeneration, and at lower densities may significantly distort the species profile of recruitment (see Putman, 1996b, 2009).

Rackham (1975) and Tabor (1993) also highlight damage which may be caused to woodland ground flora at high deer densities, while Cooke has also reported comprehensively on the effects of muntjac (*Muntiacus reevesi*) at high densities on the ground flora within Monk's Wood National Nature Reserve in Cambridgeshire (summarised, for example, in Cooke, 1994, 1995, 2005, 2006).

Heavy impacts within woodlands – whether in relation to effects on ground flora, shrub layer or stand structure (establishment of canopy species), clearly have effects beyond those simply on the vegetation and, as mentioned above, may affect populations of butterflies or other invertebrates, smaller mammals and birds as well as their predators (see Section 6.2.1, and references). While there are a number of species of woodland birds which may derive positive advantage from heavy grazing, such as wood warblers (*Phylloscopus sibilatrix*), pied flycatchers (*Ficedula hypoleuca*) and redstarts (*Phoenicurus phoenicurus*), there is growing experimental evidence in English woodland that through their effects on the understorey vegetation, deer at moderate to high densities can also effect declines in the abundance and breeding success of other woodland bird species (such as nightingales *Luscinia megarhynchos*; Fuller, 2001; Gill and Fuller, 2007).

Nor is all concern about impacts on conservation focused within woodlands: in Scotland concerns are also increasingly voiced about impact on open ground habitats such as dry heathland, blanket bog, Atlantic wet heath – all protected Natural Habitats under EU law. Indeed in Scotland, the Deer Commission of Scotland (DCS) is empowered (under the provisions of the Deer Act (Scotland) of 1996) to take action in situations where deer are considered to be causing a problem to the public interest – whether this is in terms of damage to agriculture, forestry, conservation habitats (defined as 'damage to the natural heritage') or public safety; the DCS is increasingly using these powers to try and negotiate management agreements with landowners in areas where damage by deer to protected habitats is considered to be resulting in deterioration of these habitats.

Wild boar are also implicated in damage to conservation values. In Sardinia, for example, a significant decline in the abundance of orchids within the Asinara National Park has been attributed to loss of rhizomes due to extensive rooting by boar; within the Porto Conte Regional Park, Mediterranean palm (*Chamaerops humilis*) was virtually eliminated as a result of high levels of seed predation (M. Apollonio, pers. comm.)

6.3.3 Which ungulate species seem to be most significant in terms of damage?

Evidence from Apollonio *et al.* (2010b), and Tables 6.1 and 6.2 suggest that in Europe generally, wild boar, moose (where they occur), and red deer are seen as the most damaging wild ungulate species in an agricultural context. In Slovenia, in 1998–2000 the value of claims made in compensation for damage by wild boar has reached €460 000, which was about 60% of all agricultural damage claims in Slovenia in the same period. Calculated wild boar damage in 2005 was €15 per square kilometre within the entire area of Slovenia. The actual breakdown of damage reimbursed in 2005, according to crop type, was: damage to cereal crops, 52.3%; damage to pastures, 42.4%; other types of damage (orchards, vineyards), 5.3% (Adamic and Jerina, 2010). In France, wild boar is responsible for 87% of the total amount paid for big game damage; red deer may also cause high levels of crop damage (10% of the total amount paid) (Maillard *et al.*, 2010). As noted above, it has been estimated that the damage to agriculture caused by wild boar in Europe as a whole may be in excess of 80 million euros per year (F. Morimando, pers. comm.)

In forestry, moose, red deer and roe deer are most important. Locally also mouflon and sika deer have a strong impact on vegetation. In Norway, the 'average moose' has an estimated daily intake of 11–12 kg of Scots pine shoots during winter (Solbraa, 2002). Moose in Norway stay in their winter grounds between 4 and 5 months a year, which means that the average moose needs 1500–1700 kg of browse during winter. The impact this browsing has on timber production of Scots pine is in reduction of number of trees per area, reduction in growth rate and also in timber quality. It may also result in changes in tree species dominance, as spruce may be planted instead of Scots pine to reduce impact of browsing (Solbraa, 2002). In the last two decades prices for forestry products have been relatively low, and in most areas landowners have been willing to trade the costs of forest damage for a high density of moose.

In steep mountain forests of the alpine region also, chamois can play a decisive role in delaying or preventing forest regeneration by twig browsing. Roe deer, the most abundant ungulate species in Europe, is often underestimated in its impact on tree and shrub composition particularly in mixed forests. Roe deer is the most selective browser of European ungulates.

6.4 Management options and management practice

In general it would seem that the most common approach adopted in all countries to try to reduce ungulate impacts (at least in relation to grazing and browsing impacts on forestry, agriculture or conservation habitats) is through reducing regional or local ungulate densities by culling.

However, it is apparent that such attempts are not always effective in achieving their objectives. This is in part because land-use objectives are often not sufficiently clear – or competing objectives show some conflict. Equally commonly, culling effort may be insufficient (because managers have, for example, underestimated rates of recruitment), inappropriately targeted in terms of the age or sex classes of animals culled, or may be insufficiently coordinated over a local area (Putman, 2004; see also Apollonio *et al.* 2010a). Further, particularly where populations may be resource limited, ungulate populations may respond to local reductions of density by increased productivity (increase in fecundity, increase in survival) or 'rebound' through immigration into areas of reduced population density, where competition is in consequence reduced (Putman, 1996b, 2004; Reimoser, 1986).

But perhaps more significantly, most published analyses of the impact of ungulates on forestry and agriculture (or indeed involvement in road traffic accidents; Chapter 8) agree that actual impact levels are only weakly related to ungulate density and that many other factors affect the actual impact even of a constant density of animals. Thus it is clear that the impact of ungulates on forest or agricultural crops is affected by factors such as the availability of alternative forage, overall landscape structure, the proximity of shelter habitats to foraging habitats, and quality of shelter habitats (see, for example, Gill, 1992a, 1992b; Kay, 1993; Putman, 1994; Reimoser and Gossow, 1996), and that in consequence, impact is not closely related to density per se – or certainly not related to density in any linear way. Even in wild boar, where levels of agricultural damage may be more directly related to local density (Bouldoire and Havet, 1982; Labudzki and Wlazelko, 1991; Spitz and Sovan, 1999; Schley *et al.*, 2008), habitat structure, and specifically proximity of agricultural areas to woods or dense shrubs, has also been shown to have a significant effect on actual damage levels sustained (Janeau and Gallo Orsi, 1992; Spitz and Sovan, 1999; Wilson, 2004).

In consequence, at least for most species, rather than experiencing a progressive increase in levels of damage suffered as population density increases, it would appear that significant damage may be noted after the number of ungulates exceeds some minimum threshold damage – and that further variation in density has very little relationship to actual damage levels sustained. Thus, damage levels tend to remain low – and relatively constant, until the population density passes a certain breakpoint, when impact suddenly and dramatically increases. Yet the truth is we know very little of such thresholds (although see recent review by Watson *et al.*, 2009; Putman *et al.*, 2011).

Following Holloway (1967), Ratcliffe (1987, 1989) has suggested that threshold densities below which damage levels to regenerating woodland are broadly tolerable (primarily in relation to the impacts of red and sika deer in commercial forests in Scotland) are 4 deer per 100 ha (see also Wagenknecht, 1986). Threshold densities for roe deer have been suggested at between 4 to 12 roe deer per 100 ha depending on habitat quality (Raesfeld *et al.*, 1985) – but such figures are largely untested and it is clear that different thresholds obtain for different types of damage as well as for different sites (which may be more or less productive).

Within the same woodland context, population densities at which natural regeneration is suppressed may differ significantly from those at which browsing damage to established or planted trees, or bark-stripping damage, reach economic significance. Densities of deer which may be tolerated in planted forests may thus be very different from those which would be acceptable in ancient woodlands managed for conservation or amenity value, or commercial forests replenished by natural regeneration (as is the case in many continental European forestry systems). Similarly, tolerable densities in a woodland context may be markedly different from densities which might be acceptable in other contexts – where different thresholds might be recorded for tolerable impacts in agriculture, for example, or in relation to increased risk of road traffic accidents (see, for example, Putman *et al.*, 2011).

Perhaps more importantly, even at a given density, damage levels caused by deer show very substantial variation depending on a number of environmental and cultural factors. These include (*inter alia*) crop type, distance of sensitive crop from cover, size of planted area, distance of sensitive crop from alternative preferred forages, habitat structure and cultural system. Reimoser and Gossow (1996), for example, suggest that levels of deer damage to forestry or agricultural crops relate less to deer density per se but more to the effective balance between (food-independent) 'attraction factors' for deer (factors such as extent of woodland edge and amount of thermal cover) and natural food supply. Where habitat structure is very attractive to deer yet the natural food supply is sparse, more damage may be anticipated than where the 'attractiveness' of an area is low in relation to the forage availability.

In relation to this Reimoser (2003) has shown that the most susceptible silvicultural systems are clear-cuts with afforestation, particularly small ones (<2 ha) and, with respect to browsing, also timber harvest by single-tree selection when only little light reaches the forest floor. Least susceptible are combinations of shelterwood felling and group selection systems with natural regeneration (Reimoser and Gossow, 1996). Völk (1998, 1999) presented results of a large-scale, long-term study for the eastern Alps, confirming that the type of forest management was the most important factor in determining bark-stripping damage by red deer. This factor was far more important than other factors such as deer density, hunting intensity, supplementary feeding, and disturbance by tourism.

As with forestry, numerous factors other than density would appear to affect vulnerability of agricultural crops, and degree of damage sustained. In practice, damage appears to be related once again to juxtaposition of cover (harbourage) adjacent to vulnerable fields, and availability of alternative, natural, forage (Doney and Packer, 1998; Watson *et al.*, 2009; Putman *et al.*, in prep.).

In much the same way, both within the UK and elsewhere it has been clearly established that the frequency of deer–vehicle collisions (DVCs) is not simply related to deer density but also road density, traffic volume and traffic speed (see, for example, Langbein, 2007; Chapter 8, this volume) as well as a number of other environmental factors (e.g. Bashore *et al.*, 1985; Finder *et al.*, 1999; Hubbard *et al.*, 2000; Malo *et al.*, 2004; Seiler, 2004; Putman *et al.*, 2004). In all these studies certain consistent features emerge as characteristic of sites likely to suffer a high frequency of deer-related road traffic accidents (Putman *et al.*, 2004; and again see Chapter 8, this volume), namely: number of lanes of traffic (width of road); traffic volume and speed; presence or absence of a central barrier; close association with woodland or forest cover near to the carriageway; landscape diversity (variability and patch size); the presence of obvious travel corridors across the roadway, such as rivers, dry gullies or other linear structures leading down at an angle to, or perpendicular to, the roadway.

Such conclusions – in indicating that damage levels are at best only weakly related to prevailing ungulate densities – suggest that direct population reduction alone, while it may alleviate the problem in the short term, is unlikely to have any marked effect unless ungulate numbers are reduced very substantially to a minimum presence. Thus (unless populations are reduced to very low levels indeed) management efforts based exclusively or primarily on attempted reduction of ungulate population numbers in an area may not have any significant impact in reducing damage levels.

All this is not intended to suggest that control of a perceived pest problem by direct population reduction is never appropriate. In many situations it may be the only option available, and if carried out carefully and with full understanding of the underlying dynamics of the species concerned it may prove an effective method of control. However, while direct control of ungulate populations may have a part to play in reducing levels of damage sustained, the lack of a clear relationship between severity of damage and actual animal density suggests *that control of ungulate numbers alone may not be entirely effective in delivering a reduction in impact.*

Indeed it is suggested that we might expect to decouple control of herbivore populations and control of damage, deploying different, though complementary, approaches to achieve essentially separate goals. While the regulation of numbers of ungulates within any area may contribute partially to regulation of impact, it is perhaps best seen as directed primarily to regulating the numbers of the deer themselves in relation to the land's capacity to support healthy stocks – while separate consideration may need to be given to complementary strategies which will help to control their impact on conservation, forestry or agricultural interests.

Putman (1996a, 2004) and Reimoser and Gossow (1996) thus both emphasise a need to distinguish between management approaches which may be considered appropriate in attempts to control or regulate ungulate populations, and (a different set of) management approaches which may be effective in management to control damage.

While the latter (attempts to control damage) may indeed include elements of population control, they should also explore alternative approaches such as fencing and other physical barriers; habitat manipulations to increase the availability of alternative forages or alterations to forest management and culture methods which may change the balance between forage availability and the availability of cover habitats; diversionary feeding, etc. (Putman, 1998, 2004; Reimoser, 2001a, 2001b, 2003).

We fully accept that none of these approaches is likely to be entirely effective on its own (any more than are simple attempts at population control). Kuiters *et al.* (1996), for example, point out that ungulate impact on forest development and its dependence on spatial and temporal patterns of forest regeneration are still poorly understood. This makes it very difficult to control ungulate herbivory effectively in commercial and conservation forests by habitat manipulation alone. Much further work is required on all the factors 'predisposing' different cultural ecosystems to ungulate damage, if such manipulations are to be effective, and Putman (1996a, 2004) recognised that such methods would usually need to be deployed within a wider management package involving at least some element of direct population control. Nonetheless, he urged a greater focus on such methods as likely to be of much greater *long-term* effectiveness in controlling impacts in many situations.

It is equally important that likely future impacts of ungulates should be taken into account in early stages of planning of any *new* environmental enterprise (forestry, agricultural activities, traffic routes – or whatever else). While it has taken a long time for the recognition to grow that insensitive planning towards some given, single objective, might in itself have contributed significantly to an increase in ungulate densities, or have caused some environmental imbalance resulting in an increase in damaging impacts, it has more recently become evident that the extent of damaging impacts from ungulate populations is strongly influenced by the management methods employed in trying to achieve those goals in the first place. While subsequent changes in cultural methods may have more limited efficacy in reducing damage (above), it is urged that design of any proposed new enterprise and cultural methods to be adopted in delivering the objectives set should be selected from the outset to minimise likely impacts and facilitate effective management.

All these considerations imply that the most effective management strategies for reducing ungulate impacts in the future will require integration of a number of different approaches of both population control and habitat management. We believe that there is a need for a much more holistic approach to the integration of ungulate species into cultivated landscapes, with proper, landscape-level planning to ensure adequate habitat structures available for plants and animals, thereby reducing conflicts. It is also clear that management for control or limitation of damaging impacts must also be integrated with wider management objectives which may include sustainable exploitation of those same ungulate populations as a positive resource.

6.5 Conclusions

An analysis of methods and systems for assessment of ungulate damage in different European countries (as reported in Apollonio *et al.*, 2010b and reviewed here in Tables 6.1 and 6.2) reveals enormous diversity both in methods used to estimate damage and in the extent to which different countries have routine, nationwide monitoring systems in place to record the actual extent of ungulate damage overall. Also the sensitivities to the various types of ungulate damage are significantly different between the countries. As a result of this it is not possible to come up with any meaningful estimates on what may be the distribution, or economic significance, of game damage to agriculture, forestry or conservation within Europe as a whole.

In contrast to negative impacts, possible positive impacts of ungulates on vegetation structure are very rarely searched for and investigated to recognise a balance of both. Particularly on habitats of conservation significance, both kinds of impact were recorded only in a few countries. It would appear that awareness of this lack is increasing, and this form of thinking may be more widely promoted in the future.

It is further clear that in many countries, management is not specifically directed towards reduction of damaging impacts. In general it would appear that management is more typically primarily directed towards hunting per se, and it is 'hoped' that regulation of animal populations may be successful in containing damage within tolerable levels. It is equally apparent that this is not the case: lower ungulate densities need not necessarily be associated with less damage, nor higher densities with more damage. As we have noted, *effective* control of population numbers may have some role to play in reducing or controlling impacts of ungulates on vegetational communities, but is in general terms unlikely to be entirely effective in the long term. Further, inappropriate culling, or inappropriate selection of age and sex structure to be culled, are likely to result in an increase rather than decrease in impact (examples in Putman, 2004), and may even be responsible for an increase in damage suffered. Unacceptable ungulate damage may thus be directly promoted by poor management practices.

It becomes clear that the actual impact experienced in any given situation is not simply related to ungulate species, numbers or density, but that the extent of damage caused by animals even at constant density may be significantly influenced by landscape, availability of shelter and alternative foraging opportunities, cultural systems, etc. This suggests to us that for management to be truly effective, the managers (landowner, farmer, forester, landscape planner, etc.) will have to adopt a more holistic approach and recognise that the interplay between ungulates and their natural and man-made habitats must be taken into account more consciously and more actively if a better balance between wild ungulates and vegetation is to be achieved.

It is our belief that this more integrated approach to management of wildlife impacts is crucial. No single approach can solve in the long term the different problems relating to wildlife management (avoidance of ungulate damage, protection of endangered species); they require complementary inputs from all stakeholders – foresters, hunters, farmers, tourist authorities, conservationists, regional planning authorities and local communities – with plans coordinated over large enough regions to be relevant for the ungulate species of interest.

We believe that appropriate expert systems for integrative ungulate management should be developed more widely. We also recommend that European guidelines should be developed on methodologies to be adopted for the formal assessment of damage (minimum standards), together with a requirement for member states to implement some scheme of regular monitoring on a national scale, as a basis for understanding the extent of conflicts that may exist and monitoring the success (or failure) of management measures adopted to address the problems identified. Our analyses in these pages surely serve to highlight the remarkable lack of objective information currently available on which to base appropriate management policies in Europe.

References

Adamic, M. and Jerina, K. (2010) Ungulates and their management in Slovenia. In M. Apollonio, R. Andersen and R. Putman (eds.) *European Ungulates and their Management in the 21st Century*. Cambridge, UK: Cambridge University Press, pp. 507–526.

Andersen, R. and Holthe, V. (2010) Ungulates and their management in Denmark. In M. Apollonio, R. Andersen, and R. Putman (eds.) *European Ungulates and their Management in the 21st Century*. Cambridge, UK: Cambridge University Press, pp. 71–85.

Andersen, R. Lund, E., Solberg, E. and Sæther, B.-E. (2010) Ungulates and their management in Norway. In M. Apollonio, R. Andersen, and R. Putman (eds.) *European Ungulates and their Management in the 21st Century*. Cambridge, UK: Cambridge University Press, pp. 14–36.

Andersone-Lilley, Z., Balčiauskas, L., Ozoliņš, J., Randveer, T. and Tõnisson, J. (2010) Ungulates and their management in the Baltics (Estonia, Latvia and Lithuania). In M. Apollonio, R. Andersen, and R. Putman (eds.) *European*

Ungulates and their Management in the 21st Century. Cambridge, UK: Cambridge University Press, pp. 103–128.

Apollonio, M., Andersen, R. and Putman, R. (2010a) Present status and future challenges for European ungulate management. In M. Apollonio, R. Andersen, and R. Putman (eds.) *European Ungulates and their Management in the 21st Century*. Cambridge, UK: Cambridge University Press, pp. 578–604.

Apollonio, M., Andersen, R., and Putman, R. (eds.) (2010b) *European Ungulates and their Management in the 21st Century*. Cambridge, UK: Cambridge University Press, 604 pp.

Apollonio, M., Ciuti, S., Pedrotti, L. and Banti, P. (2010c) Ungulates and their management in Italy. In M. Apollonio, R. Andersen and R. Putman (eds.) *European Ungulates and their Management in the 21st Century*. Cambridge, UK: Cambridge University Press, pp. 475–506.

Arnold, W. (2005) Schwarzwild: Hintergründe einer Explosion. *Österreichs Weidwerk* **1** (supplement), 8–11.

Bakker, J.P. (1989) *Nature Management by Grazing and Cutting*. Dordrecht, Netherlands: Kluwer.

Bakker, J.P., de Bie, S., Dallinga, J.H., Tjaden, P. and de Vries, Y. (1983) Sheep-grazing as a management tool for heathland conservation and regeneration in the Netherlands. *Journal of Applied Ecology* **20**, 541–560.

Bartoš, L., Kotrba, R. and Pintíř, J. (2010) Ungulates and their management in the Czech Republic. In M. Apollonio, R. Andersen and R. Putman (eds.) *European Ungulates and their Management in the 21st Century*. Cambridge, UK: Cambridge University Press, pp. 243–261.

Bashore, T.L., Tzilkowski, W.M. and Bellis, E.D. (1985) Analysis of deer-vehicle collision sites in Pennsylvania. *Journal of Wildlife Management* **49**, 769–774.

Bergström, R., Jernelid, H., Lavsund, S., Lundberg, K. and Wallin, K. (1995) *Älgtäthet – betestryck – fodertillgång – skogstillstånd – skadenivåer – skaderisker*. Slutrapport från projektet Balanserad älgstam. [Moose densities – browsing and damage; Final report from the research project 'Balanced moose populations'] (In Swedish) Uppsala, Sweden: Department of Wildlife Ecology, University of Agricultural Sciences.

Bokdam, J. (2003) Nature conservation and grazing management. PhD thesis, University of Wageningen, the Netherlands.

Bouldoire, J.-L. and Havet, P. (1982) Nature et importance des dégats aux cultures causés par les grand gibiers et les sangliers. *Bulletin Mensuel Office National de la Chasse* **48**, 10–16.

Büchsenmeister, R. and Gugganig, H. (2004) Stammschäden im Österreichischen Wald. Bundesamt und Forschungszentrum für Wald, Vienna. *BFW-Praxisinformationen* **3**, 12–14.

Callander, R.F. and Mackenzie, N.A. (1991) *The Management of Wild Red Deer in Scotland*. Perth, UK: Scottish Rural Forum.

Carranza, J. (2010) Ungulates and their management in Spain. In M. Apollonio, R. Andersen and R. Putman (eds.) *European Ungulates and their Management in the 21st Century*. Cambridge, UK: Cambridge University Press, pp. 419–440.

Casaer, J. and Licoppe, A. (2010) Ungulates and their management in Belgium. In M. Apollonio, R. Andersen and R. Putman (eds.) *European Ungulates and their Management in the 21st Century*. Cambridge, UK: Cambridge University Press, pp. 184–200.

Cerkal, R., Dvorak, J., Kamler, J. and Vejrazka, K. (2006) Poskozovani porostu jecmene bylozravei. [Game damage to barley] (In Czech) In J. Zimolka (ed.) *Jecmen – formy a uzitkove smery v Ceske Republice*. Prague: Profi Press, pp. 120–125.

Cooke, A.S. (1994) Colonisation by muntjac deer *Muntiacus reevesi* and their impact on vegetation. In M.S. Massey and R.C. Welch (eds.) *Monks Wood National Nature Reserve, the Experience of 40 Years: 1953–93*. Peterborough, UK: English Nature, pp. 45–61.

Cooke, A.S. (1995) Muntjac damage in woodland. *Enact* **3**,12–14.

Cooke, A.S. (2005) Muntjac deer *Muntiacus reevesi* in Monks Wood NNR: their management and changing impacts. In C. Gardiner and T. Sparks (eds.) *Ten Years of Change: Woodland Research at Monks Wood NNR, 1993–2003*, Research Report 613. Peterborough, UK: English Nature, pp. 65–74.

Cooke, A.S. (2006) *Monitoring Muntjac Deer* Muntiacus reevesi *and their Impacts in Monks Wood National Nature Reserve*. Research Report 681. Peterborough, UK: English Nature,172 pp.

Cousins, M. (1987) An investigation of roe deer habitat usage and associated tree damage in a Scottish estate of mixed woodland type. BSc (Hons) thesis (Environmental Sciences), University of Southampton, UK.

Csányi, S. and Lehoczki, R. (2010) Ungulates and their management in Hungary. In M. Apollonio, R. Andersen and R. Putman (eds.) *European Ungulates and their Management in the 21st Century*. Cambridge, UK: Cambridge University Press, pp. 291–318.

Donaubauer, E. (1994) Zur Wildschadenssituation in Europa. Salzburg, Austria: CIC Tagung. Unpublished typescript, 11 pp.

Doney, J. (1999) Validation of responses from the agricultural sector targeted by the lowland deer questionnaire. Appendix 4 of J. Packer *et al.*: *Field and Desk Studies to Assess Tolerable Damage Levels for Different Habitats and Species of Deer*. Report to Ministry of Agriculture, Fisheries and Foods, London, on contract VC 0315.

Doney, J. and Packer, J. (1998) An assessment of the impact of deer on agriculture. In C.R. Goldspink, S. King and R.J. Putman (eds.) *Population Ecology, Management and Welfare of Deer*. Manchester, UK: British Deer Society/Universities' Federation for Animal Welfare, pp. 38–43.

Eiberle, K. and Nigg, H. (1983) Über die Folgen des Wildverbisses an Fichte und Weisstanne im montaner Lage. *Schweizerische Zeitschrift für Forstwesen* **134**, 361–372.

Eiberle, K. and Nigg, H. (1987) Gundlagen zur Beurteilung des Wildverbisses im Gebirgswald. *Schweizerische Zeitschrift für Forstwesen* **138**, 747–785.

Ellenberg, H. (1988) Eutrophierung: Veränderungen der Waldvegetation: Folgen für den Rehwildverbiss und dessen Rückwirkungen auf die Vegetation. *Schweizerische Zeitschrift für Forstwesen* **139**, 171–186.

Feber, R.E., Brereton, T.M., Warren, M.S. and Oates, M. (2001) The impacts of deer on woodland butterflies: the good, the bad and the complex. *Forestry* **74**, 271–276.

Filli, F. and Suter, W. (ed.) (2006) *Ungulate Research in the Swiss National Park*. Nationalpark-Forschung in der Schweiz 93. Zernez, Switzerland: Schweizerischer Nationalpark.

Finder, R.A., Roseberry, J.L. and Woolf, A. (1999) Site and landscape conditions at white-tailed deer collision locations in Illinois. *Landscape and Urban Planning* **44**, 77–85.

Findo, S. and Skuban, M. (2010) Ungulates and their management in Slovakia. In M. Apollonio, R. Andersen and R. Putman (eds.) *European Ungulates and their Management in the 21st Century*. Cambridge, UK: Cambridge University Press, pp. 262–290.

Flowerdew, J.R. and Ellwood, S.A. (2001) Impacts of woodland deer on small mammal ecology. *Forestry* **74**, 277–288.

Fonseca, C., Alves, J. and Silva, A. (2007) Plano Global de Gestão para a população de veados (*Cervus elaphus*) da Serra da Lousã. Aveiro, Portugal: Direcção-Geral dos Recursos Florestais/Universidade de Aveiro. 87 pp.

Fuller, R.J. (2001) Responses of woodland birds to increasing numbers of deer: a review of evidence and mechanisms. *Forestry* **74**, 289–298.

Fuller, R.J. and Gill, R.M.A. (2001) Ecological impacts of increasing numbers of deer in British woodlands. *Forestry* **74**, 193–200.

Gill, R.M.A. (1992a) A review of damage by mammals in north temperate forests: 1. Deer. *Forestry* **65**, 145–169.

Gill, R.M.A. (1992b) A review of damage by mammals in north temperate forests: 3. Impact on trees and forests. *Forestry* **65**, 363–388.

Gill, R.M.A. and Fuller, R.J. (2007) The effects of deer browsing on woodland structure and songbirds in lowland Britain. *Ibis* **149**(2), Suppl., 119–127.

Gossow, H. (1983) Zur geschichtlichen Entwicklung der Beziehungen zwischen Jagd und Waldwirtschaft. *Centralblatt für das gesamte Forstwesen* **100**(2/3), 191–207.

Gossow, H. and Reimoser, F. (1985) Anmerkungen zum Zielkonflikt Wald Wild-Weide-Tourismus. *Schweizerische Zeitschrift für Forstwesen* **136**, 913–929.

Groot Bruinderink, G.W.T.A. and Hazebroek, E. (1996) Ungulate traffic collisions in Europe. *Conservation Biology* **10**, 1059–1067.

Herrero, J., Irizar, I., Laskurain, N.A., García-Serrano, A. and García-González, R. (2005) Fruits and roots: wild boar foods during the cold season in the southwestern Pyrenees. *Italian Journal of Zoology* **72**(1), 49–52.

Herrero, J., García-Serrano, A., Couto, S., Ortuño, V.M. and García-González, R. (2006) Diet of wild boar *Sus scrofa* L. and crop damage in an intensive agroecosystem. *European Journal of Wildlife Research* **52**(4), 245–250.

Hill, S.D. (1985) Influences of large herbivores on small rodents in the New Forest. PhD thesis, University of Southampton, UK.

Hirons, G.J.M. (1984) The diet of tawny owls (*Strix aluco*) and kestrels (*Falco tinnunculus*) in the New Forest, Hampshire. *Proceedings of the Hampshire Field Club and Archaeological Society* **40**, 21–26.

Holloway, C.F. (1967) The effects of red deer and other animals on natural regeneration of Scots pine. PhD thesis, University of Aberdeen, UK.

Hubbard, M.W., Danielson, B.J. and Schmitz, R.A. (2000) Factors influencing the location of deer-vehicle accidents in Iowa. *Journal of Wildlife Management* **64**, 707–713.

Hunt, J. (2000) Habitats and deer in Scotland – how fares the public interest? Report prepared for the Royal Society for the Protection of Birds, and Worldwide Fund for Nature, Scotland.

Imesch-Bebié, N., Gander, H. and Schnidrig-Petrig, R. (2010) Ungulates and their management in Switzerland. In M. Apollonio, R. Andersen and R. Putman (eds.) *European Ungulates and their Management in the 21st Century*. Cambridge, UK: Cambridge University Press, pp. 357–391.

Janeau, G. and Gallo-Orsi, U. (1992) Wild boar session and workshop: comments and conclusions. In F. Spitz, G. Jeaneau, G. Gonzales, and S. Aulagner (eds.)

Ongules/Ungulates 91. Proceedings of the international symposium SFEPM-IRGM held at Toulouse, France, pp. 383–384.

Jantz, K. (1982) What does elk damage cost? *Sveriges Skogsvardsforbund Tidskrift* **4**, 41–42.

Kaluzinski, J. (1982) Composition of the food of roe deer living in field and the effects of their feeding on plant production. *Acta Theriologica* **27**, 457–470.

Kamler, J., Homolka, M., Heroldova, M. and Dvorak, J. (2005) Volne zijici bylozravei a poini plodiny. [Free living ungulates and field crops] (In Czech) *Folia Venatoria* **35**, 205–210.

Kay, S. (1993) Factors affecting severity of deer browsing damage within coppiced woodlands in the south of England. *Biological Conservation* **63**, 217–222.

Kuiters, A.T., Mohren, G.M.J. and Van Wieren, S.E. (1996) Ungulates in temperate forest ecosystems. *Forest Ecology and Management* **88**, 1–5.

Kusak, J. and Krapinec, K. (2010) Ungulates and their managment in Croatia. In M. Apollonio, R. Andersen and R. Putman (eds.) *European Ungulates and their Management in the 21st Century*. Cambridge, UK: Cambridge University Press, pp. 527–539.

Labudzki, L. and Wlazelko, M. (1991) Saisonale Dynamik der vom Schwarzwild im Feldanbau verursachten Schaden im Forschungsgebiet Zielonka. [Seasonal dynamics of crop damage caused by wild boar in the research area Zielonka] (In German) *Zeitschrift fur Jagdwissenschaft* **37**(4), 250–257.

Langbein, J. (2007) *National Deer-Vehicle Collisions Project (England and Wales) 2003–2005*. Final Report to the Highways Agency, London, UK.

Langbein, J. and Putman, R.J. (2006) *National Deer-Vehicle Collisions Project (Scotland) 2003–2005*. Final Report to the Scottish Executive, June 2006.

Liberg, O., Bergström, R., Kindberg, J. and von Essen, H. (2010) Ungulates and their management in Sweden. In M. Apollonio, R. Andersen and R. Putman (eds.) *European Ungulates and their Management in the 21st Century*. Cambridge, UK: Cambridge University Press, pp. 37–70.

Maillard, D., Gaillard, J-M., Hewison, M., *et al.* (2010) Ungulates and their management in France. In M. Apollonio, R. Andersen and R. Putman (eds.) *European Ungulates and their Management in the 21st Century*. Cambridge, UK: Cambridge University Press, pp. 441–474.

Malo, J.E, Suarez, F. and Diaz, A. (2004) Can we mitigate animal-vehicle accidents using predictive models? *Journal of Applied Ecology* **41**, 701–710.

Mayer, H. and Ott, E. (1991) *Gebirgswaldbau – Schutzwaldpflege*. Stuttgart, Germany: Gustav Fischer-Verlag, 587 pp.

Micu, I., Náhlik, A., Neguş, S., Mihalache, I. and Szabó, I. (2010) Ungulates and their management in Romania. In M. Apollonio, R. Andersen and R. Putman (eds.) *European Ungulates and their Management in the 21st Century*. Cambridge, UK: Cambridge University Press, pp. 319–337.

Mitchell, B., Staines, B.W. and Welch, D. (1977) *Ecology of Red Deer: A Research Review Relevant to their Management*. Cambridge, UK: Institute of Terrestrial Ecology.

Mitchell, F.J.G. and Kirby, K. (1990) The impact of large herbivores on the conservation of semi-natural woods in the British uplands. *Forestry* **63**, 333–353.

Obrtel, R. and Holisova, V. (1983) Assessment of the damage done to a crop of maize (*Zea mays*) by roe deer (*Capreolus capreolus*). *Folia Zoologica (Brno)* **32**, 109–118.

Obrtel, R., Holisova, V. and Kozena, I. (1984) Deer damage to sugar beet leaves. *Folia Zoologica (Brno)* **33**, 99–108.

Packer, J.J., Doney, J. Mayle, B.A., Palmer, C.F. and Cope, M. (1999) *Field and Desk Studies to Assess Tolerable Damage Levels for Different Habitats and Species of Deer*. Report to the Ministry of Agriculture, Fisheries and Foods, on project VC 0315. London: MAFF.

Papaioannou, H. (2010) Ungulates and their management in Greece. In M. Apollonio, R. Andersen and R. Putman (eds.) *European Ungulates and their Management in the 21st Century*. Cambridge, UK: Cambridge University Press, pp. 540–562.

Petley-Jones, R. (1995) Deer or butterflies? A woodland dilemma. *Enact* **3**(3), 8–10.

Petty, S.J. and Avery, M.I. (1990) *Forest Bird Communities*. Occasional Paper 26. Edinburgh, UK: Forestry Commission.

Pollanschütz, J. (1995) Bewertung von Verbiss- und Fegeschäden, Hilfsmittel und Materialien. *Mitteilungen der Forstlichen Bundesversuchsanstalt Wien* **169**, 99 pp.

Pollard, E. and Cooke, A.S. (1994) Impact of muntjac deer *Muntiacus reevesi* on egg-laying sites of the white admiral butterfly *Ladoga camilla* in a Cambridgeshire wood. *Biological Conservation* **70**, 189–191.

Putman, R.J. (1986a) *Grazing in Temperate Ecosystems: Large Herbivores and their Effects on the Ecology of the New Forest*. London: Croom Helm. 200 pp.

Putman, R.J. (1986b) Foraging by roe deer in agricultural areas and impact on arable crops. *Journal of Applied Ecology* **23**, 91–99.

Putman, R.J. (1989) Mammals as pests. In R.J. Putman (ed.) *Mammals as Pests*. London: Chapman and Hall, pp. 1–12.

Putman, R.J. (1994) Damage by deer in coppice woodlands: an analysis of factors affecting the severity of damage and options for management. *Quarterly Journal of Forestry* **88**, 45–54.

Putman, R.J. (1996a) Ungulates in forest ecosystems: perspectives and recommendations for future research. *Forest Ecology and Management* **88**, 205–214.

Putman, R.J. (1996b) *Deer on National Nature Reserves: Problems and Practices*. Research Report 173. Peterborough, UK: English Nature, 50 pp.

Putman, R.J. (1998) The potential role of habitat manipulation in reducing deer impact. In C.R. Goldspink, S. King and R.J. Putman (eds.) *Population Ecology, Management and Welfare of Deer*. Manchester, UK: British Deer Society/Universities' Federation for Animal Welfare, pp. 95–101.

Putman, R.J. (2004) *The Deer Manager's Companion: A Guide to Deer Management in the Wild and in Parks*. Shrewsbury, UK: Swan Hill Press.

Putman, R.J. (2009) The effects of reducing grazing pressure from red deer and sheep on vegetation within the Creag Meagaidh National Nature Reserve. In W. Cresswell *et al.* (eds.) *Grazing Management of Upland Habitats for Nature Conservation*. Stroud, UK: Cresswell Associates/The Rannoch Trust.

Putman, R.J. (2010) Ungulates and their management in Great Britain and Ireland. In M. Apollonio, R. Andersen and R. Putman (eds.) *European Ungulates and their Management in the 21st Century*. Cambridge, UK: Cambridge University Press, pp. 129–164.

Putman, R.J. and Moore, N. (1998) Impact of deer in lowland Britain on agriculture, forestry and conservation habitats. *Mammal Review* **28**, 141–164.

Putman, R.J. and Kjellander, P. (2002) Deer damage to cereals: economic significance and predisposing factors. In F. Tattersall and W. Manley (eds.)

Conservation and Conflict: Mammals and Farming in Britain. Linnean Society Occasional Publications. Otley, UK: Westbury Academic & Scientific Publishing, pp. 186–197.

Putman, R.J., Edwards, P.J., Mann, J.C.E, How, R.C. and Hill, S.D. (1989) Vegetational and faunal changes in an area of heavily grazed woodland following relief of grazing. *Biological Conservation* **47**, 13–32.

Putman, R.J., Langbein, J. and Staines, B.W. (2004) *Deer and Road Traffic Accidents: A Review of Mitigation Measures: Costs and Cost-effectiveness*. Contract report RP23A. Inverness, UK: Deer Commission, Inverness, Scotland.

Putman, R.J., Langbein, J., Green, P. and Watson, P. (2011) Identifying threshold densities for wild deer in the UK above which negative impacts may occur. *Mammal Review*, in press.

Rackham, O. (1975) *Hayley Wood: Its History and Ecology*. Cambridge, UK: Cambridge and Isle of Ely Naturalists' Trust.

Raesfeld, F., Neuhaus, A.H. and Schaich, K. (1985) *Das Rehwild: Naturgeschichte, Hege und Jagd*. Hamburg, Germany: Paul Parey Verlag, 453 pp.

Ratcliffe, P.R. (1987) *Management of Red Deer in Upland Forests*. Forestry Commission Bulletin 71. London: HMSO.

Ratcliffe, P.R. (1989) The control of red and sika deer populations in commercial forests. In R.J. Putman (ed.) *Mammals as Pests*. London: Chapman and Hall, pp. 98–115.

Reimoser, F. (1986) Wechselwirkungen zwischen Waldstruktur, Rehwildverteilung und Rehwildbejagbarkeit in Abhängigkeit von der waldbaulichen Betriebsform. Vienna: VWGÖ-Verlag. *Dissertationen der Universität für Bodenkultur* **28**, 1–318.

Reimoser, F. (1993) Funktionen des Schalenwildes (Reh, Hirsch, Gemse) im Waldökosystem und Funktionsstörungen in Mitteleuropa unter besonderer Berücksichtigung von Urwald- und Naturwaldreservaten. In M. Saniga and S. Korpel (eds.) *Symposium über die Urwälder*. Zvolen, Slovakia: Forstliche Fakultät der Technischen Universität, pp. 87–95.

Reimoser, F. (2000) Income from hunting in mountain forests of the Alps. In M.F. Price and N. Butt (eds.) *Forests in Sustainable Mountain Development: A State-of-Knowledge Report for 2000*. IUFRO Research Series 5. New York: CABI Publishing, pp. 346–353.

Reimoser, F. (2001a) The role of silviculture in wildlife management. *Fauna e Selvicoltura* **1**, 3–10.

Reimoser, F. (2001b) Forest ungulate damage: integrating wildlife and forest management practices. In R. Field, R.J. Warren, H. Okarma and P.R. Sievert (eds.) *Wildlife, Land, and People: Priorities for the 21st Century*. Bethesda, MD: The Wildlife Society, pp. 172–176.

Reimoser, F. (2002) Schutz des Waldes und Wildtiermanagement in Naturschutzgebieten. *Revue de Geographie Alpine* **90**(2), 73–85.

Reimoser, F. (2003) Steering the impacts of ungulates on temperate forests. *Journal for Nature Conservation* **10**(4), 243–252.

Reimoser, F. and Gossow, H. (1996) Impact of ungulates on forest vegetation and its dependence on the silvicultural system. *Forest Ecology and Management* **88**, 107–119.

Reimoser, F. and Reimoser, S. (1997) Wildschaden und Wildnutzen: Objektive Beurteilung des Einflusses von Schalenwild auf die Waldvegetation. *Zeitschrift für Jagdwissenschaft* **43**, 186–196.

Reimoser, F. and Ellenberg, H. (1999) Forest management systems as a component of ungulate-game pest management with special reference to roe deer and edge effects. In P.D. Cowan and C.J. Feare (eds.) *Advances in Vertebrate Pest Management*. Fürth, Germany: Filander-Verlag, pp. 219–239.

Reimoser, F. and Reimoser, S. (2010) Ungulates and their management in Austria. In M. Apollonio, R. Andersen and R. Putman (eds.) *European Ungulates and their Management in the 21st Century*. Cambridge, UK: Cambridge University Press, pp. 338–356.

Reimoser, F., Odermatt, O., Roth, R. and Suchant, R. (1997) Die Beurteilung von Wildverbiß durch SOLL-IST-Vergleich. *Allgemeine Forst- und Jagdzeitung* **168**, 214–227.

Reimoser, F., Armstrong, H. and Suchant, R. (1999) Measuring forest damage of ungulates: what should be considered. *Forest Ecology and Management* **120**, 47–58.

Ruusila, V. and Kojola, I. (2010) Ungulates and their management in Finland. In M. Apollonio, R. Andersen and R. Putman (eds.) *European Ungulates and their Management in the 21st Century*. Cambridge, UK: Cambridge University Press, pp. 86–102.

Schley, L. and Roper, T.J. (2003) Diet of wild boar *Sus scrofa* in Western Europe, with particular reference to consumption of agricultural crops. *Mammal Review* **33**, 43–56.

Schley, L., Dufresne, M., Krier, A. and Frantz, A.C. (2008) Patterns of crop damage by wild boar (*Sus scrofa*) in Luxembourg over a 10-year period. *European Journal of Wildlife Research* **54**(4), 589–599.

Schodterer, H. (2004) Die Verjüngung des Österreichischen Waldes. Vienna: Bundesamt und Forschungszentrum für Wald. *BFW-Praxisinformationen* **3**, 17–20.

Schütz, M., Krüsi, B., Edwards, P.J. (eds.) (2000) *Succession Research in the Swiss National Park*. Nationalparkforschung in der Schweiz 89. Zernez, Switzerland: Schweizerischer Nationalpark, 255 pp.

Schwarzenbach, F.H. (1982) Ansätze zur Lösung des Wildschadenproblems. *Schweizerische Zeitschrift für Forstwesen* **133**, 979–984.

Seiler, A. (2004) Trends and spatial pattern in ungulate-vehicle collisions in Sweden. *Wildlife Biology* **10**, 301–313.

SNH/DCS (2002) *Wild Deer in Scotland and Damage to the Natural Heritage*. Edinburgh, UK: Deer Commission for Scotland and Scottish Natural Heritage.

Solbraa, K. (2002) Beiteskader av elg. Presentation from University course LP 300 at the Agricultural University, Ås, Norway.

Speidel, G. (1980) Methods of evaluating the economic consequences of damage in the forest, and ways of regulating it. *Forstwissenschaftliches Centralblatt* **99**, 76–85.

Spitz, F. and Sovan, L. (1999) Environmental impact prediction using neural network modelling: an example in wildlife damage. *Journal of Applied Ecology* **36**, 317–326.

Stahl, D. (1979) *Wild, lebendige Umwelt: Probleme von Jagd, Tierschutz und Ökologie geschichtlich dargestellt und dokumentiert*. Freiburg, Germany: Alber Verlag, 349 pp.

Staines, B.W. and Welch, D. (1984) Habitat selection and impact of red deer (*Cervus elaphus*) and roe deer (*Capreolus capreolus*) in a sitka spruce plantation. *Proceedings of the Royal Society of Edinburgh B*, **82**, 303–319.

Stewart, A.J.A. (2001) The impact of deer on lowland woodland invertebrates: a review of the evidence and priorities for future research. *Forestry* **74**, 259–270.

Tabor, R.C. (1993) Control of deer in a managed coppice. *Quarterly Journal of Forestry* **87**, 308–313.

Tubbs, C.R and Tubbs, J.M (1985) Buzzards (*Buteo buteo*) and land use in the New Forest, Hampshire, England. *Biological Conservation* **31**, 46–65.

van Wieren, S.E. (1991) The management of populations of large mammals. In I.F. Spellerberg, F.B. Goldsmith and M.G. Morris (eds.) *The Scientific Management of Temperate Communities for Conservation*. Oxford, UK: Blackwell, pp. 103–127.

van Wieren, S.E. and Groot Bruinderink, G.W.T.A. (2010) Ungulates and their management in the Netherlands. In M. Apollonio, R. Andersen and R. Putman (eds.) *European Ungulates and their Management in the 21st Century*. Cambridge, UK: Cambridge University Press, pp. 165–183.

Vingada, J., Fonseca, C., Cancela, J., Ferreira, J. and Eira, C. (2010) Ungulates and their management in Portugal. In M. Apollonio, R. Andersen and R. Putman (eds.) *European Ungulates and their Management in the 21st Century*. Cambridge, UK: Cambridge University Press, pp. 392–418.

Völk, F. (1998) Schälschäden und Rotwildmanagement in Relation zu Jagdgesetz und Waldaufbau in Österreich. *Beiträge zur Umweltgestaltung A* **141**. Berlin: Erich Schmidt Verlag.

Völk, F. (1999) Langjährig erfolgreiche Rotwildüberwinterung ohne gravierende Schälschäden im Ostalpenraum. *Beiträge zur Jagd- und Wildforschung* **24**, 69–86.

Wagenknecht, E. (1986): *Rotwild*. Berlin: VEB Deutscher Landwirtschaftsverlag, 484 pp.

Wallis de Vries, M.F. (1998) Large herbivores as key factors for nature conservation, In M.F. Wallis de Vries, J.P. Bakker and S.E. van Wieren (eds.) *Grazing and Conservation Management*. Dordrecht, Netherlands: Kluwer Academic Publishers, pp. 1–20.

Wallis de Vries, M.F., Bakker, M.J.P. and van Wieren, S.E. (eds.) (1998) *Grazing and Conservation Management*. Dordrecht, Netherlands: Kluwer Academic Publishers.

Wallis de Vries, M.F., Parkinson, A.E., Dulphy, J.P., Sayer, M. and Diana, E. (2007) Effects of livestock breed and grazing intensity on biodiversity and production in grazing systems: 4. Effects on animal diversity. *Grass and Forage Science* **62**, 185–197.

Watson, P., Putman, R.J., Langbein, J. and Green, P. (2009) A review of threshold densities for wild deer in England above which negative impacts may occur. Report for Defra (Department for the Environment, Food and Rural Affairs), London. Online: www.deercollisions.co.uk/pages

Wawrzyniak, P., Jędrzejewski, W., Jędrzejewska, B. and Boro, T. (2010) Ungulates and their management in Poland. In M. Apollonio, R. Andersen and R. Putman (eds.) *European Ungulates and their Management in the 21st Century*. Cambridge, UK: Cambridge University Press, pp. 223–242.

Wildauer, L. (2006) Wildschwein (*Sus scrofa*) und Reh (*Capreolus capreolus*) in den Bezirken Zentral- und Ostösterreichs seit 1950: Abschuss- und Bestandsentwicklung, mögliche Einflussfaktoren, Wildschweinschäden in der Landwirtschaft. Diploma thesis, University of Veterinary Medicine, Vienna.

Wildauer, L. and Reimoser, F. (2007a) Wild boar (*Sus scrofa*) damage in agriculture: relation between farmers and hunters in past and present. XXVIIIth International Union of Game Biologists Congress, Uppsala, Sweden, p. 225.

Wildauer, L. and Reimoser, F. (2007b) Wild boar (*Sus scrofa*) management: what can we learn from history? XXVIIIth International Union of Game Biologists Congress, Uppsala, Sweden, p. 42.

Wilson, C.J. (2004) Rooting damage to farmland in Dorset, southern England, caused by feral wild boar *Sus scrofa*. *Mammal Review* **34**, 331–335.
Wolf, G. (1988) Dauerflächen-Beobachtungen in Naturwaldzellen der Niederrheinischen Bucht: Veränderungen in der Feldschicht. *Natur und Landschaft* **63**(4), 167–172.
Wotschikowsky, U. (2010) Ungulates and their management in Germany. In M. Apollonio, R. Andersen and R. Putman (eds.) *European Ungulates and their Management in the 21st Century*. Cambridge, UK: Cambridge University Press, pp. 201–222.
Wray, S. (1994) Competition between muntjac and other herbivores in a commercial coniferous forest. *Deer* **9**, 237–242.

7

Wild ungulate diseases and the risk for livestock and public health

EZIO FERROGLIO, CHRISTIAN GORTÁZAR AND JOAQUÍN VICENTE

7.1 Introduction

The social changes occurring across Europe in the last 40 years have had a pronounced effect on the environment, creating a dynamic situation where new pathogens or new hosts emerge, changes in population density or host behaviour affect disease prevalence and, in some cases, may allow disease agents to boost their virulence and widen their host range (Figure 7.1).

Apart from the role of pathogens in the population dynamics of wild populations of ungulates (discussed here in Chapter 11), another significant issue is the risk of transmission of disease agents between wildlife and livestock or human beings. While some pathogens exclusively infect a single host species, these are usually highly coevolved parasites with limited effect on the primary host's population (Crawley, 1992; Vicente *et al.*, 2004a). In contrast, many parasites can infect multiple host species and these are primarily responsible for outbreaks of infectious diseases in humans, livestock and indeed among wildlife (Swinton *et al.*, 2002; Woolhouse, 2002).

The increased distribution and densities of wild ungulates registered all across Europe (see chapters in Apollonio *et al.*, 2010), together with a move within the livestock industry from more intensive to more extensive farming systems, or at least systems with a lower human presence on the field, have increased the risk of contact between wildlife and livestock (e.g. Laddomada *et al.*, 1994; Gortázar *et al.*, 2007). The increased chance of direct contact between wildlife and humans is partly due to the increase in outdoor activities on one hand, but, on the other hand, the quantity of wild ungulate meat consumed on our tables has greatly increased, with an associated risk for food-borne zoonosis.

Ungulate Management in Europe: Problems and Practices, eds. Rory Putman, Marco Apollonio and Reidar Andersen. Published by Cambridge University Press.

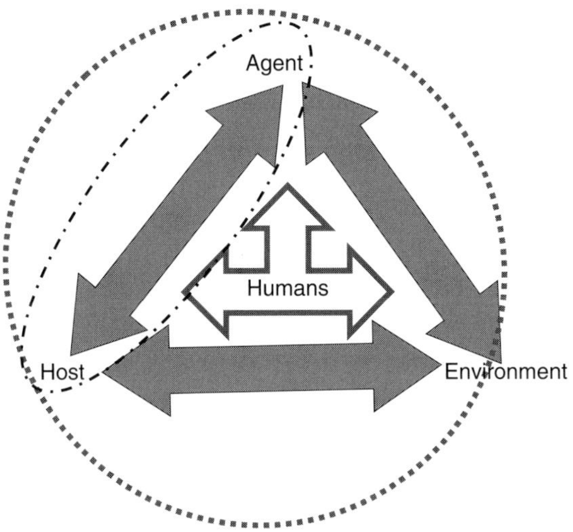

Figure 7.1 Hosts, pathogens and environment are usually strictly connected, and it is intriguing to study these links. However, if we want to manage wildlife diseases, we must consider that humans deeply influence all these factors and efforts must be directed towards managing 'human activities' as well.

When wildlife diseases attract the attention of the public, it is usually in relation to diseases that pose a threat for public health (as zoonoses which can affect humans, or as diseases with a potential for wide-scale infection of livestock), which are thus the subject of high-profile campaigns for eradication. In all such cases management has primarily been entrusted to veterinarians and intervention has usually been based on a public health approach. This is in part explained by the scarce attention paid by the veterinary profession in the past towards wildlife conservation, and also as a direct consequence of the tendency for discrete specialisation in fields such as pathology, microbiology and parasitology. As a result of this compartmentalisation in approach there is inevitably a tendency to investigate wildlife diseases in a similar way as for livestock (Ferroglio, 2003). Public health veterinary intervention usually places the main emphasis on the incriminative 'wildlife reservoir of disease' point of view, with the application of regulatory measures which usually do not consider wildlife population dynamics, welfare or conservation (Woodford, 1965). This tendency, coupled with an attitude of Governments' Veterinary authorities, which tend to treat wildlife diseases as if they fell within their own exclusive domain, has, sometimes, given rise to a conflict among public health veterinarians, wildlife veterinarians and biologists, with a perception from the last that veterinarians are arrogant towards their peers in other disciplines (Meltzer, 1995). In the past this conflict, concealed or apparent, has been detrimental for wildlife conservation, but also for the development of veterinary involvement in wildlife areas and its integration in the interdisciplinary team.

In consequence, even if in many countries wildlife diseases have been investigated for decades, it is only since the 1990s that the scientific debate became properly supported with an explosion of wildlife disease-related knowledge (Gortázar *et al.*, 2007). This change has also involved a new approach towards understanding disease agents in their fuller environmental context, and of parasite–host interactions: a whole new field styled under the title of 'disease ecology' (Hudson *et al.*, 2001). Among the most intriguing aspects of this new branch of science is the exploration of the link among wildlife pathogens, environment and human activities.

In this chapter we review what is known of the transmission of disease among wild ungulate populations and domestic livestock or humans, and the role that wild ungulates may play as reservoirs of such infection. It is not the aim of this chapter to list all the pathogens of public health concern which have been found in wild ungulates across Europe. The wildlife–domestic animal interface and the risk of pathogen transmission to livestock and humans in Europe have been recently reviewed by Frölich *et al.* (2002), Simpson (2002), Bohm *et al.* (2007) and Gortázar *et al.* (2007). We have decided to focus on reviewing some of the most relevant ungulate diseases from a European perspective, aiming moreover to discuss the available disease management options and to highlight current research and management priorities.

7.2 The wild ungulate/livestock interface

A non-exhaustive list of multi-host situations, where a disease agent might affect wild or domestic ungulates is shown in Table 7.1 (from Gortázar *et al.*, 2007). From the veterinary health perspective, multi-host situations in diseases in wildlife that are notifiable and have been eradicated in domestic livestock, or at least are almost under control in domesticates, are the worst, because a single spill-back from wildlife to livestock may have severe consequences not only on health, but also on economy. Examples include bovine TB (Phillips *et al.*, 2003) and avian influenza (Alexander, 2000). Multi-host situations are of less concern if the disease is not yet under control in domestic animals, as for example porcine circovirus type 2 (Vicente *et al.*, 2004a, 2004b) or toxoplasmosis (Gauss *et al.*, 2006), or if adequate vaccines or treatments are available and extensively used in domestics, such as for example in the case of porcine parvovirus and erysipelas (Ritzmann *et al.*,

Table 7.1 *Agent, host, current situation in Europe, relevance and main risk factors of diseases shared with European wild ungulates*

Disease	Agent	Wildlife host (domestic)	Situation in Europe	Relevance	Main risks
African swine fever	Virus	Wild boar (domestic pig)	Locally endemic in pigs; apparently no true reservoir of wild boars; apparent multi-host	Heavy economic impact	Movement of wild and domestic animals; open-air farming
Classic swine fever	Virus	Wild boar (domestic pig)	Affects domestic pig in several central and eastern European countries; wild boar acts as reservoir (oral vaccination); true multi-host	Heavy economic impact	Movement of wild and domestic animals; wildlife overabundance
Aujesky's disease	Virus	Wild boar (domestic pig)	Endemic in domestic pigs in several countries; strain differences suggest that wild boar, despite high prevalence, is no reservoir for indoor pigs; true multi-host?	Heavy economic impact; conservation concerns	Movement of wild and domestic animals, wildlife overabundance; open-air farming
Bluetongue	Virus	Wild ruminants (cattle, sheep and goats)	Expanding among domestic ruminants in Mediterranean countries; increasingly detected in wild ruminants (reservoir role unknown in Europe)	Heavy economic impact	Vector expansion; movement of wild and domestic animals

Table 7.1 (*cont.*)

Disease	Agent	Wildlife host (domestic)	Situation in Europe	Relevance	Main risks
Bovine viral diarrhoea, alphaherpesvirus, malignant catarrhal fever	Virus	Cervids (domestic ruminants)	Endemic in domestic ruminants; low prevalence in deer; apparent multi-host ?	Economic impact	Unknown
Salmonellosis	Bacteria	Wild vertebrates (poultry, all livestock)	Presence in wildlife mostly due to exposure to human/ livestock residues; huge between-country differences in prevalence in livestock; true multi-host	Heavy economic impact; zoonosis	Wildlife overabundance; open-air farming
Tuberculosis	Bacteria	Wild boar, red and fallow deer, badger, other wild mammals (cattle, goats, and extensively bred pigs)	Prevalence decreased throughout Europe, but asymptotic, endemic in badgers, wild boar and deer in several countries; strong debate on reservoir culling TBC control option; wildlife vaccination trials scheduled; true multi-host	Heavy economic impact; zoonosis	Movement of wild and domestic animal; wildlife overabundance; open-air farming

Paratuberculosis (Johnne's disease)	Bacteria	Wild ruminants, rabbit, other wild mammals (domestic ruminants)	Endemic in domestic ruminants; locally high prevalences in wild species, including deer and rabbit; increasingly reported in wildlife; true multi-host?	Heavy economic impact; zoonosis	Wildlife overabundance; movement of wild and domestic animals
Brucellosis (*B. abortus*, *B. melitensis*)	Bacteria	Wild ruminants (domestic ruminants)	Strong inter-country differences in situation; no true wildlife reservoir in Europe; spill-over	Heavy economic impact; zoonosis	Movement of wild and domestic animals
Swine brucellosis	Bacteria	Wild boar, European brown hare (domestic pig)	Present in domestic pigs, particularly in open-air system; not yet controlled; true multi-host	Economic impact; zoonosis	Open-air farming
Infectious keratoconjunctivitis	Bacteria	Wild caprinae (sheep and goat)	Endemic in sheep and goat, self-limiting in wild caprinae; apparent multi-host	Conservation concerns	Movement of wild and domestic animals
Trichinellosis	Parasite	Wild boar and other mammals (domestic pig)	Endemic in wild mammals (wild boar, carnivores) and sporadic in domestic pigs (backyard); true multi-host	Zoonosis	Open-air farming

Table 7.1 (*cont.*)

Disease	Agent	Wildlife host (domestic)	Situation in Europe	Relevance	Main risks
Sarcoptic mange	Parasite	Wild mammals (domestic mammals)	Little prevalence in domestic (treatment), but self-sustained in abundant wildlife (true multi-host); spill-over to endangered wildlife	Conservation concerns	Movement of wild and domestic animals; wildlife overabundance
Toxoplasmosis	Parasite	Wild mammals and birds (all livestock)	Widespread; true multi-host	Zoonosis	Open-air farming
Tick-borne diseases	Virus, bacteria, parasite	Wildlife (domestic animals)	Most are shared between domestic and wild animals; wildlife may act as reservoir of favouring tick numbers; true multi-host	Economic impact; some are zoonosis	Vector expansion; wildlife overabundance; movement of wild and domestic animals

2000). However, while of lesser concern it still poses potential problems since the presence of a wildlife reservoir can still be the source for a spill-over into domestic animals and is likely to contribute to increased incidence in livestock, making the goal of eradication or control more difficult.

Notwithstanding the increase in knowledge, our background of information is, in many cases, still not sufficient to decide if a given 'disease–wildlife species–livestock' triangle is of concern for animal health authorities or for wildlife managers (Simpson, 2002). This may be the case, for example, for bluetongue, a vector-borne viral disease introduced in 2004 in livestock in Europe, and rapidly spread in many European countries where serotypes 1, 2, 4, 8, 9 and 16 currently circulate (Rodríguez-Sánchez *et al.*, 2008). Since the introduction of the causal agent in Europe and the spread of the disease in many European countries is comparatively recent, we have only recently had the opportunity to study the infection in wild deer species and the potential pathways for cross-infection. Climate change has been proposed as a reason for the spread of bluetongue (Purse *et al.*, 2005) and measures to control or eradicate bluetongue virus (BTV) from Europe have been widely implemented.

However, the vaccination campaigns and movement restrictions do not currently include wild hosts of the virus, such as red deer (*Cervus elaphus*), which could play a role in the epidemiology of BTV in Europe. BTV is able to replicate in red deer after experimental infection, causing a long-lasting viraemia comparable to that of domestic ruminants (Lopez-Olvera *et al.*, 2010). Due to this last finding, red deer should be taken into account for BTV surveillance, and movement restrictions as well as vaccination schemes should be adapted to include farmed or translocated red deer.

In many cases we simply do not have data on the wild host/pathogen interaction and do not know if a wild species is susceptible to a given pathogen and, more important, can become a reservoir of this pathogen, maintaining it in an area. Many cattle viral diseases, such as infectious bovine rhinotracheitis or other bovine viral diseases, are in this status and the role of deer in their epidemiology still needs further clarification (Frölich *et al.*, 2002). The recent foot-and-mouth disease (FMD) outbreak in the UK showed that, even if themselves susceptible, at least at the densities existing in the UK, wild deer do not act as a long-term reservoir for this pathogen, as culling of infected livestock resolved the problem. Moreover, even if all deer species have been considered susceptible and able to excrete the FMD virus (Thomson *et al.*, 2001), an experimental infection study showed that there was no transmission between infected wapiti (*Cervus canadensis nelsoni*) and cattle; wapiti to wapiti transmission is also infrequent and the FMD virus cannot be found in wapiti after 28 days post infection (Rhyan *et al.*, 2008).

However, we do not know what will happen in areas with a higher deer density (and thus higher transmission rates/speeds) and with other susceptible hosts such as wild boar.

Generally speaking parasitic diseases are of low concern because helminths are usually adapted to one host species and rarely transmit to others, and moreover they are usually not notifiable. However, wild ruminants and livestock can share the same nematode, as in the case of *Haemoncus contortus*, an abomasal nematode that can cause death in juveniles (Zaffaroni *et al.*, 2000). Specific helminths shared between wildlife and domestic animals have been reviewed by Simpson (2002) and Bohm *et al.* (2007). Even if none of them alone is perhaps as significant as some of the viral diseases, in some cases nematodes can cause huge economic losses, and wildlife may complicate attempts of parasite control in domestics. The case of *Neospora caninum* infection is rather emblematic because this protozoan parasite was unknown until 1989, when it was differentiated from *Toxoplasma gondii. Neospora caninum* is nowadays considered to be one of the greatest causes of abortion in cattle (Dubey, 2003) and several researchers highlight the possible role of wild carnivores, as the definitive host, or ungulates and small mammals, which are intermediate hosts, as source of infection also for domestic animals (Barling *et al.*, 2000; Ferroglio *et al.*, 2003b, 2007b; Almeria *et al.*, 2007; Sobrino *et al.*, 2008).

Mycobacterial infections are nowadays common disease agents in wildlife and represent one of the main concerns on the wildlife–livestock interface. Bovine tuberculosis (BTb) due to *Mycobacterium tuberculosis* complex is a notifiable zoonotic infection whose importance in wild ungulates has increased since the 1990s; while it has become most significant especially among wild boar in Mediterranean Europe (Bollo *et al.*, 2000; Vicente *et al.*, 2007), the risk exists also for central and eastern Europe (Machackova *et al.*, 2003) and red deer in France (Zanella *et al.*, 2008). In the case of Mediterranean wild boar populations, artificial management, such as fencing, feeding, watering and translocations, has been suggested as a possible cause for TB re-emergence (Vicente *et al.*, 2007), but, whatever the cause, the consequential costs for farming and public health institutions are very high (Dondo *et al.*, 2006).

Even if the best known example of a European wildlife host involved in BTb maintenance and spill-over is represented by badgers in the UK and Ireland, cases are frequently found also in wild ungulates, and a recent report (Ward *et al.*, 2008) suggests that especially fallow deer can represent a reservoir for this infection, where present at very high densities (>55 head/100 ha). We should also be aware that in many cases the absence of the infection in a wild population could be due to a lack of survey, as the case of Michigan, a US state believed to be free of bovine tuberculosis since 1975, suggests. In this state BTb was discovered in white-tailed deer in 1994 and surveillance programmes have shown its presence in many counties, with foci reported in domestic and wild animals and the report of human cases among hunters (Wilkins *et al.*, 2008).

The re-emergence of BTb in many wild populations apparently is in contrast to the reduction/eradication of BTb in cattle observed in many countries. BTb cases have been reported in central Europe in deer in the 1950s (Bouvier, 1963); however, the low density of wild species was probably under the threshold for maintenance of the pathogen. The increase in wildlife densities and in wildlife management as well as changes in livestock breeding systems, with the reduction of human presence in pastures, has favoured the spill-over from livestock to wildlife which, in some cases, maintains the infection and can be the source for a spill-back to livestock (Figure 7.2).

Paratuberculosis due to *Mycobacterium avium paratuberculosis* is not a notifiable disease and there is no clear evidence of its role as a zoonosis. However, attempts to control or eradicate it in cattle are increasing in many areas. Clearly if farmers and veterinary authorities wish to control the infection in cattle, the risk that infected wildlife, contaminating soil and pasture, can cancel out their effort, must be taken into consideration in managing plans and strategies. Paratuberculosis is considered highly prevalent among wild rabbits and other wildlife in the UK (Daniels *et al.*, 2003) and in the Alps (Nebbia *et al.*, 2000; Fraquelli *et al.*, 2005) and infection has been found in red and roe deer also in areas without livestock (Robino *et al.*, 2008). In both cases re-emergence of these mycobacterial diseases constitutes a severe barrier to its eradication in livestock.

Multi-host situations of this type are equally of concern for wildlife management and conservation, as diseases can be a threat for endangered species or affect the productivity and density of wildlife populations with an economic or recreational value (Gortázar *et al.*, 2007). Among ungulates, two diseases shared between wild caprines (such as chamois and ibex) and domestic sheep and goats have important consequences on wildlife population dynamic and structure and their welfare. One is keratoconjunctivitis due to *Mycoplasma conjunctivae* in alpine chamois (Giacometti *et al.*, 2002), and the second is sarcoptic mange (*Sarcoptes scabiei*), affecting several populations of mountain herbivores in Europe (e.g. Rossi *et al.*, 2007). In both cases the disease is suspected to spread from domestic livestock to wildlife and is responsible for repeated outbreaks that affect hunting harvest, population dynamics and animal welfare.

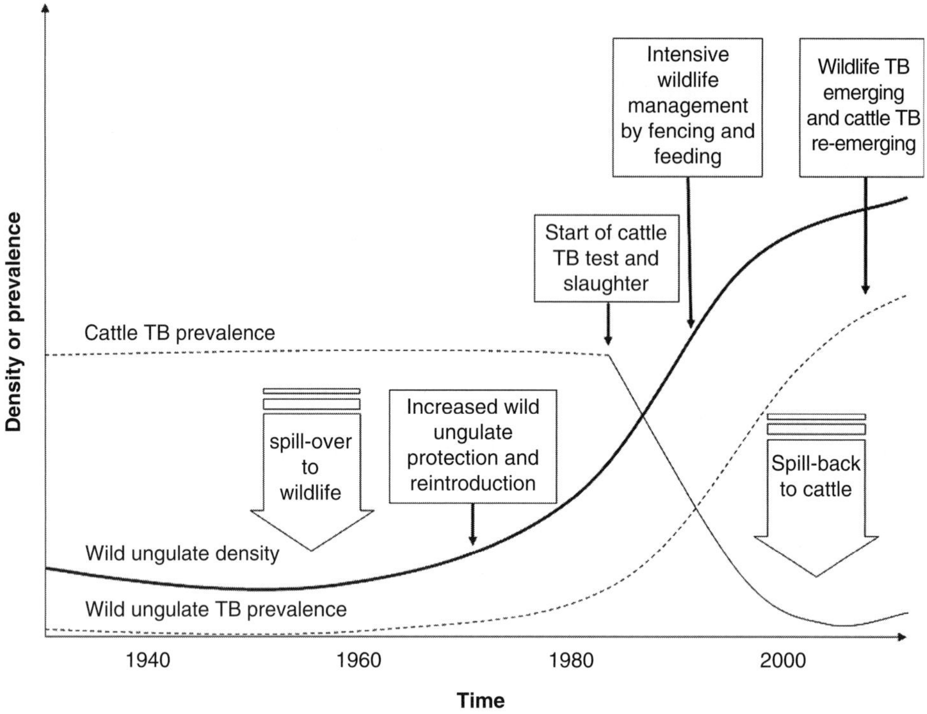

Figure 7.2 Relations among wildlife, cattle and bovine tuberculosis (BTb) transmission. Even though the spill-over of BTb to wildlife has probably also occurred in the past, BTb became endemic in many wildlife populations only with the increase in wildlife densities and management changes that have taken place in the last third of the twentieth century. Nowadays the risk of a spill-back from infected wildlife to cattle is of great concern, due to the potentially severe economical consequences to the cattle industry.

Brucellosis due to bacteria of the genus *Brucella* has been detected in alpine ibex and chamois in the Alps (Ferroglio *et al.*, 1998, 2003a) as a result of a spill-over from sheep and cattle. While there is no evidence of the persistence of the infection in chamois, the infection, due to *Brucella melitensis*, seems to be persistent in the ibex population; however, an experimental trial has suggested that, even if it cannot be completely ruled out, the transmission of *B. melitensis* from ibex to livestock appears at least to be very remote (Ferroglio *et al.*, 2007a).

In summary, actual or potential emerging diseases deserve attention, including the study of the underlying causes for the emergence of infectious diseases, which are often related to anthropogenic and environmental changes (Kuiken *et al.*, 2003; Cunningham, 2005).

7.3 Emerging diseases

Emerging infectious diseases are those where a pathogen may become self-sustaining in the new host once the initial (environment-, host- or pathogen-related) barrier to infection has been crossed. Incidence of emerging diseases has been rising in recent years, and the diseases may well occur at the fertile livestock–wildlife interface (Daszak *et al.*, 2000; Cunningham, 2005). Wild animals are the most likely source of new emerging infectious diseases that put at risk the health of human beings and livestock (Daszak *et al.*, 2000).

Chronic wasting disease (CWD) is a transmissible spongiform encephalopathy of North American cervids that has become one of the major health concerns in wapiti and white-tailed deer management in many US states. Fortunately there is no evidence of CWD in European cervids (Schettler *et al.*, 2006), even if many veterinary institutions are still stressing the risk of prion diseases in European wildlife. Obviously if infected deer from North America were to be imported into Europe, it is possible that the geographical barrier is crossed and CWD will become endemic in European ungulates.

Among other potential emerging diseases there is the Rift Valley fever, a mosquito-borne zoonotic viral disease leading to serious economic losses in livestock, particularly sheep, in Africa. Global climatic change, vector expansion or movements of domestic animals may eventually bring this disease to Europe. In addition, a number of flaviviruses that exist in tropical and subtropical America may eventually be imported through travellers or translocated animals. The host diversity of these viruses in their native range (De Thoisy *et al.*, 2004), along with the current expansion of vectors such as *Aedes albopictus* in the Mediterranean (Mitchell, 1995), may eventually cause outbreaks in Europe. Recently Krimea-Congo haemorrhagic fever, a vector-borne viral disease transmitted by *Hyalomma* ticks, has expanded westwards via Turkey, Greece and Albania (Papa *et al.*, 2008). These and other emerging diseases may become relevant threats for European wild ruminants in the coming years.

7.4 Risk assessment, and control of diseases in wildlife populations

Wildlife disease control starts usually with surveillance, knowing which diseases are present, their past and current distribution and the trends in their prevalence (Wobeser, 1994, 2002; Artois *et al.*, 2001; Artois, 2003). Disease surveillance and monitoring in European wildlife have been thoroughly reviewed and treated by Artois *et al.* (2008). At present, wildlife disease surveillance in Europe is addressed by a series of regional, national and international schemes. The status of wildlife surveillance in Europe has been analysed and reviewed by working groups in 2000 (AAVV, 2000) and the Office International of Epizootics Working Group on Wildlife Diseases reports yearly the results of the surveillance programmes carried out on wildlife. These yearly reports are available at the OIE website (www.oie.int/wildlife/eng/en_wildlife.htm) and represent a useful source of information on reported cases of wildlife diseases.

Proper implementation of a complete surveillance effort must be a priority of the veterinary authorities, as it is accepted that those countries that conduct disease surveillance of their wild animal populations are more likely to detect the presence of infectious and zoonotic diseases and to swiftly adopt countermeasures (Morner *et al.*, 2002; Artois *et al.*, 2008). The main obstacle in such schemes is represented by the difficulty of collecting samples (i.e. blood, tissues) from wildlife.

Samples can be collected when wild animals are captured and handled or found dead, but the main source of samples is represented by culled game animals. Tissues can be collected by hunters or veterinarians and sent to the laboratories fresh or frozen for analysis. Subsequent investigation may be through direct pathogen diagnosis (culture or DNA/RNA finding by polymerase chain reaction, PCR). However, probably due to economic reasons, surveillance and monitoring are mainly established on indirect diagnosis based on serological analysis searching the antibodies produced by the animals against a given pathogen. In this case sera are needed; collection of suitable sera from hunted animals must be done within a few hours from animal death and the quality of obtained sera is often not good enough. A possibility of overcoming this problem is the use of lung tissue extract, which can be easily obtained by frozen pieces of lung, instead of blood sera for serological tests (Ferroglio *et al.*, 2000).

In addition to surveillance, three basic options are available for disease management in wildlife: prevention of introduction of disease, control of existing disease or, almost impossible and very costly, eradication (Wobeser, 2002).

As a general rule the most usual risk factor is represented by the introduction of a pathogen through movements or translocations of wild or domestic animals. It is impossible to move animals without also moving their pathogen. To show how animal movements can be relevant to pathogen introduction, it has been reported that 10 new pathogens have been detected in wild ungulates in north-west Italy since 1995 (Artois *et al.*, 2008). In many reported cases the introduction of the new agents is due to the translocation of wildlife for restocking, while in some cases the pathogen was transmitted to wild ungulates by livestock sharing the same pasture. Due to social changes in many areas pastoralism practice has been changed, with a decrease in human presence with herds and flocks and a consequent increase risk of overlapping between livestock and wildlife. In turn, translocation of exotic species into Europe also conveys important risks of introduction of new pathogens, especially when foreign species share habitats with susceptible native species, potentially leading to situations where the native species become endangered (McInnes *et al.*, 2006). Introductions of North American wapiti (*Cervus canadensis*) carried into Europe the trematode *Fascioloides magna*, highly pathogenic in the red deer and in roe deer (Novobilsky *et al.*, 2006).

Fascioloides magna also offers a clear example of how difficult it is to eradicate a new pathogen after its introduction. The first description of this parasite was in a fenced hunting reserve in northern Italy during a red deer die-off which occurred in 1875, after the release of wapiti imported from North America (Balbo *et al.*, 1989). Despite several control procedures, the parasite is still present in this red deer population where it can cause disease and death in both red and fallow deer. In the Danube area the parasite has greatly increased its range, due to the active and passive movement of the intermediate host (a snail), and is now a health problem even in countries (such as Croatia) where wapiti have never been introduced (Rajkovic-Janje *et al.*, 2008). In this context a much closer collaboration is required between governmental or supragovernmental agencies devoted to animal health and to wildlife management, considering that even the movements of domestic animals can easily cause the introduction of new diseases or new vectors (e.g. rabbit haemorrhagic disease introduced from China into Germany, (Angulo and Cooke, 2002) or bluetongue virus infection in wildlife after its introduction in European livestock).

After introduction, several management tools have been deployed or proposed for the control of diseases in wildlife especially when overabundance, which can facilitate the transmission of the pathogen, represents the main risk factor. Among these tools wildlife culling has been frequently proposed as a valid option to control diseases in wildlife, giving rise to an intense scientific and social debate (e.g. in relation to badger culling for TB control, Donnelly *et al.*, 2006).

Only in the case of island populations, when geographical barriers limit animal dispersal, or in the case of introduced species (where legal and social constraints to culling are minimal), or to cope with a point-source wildlife disease outbreak (centring culling on the disease focus, plus an outer ring of vaccination), can culling to eradication generally be considered as an option. By contrast, simple population *reduction* remains a goal in many disease control efforts. However, generally, such population reduction is only temporary and has a low impact on the epidemiology of the disease, except if habitat modification is used to effect more permanent reductions of host density or to alter host distribution or exposure to disease agents (Wobeser, 2002; Acevedo *et al.*, 2007; Gortázar *et al.*, 2007). In addition, it must be noted that any risks are not necessarily directly related to actual population size because aggregation of animals, usually due to human activities (i.e. artificial feeding), can increase the risk of pathogen transmission even in populations with a not particularly high density (Acevedo *et al.*, 2007). European examples include tuberculosis in wild boar, red and fallow deer (Vicente *et al.*, 2007) and classical swine fever (Rossi *et al.*, 2005) in wild boar, and salmonellosis (Pennycott, 1998) and trichomoniasis (Höfle *et al.*, 2004) where an increase in transmission rates has been associated with artificial feeding. In many cases bans on supplementary feeding can reduce the carrying capacity for ungulates and reduce population density and, just as significant, aggregation, two key factors in infectious disease transmission (Acevedo *et al.*, 2007; Vicente *et al.*, 2007).

Rather more selective culling is usually proposed when affected individuals are readily identifiable (Wobeser, 2002). This approach is based on the apparent ability of strategies of culling of obviously infected individuals, or test and slaughter, to control and eradicate infection in domestic animals. However, the status of infection of wild animals cannot easily be determined and many infected individuals cannot be detected; the proportion of animals killed by selective culling thus may have little or no effect on the spread of the disease. Field experience, such as the one with chamois mange in the Cantabrian Mountains (Spain) (Fernandez-Moran *et al.*, 1997), or infectious keratoconjunctivitis in chamois (E. Ferroglio, unpublished data), clearly shows that, apart from some social aspect (i.e. giving to hunters the feeling of doing something), such selective culling is of little or no value in disease management of free-ranging wildlife. Indeed, it may even prove counterproductive. Disruption of social structure with increased movement and, therefore, increased contact rate (at intra- or interspecific level, Donnelly *et al.*, 2006) may lead to actual increases in disease prevalence, or spread of disease to previously uninfected population nuclei (e.g. wild boar and swine fever after high hunting pressure, Guberti *et al.*, 1998).

One option to avoid or prevent wildlife contact with domestic livestock, or vice versa, is the use of barriers. However, such barriers present cost and physical limitations (two barriers are required to avoid direct contact) which restrict such approaches to certain spatially limited risk situations. For example, it may make sense to exclude badgers from cattle farms in TB endemic areas (Garnett *et al.*, 2002) or to prevent wild boar contact with open-air bred pigs or other livestock (Parra *et al.*, 2005).

The possibility of controlling wildlife diseases by chemical treatment with drugs, as we are accustomed to do in domestic animals and in human medicine, is sometimes proposed by stakeholders. Wildlife treatment with drugs is increasingly frequent, especially against parasites in economically valuable game species. Spanish ibex (*Capra pyrenaica*) have been treated against sarcoptic mange with ivermectin (Leon-Vizcaino *et al.*, 2001), and wood pigeons (*Columba palumbus*) have been treated against trichomonas with dimetridazole (Höfle *et al.*, 2004). Also anthelminthic treatments are frequent in ungulates and in game birds (Fernandez de Mera *et al.*, 2004; Rodriguez *et al.*, 2006; Villanúa *et al.*, 2006a, 2006b). However, in many cases the effectiveness of these treatments is unclear, and ethical and public health issues need to be addressed because, for example, the use of antibiotics in game species may affect meat hygiene or non-target species can take the drugs. This latter risk is greatly increased when baiting with drugs is not selective due to poor knowledge of animals' feeding preferences.

Vaccination is widely used in disease control; however, wildlife vaccination is only exceptionally undertaken, and it is normally limited to the most significant diseases (those that cause serious economic losses, those that are almost under control in domestics, and those where wildlife reservoirs are of paramount importance in maintaining the disease). In Europe this is the case for fox rabies (Artois *et al.*, 1993), classical swine fever (Kaden *et al.*, 2002) and, probably soon, bovine TB. In contrast to culling, oral vaccination has the advantages of being painless, thus avoiding animal welfare problems, and does not cause behavioural problems such as increased dispersal or immigration. Vaccination makes sense if the huge investment is the only way to control a disease in its wildlife reservoir, if the costs are clearly outbalanced by the costs of taking no action at all and provided that the effectiveness and safety of the vaccine have been tested in captivity. In most cases, vaccination needs to be combined with other management measures, and the ecology of the host species needs to be carefully considered.

In summary, management of diseases of wild animals cannot be done without a change in human activities (Wobeser, 2002), and a sound scientific basis is strongly needed before suggesting any corrective measures that can create or increase conflicts among the different stakeholders: veterinary authorities, hunters, conservationists, livestock breeders and the general public. This will require the need for multidisciplinary teams including not only ecologists, wildlife managers, foresters, wildlife veterinarians, public health authorities, breeders and agriculture officers, but also, considering the importance of the human factor, sociologists.

7.5 Conclusions

Among the diseases listed in Table 7.1 not all have the same economic impact, and for only a few of them is control of the disease in wildlife reservoirs actually of paramount importance for the success of general control schemes in livestock. These diseases (rabies, classical swine fever, swine brucellosis and bovine TB) are the most crucial, and European research efforts are clearly biased towards them.

Nonetheless, even more research is needed into the remaining diseases, where current knowledge is limited regarding their management. In a number of cases scientific evidence is currently inadequate to determine if European wildlife has a high probability of substantially affecting regional disease status or not. This is the case, for example, of bluetongue, bovine viral diarrhoea, alphaherpesvirus infections, malignant catarrhal fever, and brucellosis (*B. abortus* and *B. melitensis*). These diseases deserve a greater research effort to determine their relevance in relation to wildlife. Finally, there are many other diseases that have either a more limited impact on economy and public health, or the role of wildlife in their control is not considered relevant. This is the case, for example, of many parasitic diseases. Research in these areas exists, in part, because macroparasites are easy to handle in experimental studies, and it provides a valuable source of basic scientific knowledge on disease ecology, which may be applied later to more relevant diseases (e.g. Hudson *et al.*, 2001; Vicente *et al.*, 2004a, 2004b).

Regarding the kind of research needed in the wildlife disease field, surveillance and descriptive studies, in general, are still valuable, especially in regions, species or diseases that have received less attention. Nonetheless, limiting the effort to the mere reporting of wildlife disease outbreaks is of limited value if management recommendations are not given. Moreover wildlife managers and the public may perceive animal health authorities as purveyors of bad news ('you got this disease') with no positive counterpart ('you can do this') (Ferroglio, 2003). Therefore, more experimental approaches are strongly needed if the aim is to produce substantial knowledge that enables researchers to make targeted management recommendations. Experimental studies should ideally combine indoor experiments, such as experimental infections or vaccination trials, with field experiments testing hypotheses regarding, for example, the effects of host aggregation and density on disease prevalence (e.g. Donnelly *et al.*, 2006; Acevedo *et al.*, 2007) as well as pathogen persistence. Mathematical modelling may help to identify those knowledge gaps that most urgently need experimental research (Morgan *et al.*, 2006), but research effort must also be focused on the development of diagnostic tools appropriate for wildlife species (Simpson, 2002).

References

A.A.V.V. (2000) Reporting system of wildlife diseases in Europe: European Network on wildlife as reservoirs of pathogens including zoonoses. FAIR-CT 98–4361, Madrid June 2000.

Acevedo, P., Vicente, J., Höfle, U., *et al.* (2007) Estimation of European wild boar relative abundance and aggregation: a novel method in epidemiological risk assessment. *Epidemiology and Infection* **135**, 519–527.

Alexander, D.J. (2000) A review of avian influenza in different bird species. *Veterinary Microbiology* **74**, 3–13.

Almeria, S., Vidal, D., Ferrer, D., *et al.* (2007) Seroprevalence of *Neospora caninum* in non-carnivorous wildlife from Spain. *Veterinary Parasitology* **143**, 21–28.

Angulo, E. and Cooke, B. (2002) First synthesize new viruses then regulate their release? The case of the wild rabbit. *Molecular Ecology* **11**, 2703–2709.

Apollonio, M., Andersen, R. and Putman, R. (eds.) (2010) *European Ungulates and their Management in the 21st Century*. Cambridge, UK: Cambridge University Press.

Artois, M. (2003) Wildlife infectious disease control in Europe. *Journal of Mountain Ecology* **7**, 89–97.

Artois, M., Masson, E., Barrat, J. and Aubert, M.F.A. (1993) Efficacy of 3 oral rabies vaccine-baits in the red fox: a comparison. *Veterinary Microbiology* **38**, 167–172.

Artois, M., Delahay, R., Guberti, V. and Cheeseman, C. (2001) Control of infectious diseases of wildlife in Europe. *Veterinary Journal* **162**, 141–152.

Artois, M., Bengis, R., Delahay, R., *et al.* (2008) Wildlife disease surveillance and monitoring. In R.J. Delahay, G.C. Smith and M.R. Hutchings (eds.) *Management of Diseases in Wild Mammals*. Tokyo: Springer, pp.187–214.

Balbo, T., Rossi, L. and Meneguz, P.G. (1989) Integrated control of *Fascioloides magna* infection in northern Italy. *Parassitologia* **31**, 137–144.

Barling, K.S., Sherman, M., Peterson, M.J., *et al.* (2000) Spatial associations among density of cattle, abundance of wild canids, and seroprevalence to *Neospora caninum* in a population of beef calves. *Journal of American Veterinary Medicine Association* **217**, 1361–1365.

Bohm, M., White, P.C.L., Chambers, J., Smith, L. and Hutchings, M.R. (2007) Wild deer as a source of infection for livestock and humans in the UK. *Veterinary Journal* **174**, 260–276.

Bollo, E., Ferroglio, E., Dini, V., *et al.* (2000) Detection of *Mycobacterium tuberculosis* complex in lymph nodes of wild boar (*Sus scrofa*) by a target-amplified test system. *Journal of Veterinary Medicine B* **47**, 337–342.

Bouvier, G. (1963) Transmission possible de la tuberculose et de la brucellosi du tibie à l'homme et aux animaux domestiques et sauvages. *Bulletin de l'Office International des Epizooties* **59**, 433–436.

Crawley, M.J. (1992) *Natural Enemies: The Population Biology of Predators, Parasites and Diseases*. London: Blackwell.

Cunningham, A.A. (2005) A walk on the wild side – emerging wildlife diseases. They increasingly threaten human and animal health. *British Medical Journal* **331**, 1214–1215.

Daniels, M.J., Henderson, D., Greig, A., *et al.* (2003) The potential role of wild rabbits *Oryctolagus cuniculus* in the epidemiology of paratuberculosis in domestic ruminants. *Epidemiology and Infection* **130**, 553–559.

Daszak, P., Cunningham, A.A. and Hyatt, A.D. (2000) Emerging infectious diseases of wildlife: threats to biodiversity and human health. *Science* **287**, 443–449.

De Thoisy, B., Dussart, P. and Kazanji, M. (2004) Wild terrestrial rainforest mammals as potential reservoirs for flaviviruses (yellow fever, dengue 2 and St Louis encephalitis viruses) in French Guiana. *Transaction of the Royal Society of Tropical Medicine and Hygiene* **98**, 409–412.

Dondo, A., Ferroglio, E., Goria, M., *et al.* (2006) Economic significance of bovine tuberculosis in Italy and effect of *M. bovis* infection in wild swine. In C.O. Thoen, J.H. Steele and M.J. Gilsdorf (eds.) *Mycobacterium bovis Infection in Animals and Humans* (2nd edn.). Oxford, UK: Blackwell Publishing, pp. 117–123.

Donnelly, C.A., Woodroffe, R., Cox, D.R., *et al.* (2006) Positive and negative effects of widespread badger culling on tuberculosis in cattle. *Nature* **439**, 843–846.

Dubey, J.P. (2003) Neosporosis in cattle. *Journal of Parasitology* **89**, 42–56.

Fernandez de Mera, I.G., Vicente, J., Gortázar, C., Höfle, U., Fierro, Y. (2004) Efficacy of an in-feed preparation of ivermectin against helminths in the European wild boar. *Parasitology Research* **92**, 133–136.

Fernandez-Moran, J., Gomez, S., Ballesteros, F., *et al.* (1997) Epizootiology of sarcoptic mange in a population of Cantabrian chamois (*Rupicapra pyrenaica parva*) in northwestern Spain. *Veterinary Parasitology* **73**, 163–171.

Ferroglio, E. (2003) Wildlife veterinarian: a necessary contribution to wildlife conservation?. In J.M. Perez (ed.) *In Memoriam al Prof. Isidoro Ruiz Martinez*. Jaen, Spain: Universidad de Jaen (Spain).

Ferroglio, E., Tolari, F., Bollo, E. and Bassano, B. (1998) First isolation of *Brucella melitensis* from Alpine ibex. *Journal of Wildlife Diseases* **34**, 400–402.

Ferroglio, E., Rossi, L. and Gennero, S. (2000) Lung-tissue extract as an alternative to serum for surveillance for brucellosis in chamois. *Preventive Veterinary Medicine* **43**, 117–122.

Ferroglio, E., Gennero, S., Rossi, L. and Tolari, F. (2003a). Epidemiologia della brucellosi nel camoscio alpino. *Journal of Mountain Ecology* **7**, 229–232.

Ferroglio, E., Wamba, E., Trisciuoglio, A. and De Meneghi, D. (2003b). Antibodies to *Neospora caninum* in wild mammals from Kenya, East Africa. *Veterinary Parasitology* **118**, 43–49.

Ferroglio, E., Gennero, M.S., Pasino, M., *et al.* (2007a) Cohabitation of a *Brucella melitensis* infected Alpine ibex (*Capra ibex*) with domestic small ruminants in an enclosure in Gran Paradiso National Park, in western Italian Alps. *European Journal of Wildlife Research* **53**, 158–160.

Ferroglio, E., Pasino, M., Romano, A., *et al.* (2007b) Evidence of *Neospora caninum* DNA in wild rodents. *Veterinary Parasitology* **148**, 346–349.

Fraquelli, C., Carpi, G., Bregoli, M., *et al.* (2005) Epidemiology of paratuberculosis in two red deer (*Cervus elaphus*) populations of Trentino (northern Italy). In E.J.B. Manning and S.S. Nielsen (eds.) *Proceedings of the 8th International Colloquium on Paratuberculosis*, held Copenhagen Denmark. Madison, WI: International Association for Paratuberculosis.

Frölich, K., Thiede, S., Kozikowski, T. and Jakob, W. (2002) A review of mutual transmission of important infectious diseases between livestock and wildlife in Europe. *Annals of the New York Academy of Sciences* **969**, 4–13.

Garnett, B.T., Delahay, R.J. and Roper, T.J. (2002) Use of cattle farm resources by badgers (*Meles meles*) and risk of bovine tuberculosis (*Mycobacterium bovis*) transmission to cattle. *Proceedings of the Royal Society B–Biological Sciences* **269**, 1487–1491.

Gauss, C.B.L., Dubey, J.P., Vidal, D., *et al.* (2006) Prevalence of *Toxoplasma gondii* antibodies in red deer (*Cervus elaphus*) and other wild ruminants from Spain. *Veterinary Parasitology* **136**, 193–200.

Giacometti, M., Janovsky, M., Belloy, L. and Frey, J. (2002) Infectious keratoconjunctivitis of ibex, chamois and other Caprinae. *Revue Scientifique et Technique de l'Office International des Epizooties* **21**, 335–345.

Gortázar, C., Ferroglio, E., Höfle, U., Froelich, K. and Vicente, J. (2007) Diseases shared between wildlife and livestock: a European perspective. *European Journal of Wildlife Research* **53**, 241–256.

Guberti, V., Rutili, D., Ferrari, G., Patta, C. and Oggaino, A. (1998) Estimate the threshold abundance for the persistence of the classical swine fever in the wild boar population of the eastern Sardinia. In *Report on Measures to Control Classical Swine Fever in European Wild Boar*. Document VI/7196/98-AL. Perugia, Italy: Commission of the European Communities, Directorate General VI for Agriculture.

Höfle, U., Gortázar, C., Ortiz, J.A., Knispel, B. and Kaleta, E.F. (2004) Outbreak of trichomoniasis in a woodpigeon (*Columba palumbus*) wintering roost. *European Journal of Wildlife Research* **50**, 73–77.

Hudson, P.J., Rizzoli, A., Grenfell, B.T., Heesterbeek, H. and Dobson, A.P. (2001) *The Ecology of Wildlife Diseases*. Oxford, UK: Oxford University Press.

Kaden, V., Heyne, H., Kiupel, H., *et al.* (2002) Oral immunisation of wild boar against classical swine fever: concluding analysis of the recent field trials in Germany. *Berliner Muncher Tierarztliche Wochenschrift* **115**, 179–185.

Kuiken, T., Fouchier, R., Rimmelzwaan, G. and Osterhaus, A. (2003) Emerging viral infections in a rapidly changing world. *Current Opinion in Biotechnology* **14**, 641–646.

Laddomada, A., Patta, C., Oggiano, A., *et al.* (1994) Epidemiology of classical swine fever in Sardinia: a serological survey of wild boar and comparison with African swine fever. *Veterinary Record* **134**, 183–187.

Leon-Vizcaino, L., Cubero, M.J., Gonzalez-Capitel, E., *et al.* (2001) Experimental ivermectin treatment of sarcoptic mange and establishment of a mange-free population of Spanish ibex. *Journal of Wildlife Diseases* **37**, 775–785.

Lopez-Olvera, J.R., Falconi, C., Fernandez-Pacheco, P., *et al.* (2010) Experimental infection of European red deer (*Cervus elaphus*) with bluetongue serotypes 1 and 8. *Veterinary Microbiology* **145**, 148–152.

Machackova, M., Matlova, L., Lamka, J., *et al.* (2003) Wild boar (*Sus scrofa*) as a possible vector of mycobacterial infections: reviews of literature and critical analysis of data from Central Europe between 1983 to 2001. *Veterinary Medicine* **48**, 51–65.

McInnes, C.J., Wood, A.R., Thomas, K., *et al.* (2006) Genomic characterization of a novel poxvirus contributing to the decline of the red squirrel (*Sciurus vulgaris*) in the UK. *Journal of General Virology* **87**, 2115–2125.

Meltzer, D.G.A. (1995) Veterinary wildlife reseach and its role in community development. *Journal of the South African Veterinary Association* **66**, 187–189.

Mitchell, C.J. (1995) Geographic spread of *Aedes albopictus* and potential for involvement in arbovirus cycles in the Mediterranean Basin. *Journal of Vector Ecology* **20**, 44–58.

Morgan, E.R., Lundervold, M., Medley, G.F., *et al.* (2006) Assessing risks of disease transmission between wildlife and livestock: the Saiga antelope as a case study. *Biological Conservation* **131**, 244–254.

Morner, T., Obendorf, D.L., Artois, M. and Woodford, M.H. (2002) Surveillance and monitoring of wildlife diseases. *Revue Scientifique et Technique de l'Office International des Epizooties* **21**, 67–76.

Nebbia, P., Robino, P., Ferroglio, E., *et al.* (2000) Paratuberculosis in red deer (*Cervus elaphus hippelaphus*) in the Western Alps. *Veterinary Research Communication* **24**, 435–443.

Novobilsky, A., Horackova, E., Hirtova, L., Modry, D. and Koudela, B. (2006) The giant liver fluke *Fascioloides magna* (Bassi 1875) in cervids in the Czech Republic and potential of its spreading to Germany. *Parasitology Research* **100**, 549–553.

Papa, A., Bino, S., Papadimitriou, E., *et al.* (2008) Suspected Crimean Congo haemorrhagic fever cases in Albania. *Scandinavian Journal of Infectious Diseases* **3**, 1–3.

Parra, A., Larrasa, J., Garcia, A., Alonso, J.M. and de Mendoza, J.H. (2005) Molecular epidemiology of bovine tuberculosis in wild animals in Spain: a first approach to risk factor analysis. *Veterinary Microbiology* **110**, 293–300.

Pennycott, T.W. (1998) Population density and infectious disease at bird tables. *Veterinary Record* **142**, 523.

Phillips, C.J., Foster, C.R., Morris, P.A. and Teverson, R. (2003) The transmission of *Mycobacterium bovis* infection to cattle. *Research in Veterinary Science* **74**, 1–15.

Purse, B.V., Mellor, P.S., Rogers, D.J., *et al.* (2005) Climate change and the recent emergence of bluetongue in Europe. *Nature Reviews Microbiology* **3**, 171–181.

Rajkovic-Janje, R., Bosnic, S., Rimac, D. and Gojmerac, T. (2008) The prevalence of American liver fluke *Fascioloides magna* in red deer from Croatian hunting grounds. *European Journal of Wildlife Research* **54**, 525–528.

Rhyan, J., Deng, M., Wang, H., *et al.* (2008) Foot and mouth diseases in North American bison and elk: susceptibility, intra- and interspecific transmission, clinical signs and lesions. *Journal of Wildlife Diseases* **44**, 269–279.

Ritzmann, M., Gyra, H., Johannes, S., Hausleithner, D. and Heinritzi, K. (2000) Comparative investigations of a combined vaccine against porcine parvovirus and swine erysipelas and corresponding monovalent vaccines in different vaccination schedules. *Tierarztl Pax* **28**, 23–27.

Robino, P., Nebbia, P., Tramuta, C., *et al.* (2008) Identification of *Mycobacterium avium* subsp. *paratuberculosis* in wild cervids (*Cervus elaphus hippelaphus* and *Capreolus capreolus*) from north-western Italy. *European Journal of Wildlife Research* **54**, 357–360.

Rodriguez, O., Fernandez de Mera, I.G., Vicente, J., Peña, A. and Gortázar, C. (2006) Efficacy of in-feed-administered ivermectin on *Elaphostrongylus cervi* first-stage excretion in red deer (*Cervus elaphus*). *Parasitology Research* **98**, 176–178.

Rodríguez-Sánchez, B., Iglesias-Martín, I., Martínez-Avilés, M. and Sánchez-Vizcaíno, J.M. (2008) Orbiviruses in the Mediterranean basin: updated epidemiological situation of bluetongue and new methods for the detection of BTV serotype 4. *Transboundary and Emerging Diseases* **55**, 205–214.

Rossi, L., Fraquelli, C., Vesco, U., *et al.* (2007) Descriptive epidemiology of a scabies epidemic in chamois in the Dolomite Alps, Italy. *European Journal of Wildlife Research* **53**, 131–141.

Rossi, S., Fromont, E., Pontier, D., *et al.* (2005) Incidence and persistence of classical swine fever in free-ranging wild boar (*Sus scrofa*). *Epidemiology and Infection* **133**, 559–568.

Schettler, E., Steinbach, F., Eschenbacher-Kaps, I., *et al.* (2006) Surveillance for prion disease in cervids, Germany. *Emerging Infectious Diseases* **12**, 319–322.

Simpson, V.R. (2002) Wild animals as reservoirs of infectious diseases in the UK. *Veterinary Journal* **163**, 128–146.

Sobrino, R., Dubey, J.P., Pabon, M., *et al.* (2008) *Neospora caninum* antibodies in wild carnivores from Spain. *Veterinary Parasitology* **155**, 190–197.

Swinton, J., Woolhouse, M.E.J., Begon, M.E., *et al.* (2002) Microparasite transmission and persistence. In P.J. Hudson, A. Rizzoli, B.T. Grenfell, H. Heesterbeek and A.P. Dobson (eds.) *The Ecology of Wildlife Diseases*. Oxford, UK: Oxford University Press, pp.83–101.

Thomson, G.R., Bengis, R.G. and Brown, C.C. (2001) Picornavirus infections. In E.S. Williams and I.K. Barker (eds.) *Infectious Diseases of Wild Mammals* (3rd edn.). London: Manson Publishing, pp.119–130.

Vicente, J., Fierro, Y., Martinez, M. and Gortázar, C. (2004a) Long-term epidemiology, effect on body condition and interspecific interactions of concomitant infection by nasopharyngeal bot fly larvae (*Cephenemyia auribarbis* and *Pharyngomyia picta*, Oestridae) in a population of Iberian red deer (*Cervus elaphus hispanicus*). *Parasitology* **129**, 349–361.

Vicente, J., Segalés, J., Höfle, U., *et al.* (2004b) Epidemiological study on porcine circo virus type 2 (PCV2) infection in the European wild boar (*Sus scrofa*). *Veterinary Research* **35**, 243–253.

Vicente, J., Höfle, U., Garrido, J.M., *et al.* (2007) Risk factors associated with prevalence of tuberculosis-like lesions in wild boar and red deer in south central Spain. *Veterinary Research* **38**, 451–464.

Villanúa, D., Acevedo, P., Höfle, U., Rodríguez, O. and Gortázar, C. (2006a) Changes in transmission stage excretion after pheasant release. *Journal of Helminthology* **80**, 313–318.

Villanúa, D., Höfle, U., Pérez-Rodríguez, L. and Gortázar, C. (2006b) *Trichomonas gallinae* in wintering common wood pigeons *Columba palumbus* in Spain. *Ibis* **148**, 641–648.

Ward, A.I., Etherington, T.R. and Smith, G.C. (2008) Exposure of cattle to *Mycobacterium bovis* excreted by deer in southwest England: a quantitative risk assessment. Consultancy report to TB Programme, Food and Farming Group, Defra, London.

Wilkins, M.J., Meyerson, J., Barlett, P.C., *et al.* (2008) Human *Mycobacterium bovis* infection and bovine tuberculosis outbreak, Michigan, 1994–2007. *Emerging Infectious Diseases* **14**, 657–660.

Wobeser, G.A. (1994) *Investigation and Management of Disease in Wild Animals*. New York: Plenum.

Wobeser, G.A. (2002) Disease management strategies for wildlife. *Revue Scientifique et Technique de l'Office International des Epizooties* **21**, 159–178.

Woodford, M. (1965) The role of the veterinarian in wildlife conservation. *Veterinary Record* **77**, 1311–1317.

Woolhouse, M.E. (2002) Population biology of emerging and reemerging pathogens. *Trends in Microbiology* **10**, 3–7.

Zaffaroni, E., Manfredi, M.T., Citterio, C., *et al.* (2000) Host specificity of abomasal nematodes in free ranging alpine ruminants. *Veterinary Parasitology* **90**, 221–230.

Zanella, G., Durand, B., Hars, J., *et al.* (2008) *Mycobacterium bovis* in wildlife in France. *Journal of Wildlife Diseases* **44**, 99–108.

8

Traffic collisions involving deer and other ungulates in Europe and available measures for mitigation

JOCHEN LANGBEIN, RORY PUTMAN AND BOSTJAN POKORNY

8.1 Introduction: the scale of the problem

As road infrastructures proliferate, traffic volumes and speeds rise, and ungulate densities also increase throughout Europe (Gill, 1990; Apollonio *et al.*, 2010), so the frequency of road traffic accidents involving wildlife also escalates throughout Europe.

In 1982 some 10 000 road accidents were recorded in Sweden due to collisions with moose, red deer and roe deer; by 1993 that number had risen to 55 000, with mortality of roe deer alone in excess of 50 000 (Groot Bruinderink and Hazebroek, 1996). Statistics presented by Groot Bruinderink and Hazebroek showed this to be a general trend throughout Europe, and suggested that, at that time, vehicle–ungulate collisions in Europe as whole may have been of the order of 500 000. Estimates offered by Apollonio *et al.* (2010) indicate that, at least in those countries where estimates are attempted, numbers had risen substantially by 2005 (Table 8.1). Formal records are only maintained in a small proportion of countries, whereas in many others comparable data are not available. It is therefore not possible to offer an accurate estimate for the total number of collisions occurring in Europe as a whole, although we may note that totals recorded in the table (for fewer than half the countries of Europe) already approximate to 400 000, with the full toll of ungulates killed annually on European roads likely to be closer to 1 million.

The scale and recent escalation of wildlife collisions in Europe are mirrored by figures from North America. The number of deer–vehicle collisions (DVCs) in the United States during the early 1990s was estimated as 538 000 per annum by Romin and Bissonette (1996) and 726 000 by Conover *et al.* (1995),

Ungulate Management in Europe: Problems and Practices, eds. Rory Putman, Marco Apollonio and Reidar Andersen. Published by Cambridge University Press.

Table 8.1 *Number of traffic accidents causing death of deer (all species) in different European countries. In almost all cases numbers are dominated by roe deer*

Country	Years	Average numbers of ungulates killed per year	Source
Austria	2000–2006	40 500	Austrian national statistics
Switzerland	2000–2006	8000–10 000	Imesch-Bebié *et al.*, 2010
Slovenia	2001–2006	5970	Slovene Hunters Association
Croatia	2002–2005	960	Official Croatian statistics
Hungary	2000–2005	3670	Official Hungarian hunting statistics
Finland[a]	2000–2005	5000	Ruusila and Kojola, 2010
Denmark	2003–2006	6000	Andersen and Madsen, 2007
Norway[a]	2000–2005	8870	Andersen *et al.*, 2010
Sweden[a]	2005	61 000	Seiler, 2004
Germany	2005	227 000	Kerzel, 2005
Netherlands	2000–2004	5400	van Wieren and Groot Bruinderink, 2010; S.E. van Wieren, pers. comm.
England/Wales	2000–2005	31 000–45 000	Langbein, 2007a
Scotland	2000–2005	6500–10 000	Langbein and Putman, 2006
France	2004	23 500	Maillard *et al.*, 2010
Spain	2003–2004	>4050	Carranza, 2010

[a]includes moose.

although the latter suggested that even then the true figure may well have been in excess of 1 million. Annual assessments undertaken by State Farm Insurance and the US Insurance Institute for Highway Safety (2004) confirm that DVCs in the USA had risen to over 1.5 million per annum by 2004.

Such accidents may have considerable impact. Taken in relation to estimated population size in the different countries considered, Groot Bruinderink and Hazebroek's (1996) road kill figures for roe deer equate to between 1.6% (Norway) and 6% (Germany) of the annual spring population. Recent analyses in the UK suggest that the total mortality imposed through DVCs as a proportion of national spring population size is estimated to lie between 3% to 7% for roe deer, 1% to 3% for red deer and from 7% to 13% for fallow deer, making DVCs almost certainly the major cause of annual mortality among wild deer, aside from deliberate culls taken as part of management (Langbein, 2007a). The situation is very similar in Slovenia, where for example yearly road-kill of roe deer represented between 11% and 15% of the total recorded mortality of this species in the period 2000–2006 (unpublished data of Slovene Hunters Association). In addition to recorded road casualties, many animals hit by traffic are not necessarily killed outright or found at the roadside; thus the total toll is probably significantly higher with addition of those animals that get away from the immediate scene of the accident but die later, for which in many cases the driver of the vehicle involved may not even have been aware that the object hit was an ungulate. This 'hidden proportion' also poses additional problems of welfare, since animals not killed outright may survive many hours or days in considerable pain before death (Putman, 1997).

In addition to the impact on ungulate populations themselves, collisions of vehicles with any of the different species represent a significant risk-factor to road safety, and many accidents involving ungulates result in significant damage to persons and/or property (Putman, 1997; Langbein, 2007a). A range of published literature from Europe and the USA indicate the scale of the problem (e.g. Allen and McCulloch, 1976; Hansen, 1983; Hartwig, 1991, 1993; Fehlberg, 1994; Conover *et al.*, 1995; Romin and Bissonette, 1996; Putman, 1997; Bissonette *et al.*, 2008), and some comprehensive economical analyses revealed that an average total economic loss due to DVCs reaches as much as €2000–€2500 per single collision (Danielson and Hubbard, 1998; Wu, 1998; Bissonette *et al.*, 2008).

8.1.1 Human injury and death

From such studies, it is also clear that between 1% and 5% of 'reported' road traffic accidents involving ungulates would be expected to result in human injury. Numbers of humans killed or injured in traffic accidents involving ungulates in Germany in the late 1980s/early 1990s, for example, were estimated at approximately 25 killed and 2500 injured per annum (Hartwig, 1991), with damage to property estimated in 1993 at US$280 million (Fehlberg, 1994). Based on such statistics, recent reviews have suggested that as many as 30 000 and 29 000 human injury accidents are caused each year in Europe and the United States respectively in traffic collisions with ungulates (Groot Bruinderink and Hazebroek, 1996; Romin and Bissonette, 1996; Bissonette *et al.*, 2008).

The number of human fatalities is considerably lower, but is still of concern. Hartwig (1991) reported from a survey of game-related road traffic accidents (RTAs) in Nordrhein Westfalen in Germany during 1989 that 5 out of 9395 accidents involving deer led to human fatalities (the equivalent of one fatality for every 2000 accidents). Statistics available for the UK indicate that between the years 2000 and 2005, on average 12 human fatalities are known to have occurred per year as the result of road accidents involving deer, with an additional 100 incidents resulting in serious injuries, and 450 slight injuries (Langbein and Putman, 2006). Once again across Europe comparable data are limited to only a few countries, but ungulate collisions in Finland are known to have resulted in 3 fatalities and 215 injuries in 2006 (Ruusila and Kojola, 2010); in Spain (2004) collisions resulted in 17 fatalities, 76 serious injuries and around 400 slight injuries (Carranza, 2010), and in France (again in 2004) vehicle collisions with ungulates resulted in 20 fatalities and 340 injuries (Maillard *et al.*, 2010).

In the USA the Insurance Institute for Highway Safety (2004; 2008) found that on average 119 human fatalities resulted from vehicle collisions involving animals between 1993 and 1997. That death toll rose to 155 per year between 1998 and 2002, and to 205 per year between 2003 and 2007; overall, ungulates were found to be the animal taxon involved in 77% of these fatal animal-related RTAs (IIHS, 2008).

8.1.2 Damage to property

The level of detailed information recorded for ungulate vehicle collisions varies within, and between, different countries. For major accidents, and particularly those involving human injury, details of whether an animal was involved may be recorded, but accidents of a minor nature are often recorded by the authorities in terms of overall numbers only. To overcome this problem, a special investigation was commissioned in one of Germany's federal states (Nordrhein-Westfalen) in conjunction with the state police forces (Hartwig, 1991, 1993). This set out to determine the numbers of all game-related traffic accidents occurring during 1989, including those causing only minor damage (up to DM3000, or *c.* £1200) to vehicles, which are normally not recorded in any detail. The study logged a total of 557 000 traffic accidents, of which 348 942 caused only 'minor' damage (<DM3000), and 208 176 involved either human fatalities or injury and/or major damage (>DM3000) to vehicles.

Estimates of the economic cost of such incidents are complex, because in some cases only the direct costs of damage to vehicles and property is presented (*c.* 25 million euros in the UK; Langbein and Putman, 2006), while in other estimates a component is added in respect of human injury and associated cost of emergency services, or 'loss of earnings' (e.g. in Utah among total costs of *c.* $7.5 million, vehicle damage costs represent only 39% of total costs, while human fatality costs are assessed as 53%, human injury costs as 2%, and loss of deer as 6% of total costs; Bissonette *et al.*, 2008). Total costs of DVC in Slovenia are assessed at *c.* €15 million per annum (Pokorny *et al.*, 2008).

Costs of ungulate–vehicle collisions in Finland are estimated at €163 million per annum, in Sweden and France at around €100 million (per annum) and in Germany at €447 million (values presented in Apollonio *et al.*, 2010), but it is difficult to interpret such global estimates.

More directly, the average vehicle repair costs arising from each ungulate–vehicle collision has been assessed in the early 1990s at close to US$1500 in both Europe (Hartwig, 1993; Fehlberg, 1994) and the United States (Decker *et al.*, 1990; Witmer and de Calesta, 1992; Conover *et al.*, 1995). The annual repair costs arising (excluding human injury compensation) through ungulate accidents in Europe alone are thus estimated to lie by now well in excess of one billion euros.

8.2 Factors affecting accident frequency

Collisions with ungulates are not distributed randomly in space and time, and there are a number of environmental factors which affect the frequency of such accidents. These factors include road type (major/minor road) and traffic volume, habitat characteristics of the roadside and habitat mosaic in wider landscape, time of day and season. These different factors interact to affect accident risk. Furthermore, the different ungulate species themselves appear to differ in their susceptibility to being involved in traffic accidents (Langbein and Putman, 2006), while partly due to differences in size, collisions with different species of ungulates also have different implications in terms of severity of damage and likelihood of injuries caused to drivers (Ückermann, 1983; Conover *et al.*, 1995; Malo *et al.*, 2004; Langbein, 2007a).

8.2.1 Season and time of day

Ungulates of all species regularly cross minor roads (narrow roads of relatively low traffic volume) during routine daily movements within an established home range. Although surprisingly few published data are available for most species, it is common experience to encounter red, roe or fallow, for example, crossing such roads in the course of regular movements to or from foraging areas within their range (Putman, 1997; Langbein, 2007b). Primary roadways (motorways, dual carriageways), as also railways, appear to constitute more of a recognised barrier to movement. ‘Casual’ crossings of these major routes are less frequent and for most ungulate species, it would appear that the boundaries of home ranges commonly tend to coincide with major ‘barriers’ of this kind. Major roadways are, however, not totally impermeable: in some cases they may indeed be crossed in the context of daily movement around the home range; in addition there are likely to be peaks in road crossing coinciding with larger-scale dispersal movements (of juveniles leaving their natal range, or mature males seeking mating opportunities during the rut).

In response to these patterns of movement we might expect crossings of minor roads of low traffic volume to be frequent, regular and distributed throughout the year, since such roads will be crossed regularly during the course of normal movements around an established home range.

For crossings of more major roads, by contrast, we might expect the first strong seasonal peak in crossings in late spring and early summer as the consequence of dispersal of juveniles of all species. However, other seasonal peaks in road-crossing activity are likely to be rather more species specific, associated with mating movements of, in particular, mature males at the onset and end of the rut (e.g. in mid summer in the case of roe deer, and in autumn in the case of red and fallow deer).

These conclusions are matched by available data on patterns of actual traffic accidents involving different species of European deer (Figure 8.1a,b). For both red and fallow deer, a consistent seasonal peak in reported accidents coincides with the September–November peak in movement associated with the onset of the rut in autumn (Ückermann, 1964; Langbein, 1985; Carsignol, 1989; Desire and Recorbet, 1990; Hartwig, 1991; Groot Bruinderink and Hazebroek, 1996; Pokorny, 2006). In a study of movement patterns of fallow deer and the frequency and distribution of traffic accidents involving deer in an area of north Staffordshire (UK) between 1983 and 1985, Langbein (1985) also noted that minor roads showed a high level of deer crossings throughout the year, but crossings of major roads were notably seasonal and concentrated in autumn (October) and in the period between February and April (conclusions based on a sample of 986 recorded crossings), with deer accidents also coinciding once again with the peak of the fallow rut in that area (Langbein, 1985).

For roe deer the seasonal peaks in accident frequency appear less well defined, but the majority of road-related mortality falls within the period April–May (Figure 8.1a,b). This appears to coincide with the reduced speed of movement of females with very young fawns at heel, with the period of dispersal of yearlings from their natal range, as well as with the period of establishing new territories by adults, particularly bucks (Desire and Recorbet, 1990; Hartwig, 1991; Groot Bruinderink and Hazebroek, 1996; SGS Environment, 1998; Pokorny, 2006). Of 3826 accident reports between 2003 and 2005 in Great Britain where roe deer were positively identified as the species involved, 26% of incidents occurred in April and May (Langbein and

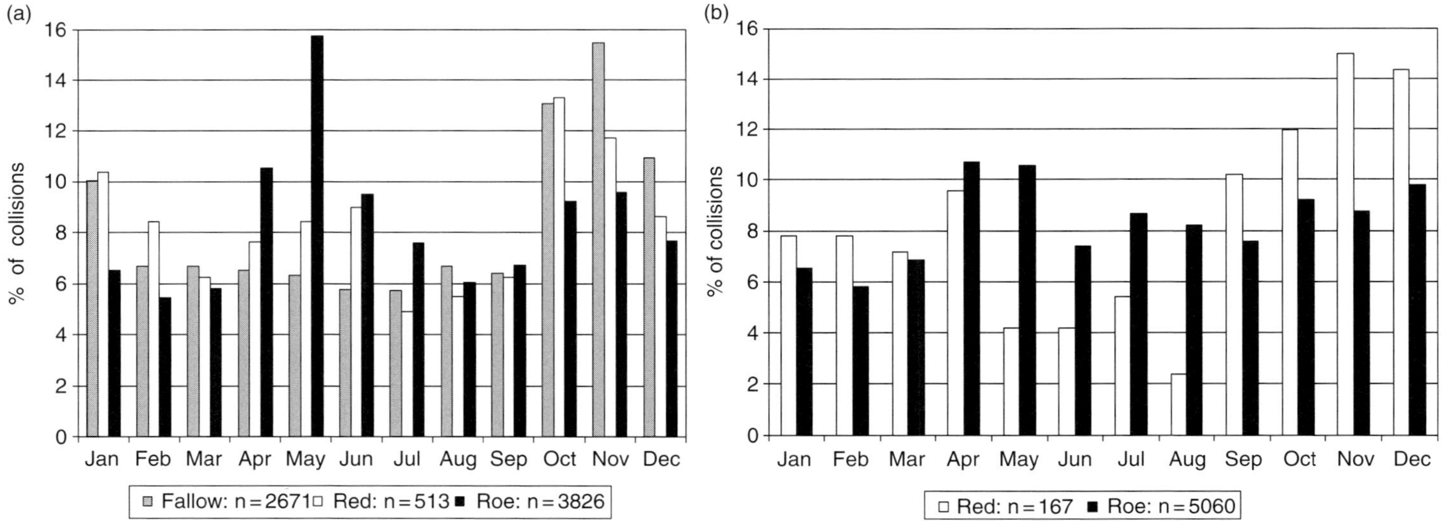

Figure 8.1 Seasonal variation in the frequency of collisions with roe, red and fallow deer. (a) Percentage of roe deer, red deer and fallow deer vehicle collisions recorded in differing months in Great Britain between 2003 and 2005 (based on data from Langbein and Putman, 2006 and Langbein, 2007a, using subsamples of 7060 reports giving reliable detail on species). (b) Percentage of roe deer and red deer vehicle collisions recorded in differing months in Slovenia (based on data provided by the Slovene Hunting Information System for roe deer killed in 2007 and for red deer during 2006 and 2007).

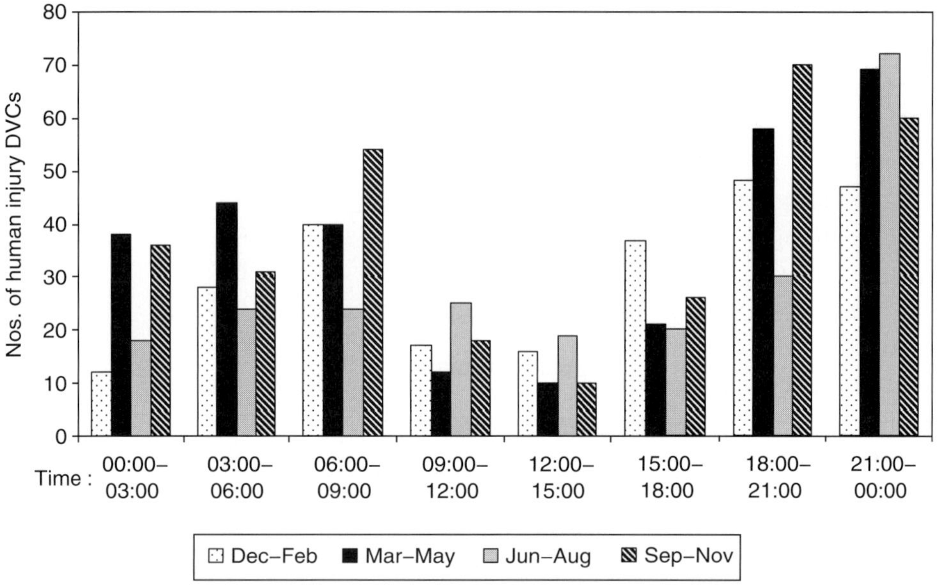

Figure 8.2 Diurnal patterns in occurrence of deer–vehicle collisions (DVCs) in differing seasons in Great Britain (from Langbein, 2007a; based on data restricted to DVCs leading to human casualties registered by police, for which times of incidents are generally most reliable).

Putman, 2006; see also Figure 8.1a). Similarly, in Slovenia 21% of all known collisions with roe deer occurred in these same two months both during 1999–2001 (Pokorny, 2006) and in 2007 (unpublished data of Slovene Hunters' Association: Figure 8.1b). However, the risk for collision with roe deer is also high during the summer (particularly in the rut period) and autumn (primarily due to the greater activity of fawns and clearance of maize fields) (Pokorny *et al.*, 2008).

As well as reporting such seasonal patterns, a number of studies have found that the majority of road-related accidents occur during the hours of darkness, and particularly at dusk or dawn (e.g. Ückermann, 1964; Langbein, 1985, 2007a; Desire and Recorbet, 1990; Lavsund and Sandegren, 1991; Hartwig, 1993; Groot Bruinderink and Hazebroek, 1996; Haikonen and Summala, 2001; Pokorny, 2006; and see Figure 8.2). While this coincides with the period of maximum deer activity, the effect may also be partly due to reduced driver visibility at these periods and accentuated when 'rush hour' periods coincide with poor light conditions of dawn and dusk in autumn (Sanders, 1985; Langbein, 1985). Langbein (1985) suggested that the coincidence of rush hour traffic peaks with twilight in autumn and spring may be important in exacerbating the seasonal peaks in traffic accidents; and that this may contribute to the fact that DVCs overall tend to peak just after, rather than at, the height of the fallow deer, red deer and sika deer rut. This may also contribute to the commonly noted escalation of accidents in autumn with those species of deer, including roe (Figure 8.1), which do not rut at this time of year. A further factor which in the USA has been proposed as contributing to the autumnal peak in DVCs is the start of the hunting season (Etter *et al.*, 2002), though this is generally considered of lesser importance than rutting activity (Puglisi *et al.*, 1974; Gleason and Jenks, 1993). The beginning of November also marks the end of the close season for females for most deer species in the UK (but more generally across Europe hunting seasons on antler-less deer start in early or mid September in most countries), but no European studies have to date investigated effects of hunting pressure on DVCs.

8.2.2 Road type, traffic volume and speed

It is self evident that frequency of ungulate–vehicle collisions will be related to the number of vehicles as well as numbers of ungulates. However, although the density of roads and the speed of traffic have already been demonstrated as the main risk factors (e.g. Romin and Bissonette, 1996; Lode, 2000), some recent studies showed no relation between average daily traffic flow (as the measure of traffic volume) or posted speed limit (as the measure of traffic speed) and rates of collisions with ungulates (Bissonette *et al.*, 2008). In contrast to these two traffic-related variables where a consensus is lacking on their effects, road density has a much more pronounced effect on collision rates. For example, by spatial regression analysis of 6031 records on exact position of deer–vehicle collisions in 9639 Slovene 1×1 km quadrants (situated only in those hunting grounds which provided exact locations of DVCs for the period 1999–2001) it has already been demonstrated that of 40 independent variables (which mainly determine characteristics of the landscape), density of roads in a single quadrant predominantly determines the number of roe deer vehicle collisions in the same quadrant (Pokorny, 2006).

While the majority of recorded collisions with ungulates occur on secondary roads, due to their greater overall length within most national road networks, accident *frequency* (per unit length of carriageway) has consistently been found to be higher on primary trunk roads or major throughways where speed of traffic and, most of all, total traffic volumes are greater (Pojar *et al.*, 1975; Bashore *et al.*, 1985; Desire and Recorbet, 1990; Hartwig, 1993; Langbein and Putman, 2006; McShea *et al.*, 2008). However, the last is not the case in some central European countries (e.g. Austria, Croatia, Slovenia) where most major highways are completely fenced, and therefore the possibility for deer to venture onto the road itself is comparatively low.

Based on their assessment of recorded accidents involving roe deer in France (Desire and Recorbet, 1990), major roads accounted for a disproportionately higher number of collisions: 5.3% of all roe deer killed were killed on motorways, despite the fact that these amounted to only 0.8% of the total road network; only 5.8% of recorded accidents were on minor roads which, by contrast, comprised 49% of recorded road length. Similar results were found by Hartwig (1993) in Germany, where motorways accounted for 21.2% of all wildlife-related road traffic accidents, even though they made up only 7% of the length of major roads in the area of study. Motorways and primary trunk routes together accounted for 37.5% of all recorded accidents in some 24% of total road length. Higher incidence of DVCs on more major routes would appear to be a function both of higher traffic volume and higher vehicle speed.

Among a sample of 14 033 DVCs reported to a study in Great Britain during 2003–2005 (Langbein and Putman, 2006; Langbein, 2007a) for which the road type was identified, 9452 (67%) occurred on A roads (A) or motorways (M) and 4579 (33%) on more minor roads (class B, C or smaller). Dividing simply by total road lengths in Britain in kilometres as 50 192 (A + M) and 387 674 (more minor roads) shows average rates of *recorded* incidents to be 14 times higher on major roads than on minor roads. However, while motorways and major roads constitute only 13% of total road length, they do carry close to 65% of the total volume of traffic in Britain; a figure roughly in line with the percentage of DVCs recorded on them. In fact as the proportion of DVCs reported to that study is likely to have been higher in the case of major roads than minor roads, incidents on minor roads are in reality likely to be somewhat more frequent than predicted by traffic volume alone. This was confirmed to be the case among that subsample of 406 DVC police reports in England (2003–2005) available to the above study which involved human injury, for which level of reporting should be least affected by road type. Of these only 49% occurred on major (A + M) roads, and 51% on minor roads; indicating that at least personal injury DVCs occur more frequently on minor roads than in relation to the traffic volume carried by them (Langbein, 2007a). This is in accord also with the findings from a number of US studies (e.g. Allen and McCullough, 1976; Grovenburg *et al.*, 2008) who both observed greater mortalities due to DVCs on two-lane paved roads than on divided highways and interstate routes (equivalent to dual carriageways and motorways in the UK).

8.2.3 Other environmental and landscape factors influencing accident risk

The type of habitats traversed, or close to a road, also has an influence on the susceptibility of a given section of road to ungulate collisions. As reviewed by Staines *et al.* (2001), one of the few consistent findings from analyses of collisions involving deer from both Europe and North America is that the majority of DVC hot spots occur within or near wooded areas, particularly where the woodland comes right down to the road edge (e.g. Ückermann, 1964; Bashore *et al.*, 1985; Romin and Bissonette, 1996; Finder *et al.*, 1999). Indeed, the risk for collision with deer is higher in fragmented landscape, where the forest edge is very long (e.g. Romin and Bissonette, 1996; Finder *et al.*, 1999; Madsen *et al.*, 2002).

Further, more recent attempts at multivariate analysis of the relative importance of the whole suite of factors that will affect incident frequency also identify proximity of woodland as a key factor often associated with higher collision rates (Hubbard *et al.*, 2000; Nielsen *et al.*, 2003; Malo *et al.*, 2004; Pokorny, 2006; Hussain *et al.*, 2007; Grovenburg *et al.*, 2008).

However, in their models of DVCs at various scales in more urban landscapes surrounding Edmonton in Canada, Ng *et al.* (2008) found areas of DVCs to be correlated foremost with high speed limits and low densities of roads, but also the combination of high road densities with non-forested vegetation of high productivity. As discussed by Ng *et al.* (2008), there are several reasons why findings from different areas may seem contradictory, not least as many studies have tended to emphasise local habitat characteristics (e.g. Puglisi *et al.*, 1974; Nielsen *et al.*, 2003), while others have focused more on effects of surrounding landscapes, patch size or their connectivity (Hubbard *et al.*, 2000; Malo *et al.*, 2004; Seiler, 2004; McShea *et al.*, 2008). Both local and spatial scales are likely to be of importance (Seiler, 2004; Ng *et al.*, 2008); and furthermore we may predict that their effects will differ depending on differences in home range pattern and habitat selection between deer species.

The original analyses of Bashore *et al.* (1985) considered a number of environmental and 'traffic-flow' characteristics associated with high recorded frequency of deer–vehicle collisions on stretches of two-lane highway in Pennsylvania between July 1979 and October 1980, concluding that the predicted probability of accidents decreases with an increasing number of homes, commercial and other buildings within the buffer area, and longer sight distance along the roadway. Their model also indicated a decrease in the 'high' DVC probability with increases in the proportion of fencing, the distance to woodlands, the ability to see a roadside object (i.e. in-line visibility), non-wooded herbs in the buffer zone, and posted speed limit. In a subsequent GIS analysis, Finder *et al.* (1999) measured topographical and habitat-related features within a 0.8 km radius of road segments in Illinois with higher than average accident rates and a series of randomly selected control sites. Once again, high accident rates for white-tailed deer (*Odocoileus virginianus*) were associated with woodland cover; a logistic regression model, developed using only landscape features derived from satellite imaging, accurately distinguished between high and low kill sites and related accident frequency to landscape diversity and (shorter) distance from adjacent woodland cover. Hubbard *et al.* (2000) published similar findings from a multiple regression analysis of land-use variables and highway characteristics in Iowa, identifying four landscape features associated with clusters of high accident frequency as: the proportional area of woodland and grass adjacent to the roadway, proportion of crop land, and the heterogeneity in size and disposition of land cover patches. The number of lanes of traffic (identifying in effect more major trunk routes or motorways) and the number of bridges across the carriageway also appeared as being two of the major predictors of high DVC locations. While this last may seem unexpected, we should note that such a finding is consistent with the earlier analyses of Finder *et al.* (1999), above, who also noted an increase in accident risk with an increasing number of gullies or other travel corridors crossing the roadway.

While many of the studies discussed above are derived from analyses of deer–vehicle incidents in North America, findings in a more specifically European context are not dissimilar. In an analysis of 115 kills of roe deer at Kalo in Denmark, between 1956 and 1985, Madsen *et al.* (2002) found no correlations between the pattern of road kills and mean daily traffic flows but noted that collision sites were strongly clumped, and sites associated with higher road kill tended to have denser vegetation (hedgerows, bushes, etc.) present on one or both sides of the road. Malo *et al.* (2004), based on analysis of the locations of 2067 DVCs occurring between 1988 and 2001 in the province of Soria (central Spain), once more identified the features characteristically associated with locations of high accident frequency as vegetation, fencing or other structures forcing the animals to cross at particular points and natural linear features perpendicular to the roadway associated with natural travel corridors.

Seiler (2004) studied trends and variation in ungulate–vehicle collisions in Sweden at varying spatial resolution in order to test the hypothesis that incidents are proportional to animal density and traffic volume. Spatial patterns were studied at the level of individual hunting areas ($N=311$), moose management districts ($N=95$), and counties ($N=22$), whereas trends in collision rates were studied at national, county and district level covering periods of 30, 16 and 12 years, respectively. During 1970–1999 the overall number of reported collisions with moose and roe deer was closely correlated with changes in annual game bags (harvest) and the increase in traffic volume. Large-scale spatial variations in ungulate–vehicle collisions also showed a strong correlation with harvest and traffic. The ratio of collision numbers to harvest, however, increased significantly over time, suggesting a growing importance of traffic over ungulate management. With increased resolution, other environmental factors such as preferred habitat, road density and the presence of normal road underpasses that can provide passages also for wildlife, gained significance over ungulate density and traffic volume.

8.3 Options for mitigation

Clearly, by any form of assessment, collisions of vehicles with ungulates represent a significant problem – and a growing one. What are the options available for mitigation, to reduce overall frequency or risk of such collisions?

The ideal goal for measures designed to reduce the risk of ungulate–vehicle collisions is not that they should seek to prevent animals from crossing a particular roadway, but that crossing should be affected more safely. Attempts to prevent crossings altogether for long lengths of roadway are likely to prove ineffective and result in animals forcing such barriers (with the added risk that they may then become trapped within the carriageway, actually increasing rather than decreasing the risk of accident). At the very least, where barriers are completely effective at preventing movement, this will result in fragmentation and isolation of previously continuous populations of ungulates and other larger wildlife (Forman *et al.*, 1997; Forman and Alexander, 1998; Mladenhoff *et al.*, 1999). Thus, the most successful measures will seek not to prevent crossings altogether but to displace these in space or time such that animals cross the road at periods of reduced (or zero) traffic flow, or in places where accident risk is reduced through enhanced visibility and/or driver awareness or provision of traffic-free wildlife passages. In our view few, if any, available mitigation measures are effective used in isolation, but become truly effective only in combination. This will be a recurring theme throughout this section.

In a review for the Deer Commission for Scotland (the Statutory Agency responsible for overall management of deer populations and their impacts on the Scottish environment and on public safety), Putman *et al.* (2004) suggested that the options available to prevent or minimise ungulate–vehicle collisions may broadly be considered as attempts to:

(i) **Prevent or control crossing**, e.g. by the use of highway fencing, roadside wildlife warning devices, reductions in local population density of particular ungulate species, chemical deterrents, or the fitting of warning whistles to vehicles.
(ii) **Increase driver awareness**, e.g. by use of various driver warning systems – whether through fixed signage, or signage responsive to driver speed or the actual presence of deer on the roadside; or removal of roadside vegetation to increase visibility.
(iii) **Provide safer crossing places** for deer (or other ungulates), e.g. by the installation of dedicated bridges or underpasses, by modification of existing passages to dual use, or by the creation of designated 'cross-walks' across the carriageway itself.

Published analyses of the efficacy of the various options, and of the circumstances in which differing approaches might be expected to be effective, are summarised below, with further detail on individual measures given by, for example, Putman *et al.* (2004), DVCIC (2003) and Mastro *et al.* (2008).

8.3.1 Preventing or controlling crossing

Roadside fencing

Our continuing review has produced no literature to change our conclusion (Staines *et al.*, 2001; Putman *et al.*, 2004) that high tensile roadside fencing is likely to remain the primary method used to try and reduce road crossings and resultant accidents at identified sites of high risk. However, we reiterate our further comments that it is essential that fencing should: 'be of adequate specification (height/mesh size) and be designed not with the expectation, or aim, of attempting to prevent road crossings altogether, but rather to channel animals towards a safer crossing point'.

Suitable specifications for the height of such fencing in relation to differing ungulate species are offered in the EU handbook *Wildlife and Traffic: A European Handbook for Identifying Conflicts and Designing Solutions* (European Commission Action 341; Iuell *et al.*, 2003) and are thus not rehearsed in detail here. We merely note that we ourselves do not necessarily agree with the specifications offered there in relation to mesh size, suggesting rather that at the base of fences this should not exceed 75 × 75 mm in areas where muntjac occur, with mesh no larger than 100 × 100 mm in roe deer areas to prevent smaller individuals squeezing through (see also Putman *et al.*, 2004).

Staines *et al.* (2001) have also emphasised that where fencing has not proved effective this has usually been related to inadequate specification of fence construction, to animals getting past the end of fencelines where insufficient length has been installed (Reed *et al.*, 1974; Ward, 1982; Clevenger *et al.*, 2001, 2002), or at road junctions where fencing is difficult. In such situations, accident risk may actually be increased at the end of fencelines or where animals become trapped in the road corridor on the wrong side of the fence (Feldhamer *et al.*, 1986; Clevenger *et al.*, 2001) and it is appropriate in any fencing scheme to incorporate means of exit from the carriageway, such as one-way gates (Reed *et al.*, 1974; Lehnert and Bissonette, 1997) or earthen escape ramps (deer leaps) (see review by Mastro *et al.*, 2008). Bissonette and Hammer (2000) found white-tailed deer used ramps 8 to 11 times more often than one-way gates.

Few studies have directly assessed the efficacy of electric fencing to reduce DVCs, although at least in the case of American moose (*Alces americanus*), Leblond *et al.* (2007) reported significant reductions in numbers of moose tracks leading onto road verges, and in numbers of collisions, after installation of electric fencing.

Complete barrier fencing attempting to prevent road crossings altogether is likely to prove ineffective in the longer term, as in the absence of any alternative route animals are eventually likely to force the fence to cross roadways (with the added risk that they may then become trapped within the carriageway, unable to escape). At the very least, where it is effective as a total barrier to movement, such fencing causes fragmentation and isolation of previously continuous populations of deer and other larger wildlife (see Iuell *et al.*, 2003). Barrier fencing is thus at its most effective when erected in short lengths and in conjunction with the provision of some alternative and safer means of crossing the carriageway, and designed so as to deflect animal movements towards these safer crossing points (see Section 8.3.3 below).

Roadside wildlife warning reflectors

With the same considerations as above, roadside reflectors are designed not to stop animal movement across roads, but to delay these at times when there is traffic in the carriageway until the roadway is clear. Working on the principle that light from approaching headlights is reflected onto the verge to provide a flash warning or continuous wall of light (depending on reflector type and deployment), they are intended to alert ungulates to approaching traffic at night or startle them to delay crossing until the road is clear.

Even if such optical reflectors are effective at all in delaying crossings, by definition they can at best only be effective at night. Research into the daily pattern of accident frequency suggests in fact that highest periods of accident risk coincide with dawn and dusk (e.g. Ückermann, 1964; Langbein, 1985;

Desire and Recorbet, 1990; Lavsund and Sandegren, 1991; Hartwig, 1993; Groot Bruinderink and Hazebroek, 1996; Haikonen and Summala, 2001; Pokorny, 2006), when approaching vehicles may not in practice have their full headlights on, and when effectiveness of the reflection is to an extent reduced by higher general ambient light conditions. Furthermore, even during darkness when more likely to be effective, such reflectors can only usefully be installed on roads of relatively low, or sporadic, traffic flow, such that there are periods of quiet between vehicles to permit safe crossing (Putman *et al.*, 2004). On roads of high or continuous night-time traffic, where reflectors are continuously activated, deer and other ungulates may be expected to more readily habituate to the reflected light. Even if they do not habituate to the reflected light or traffic in general, if intent on crossing – but not provided with interval periods of no traffic flow – animals are likely simply to force the 'barrier' and cross anyway even in the face of oncoming traffic. Despite these basic limitations there have been numerous instances of quite inappropriate deployment of wildlife warning reflectors, including on major motorways in the UK and main roads elsewhere of very high traffic flow.

Debate even over the initial effectiveness of such reflectors (immediately after installation) continues (e.g. Woodard *et al.*, 1973; Gilbert, 1982; Gladfelter, 1982; Schafer and Penland, 1985; Zacks, 1986; Waring *et al.*, 1991; Armstrong, 1992; Jared, 1992; Ford and Villa, 1993; Reeve and Anderson, 1993; Pafko and Kovach, 1996; Pepper *et al.*, 1998; D'Angelo *et al.*, 2006; Voss, 2007). From some studies it appears that when properly installed they may have some effectiveness in reducing the incidence of DVCs for a period after erection; other studies report limited effectiveness or note that any initial effect wanes over time, either due to deterioration of the reflectors themselves, lack of clearance of vegetation around the installation, or due to habituation/learning (e.g. Ujvari *et al.*, 1998).

The majority of these studies are based on analysis of changes in DVC frequency through time. Such analyses are inevitably difficult and inevitably inconclusive, given the fact that there is in any case considerable stochastic variation in incident frequency along any given stretch of roadway (Putman *et al.*, 2004; Voss, 2007). Indeed, only few studies specifically investigate the effect of reflectors (or other roadside devices) directly on deer behaviour. From one such study, D'Angelo *et al.* (2006) suggest that installation of such reflectors may even prove counter-productive. They studied the extent to which white-tailed deer altered their behaviour in relation to red, white or amber wildlife reflectors at the roadside, concluding not only that none were effective at reducing DVCs, but based on infra-red video recordings suggested that deer may in fact be more likely to be spooked and dart across the road where light reflectors were present.

In the UK, as well as many parts of Europe and North America, the most common type of optical wildlife reflectors used are red in colour, based in part on wrongful assumption that most animals, including mammals, will associate red with danger, and partly to comply with highways directives to minimise risk of glare for oncoming drivers. However, from knowledge of the limited number and characteristics of cones (colour-sensitive receptors) present in the retina of ungulates, biologists have long presumed that along with most other mammals (with exception of humans and some other primates), ungulates have comparatively limited (dichromatic) colour vision. The inability of deer and other ungulates to see red has been confirmed by a number of recent studies (VerCauteren and Pipas, 2003; Sheets and Cason, 2005; Pürstl, 2006), which leads to the inescapable conclusion that light from red reflectors will appear grey or black to them, and would therefore not produce a clear contrast in dark surroundings. Although deer are able to see better in the green to blue part of the light spectrum, Pepper (1999) found no evidence of any improved effectiveness where red reflectors were replaced with green-blue ones and VerCauteren *et al.* (2006) found green and blue lasers also to be ineffective at dispersing deer at night.

Although many deterrent manufacturers and various articles in the hunting literature continue to claim positive effects from optical wildlife reflector installations, the authors of the recent European-wide appraisal of conflicts and solutions in relation to wildlife road crossings (Iuell *et al.*, 2003) concluded that: 'These features are popular because they are cheap and easy to place. However, a thorough analysis of studies carried out over the last 40 years all over the world has found little evidence for the effectiveness of wildlife warning reflectors.'

Acoustic roadside deterrents

In order to attempt to overcome some of the drawbacks of optical-only reflectors outlined above, and to reduce likelihood of habituation to individual deterrent signals, a number of more sophisticated roadside deterrents have been developed over recent years. These generally either combine traditional white or red optical reflectors with acoustic modules to produce whistling sounds (e.g. WEGU-GFT acoustic wildlife warning reflector, Eurocontor acoustic reflector), or provide a series of signals at differing frequencies ranging from infrasound through to ultrasound (e.g. Eurocontor Ecopillars) based on the assumption that deer may habituate less readily to a mixture of signals. Another type (e.g. WIWASOL-II device) produces a series of intermittent whistling sounds readily audible to the human ear, and incorporates also a small flashing blue light. Each of the above types employ solar cells to charge integral batteries during daylight hours, with the 'deterrent' signals emitted either at intervals throughout 24 hours, or else only when triggered by bright car headlights falling on the devices at night. As has often been the case regarding optical reflectors and other wildlife deterrents in the past, various preliminary findings reported in the hunting press or other general media claim good results with such devices in reduction of ungulate collisions, mostly during the first one or two years after installation. However, firm evidence for lasting effects remains lacking in the published scientific literature and results of trials undertaken in differing countries or situations – like those for light reflectors – remain contradictory.

Positive results have been reported in Slovenia, where effectiveness of three different mitigation measures (ultrasound emitting devices; acoustic reflectors; combination of classic and acoustic reflectors) installed along 23 different road sections (22 km) was tested between July and December 2006 (Pokorny and Poličnik, 2008). In total, the number of road-killed deer (primarily roe) decreased in that first six months after the installation by 106 individuals (83%) in comparison either with the equivalent period in the year before the trial or with the average value for the equivalent periods in years 2002–2005. Differences in reductions observed among the three types were low and insignificant (e.g. 79%, 87% and 78% for ultrasound emitting devices, acoustic reflectors and combination of reflectors, respectively). On the contrary, change in DVCs calculated for an equivalent number of control sections averaged a fall of only 16% (in comparison with the equivalent period in 2005) or 3% (in comparison with average values in the period 2002–2005), and was statistically insignificant.

Results of continued monitoring over the following 18 months for a reduced set of 16 of the original road sections from January 2007 to June 2008 (Pokorny and Poličnik, 2008) showed a less pronounced but still statistically significant reduction in the number of road-killed deer, by 47% and 62%, in 2007 and first six months of 2008, respectively. From September 2007 to June 2008 effectiveness of acoustic reflectors was monitored also on 15 additional Slovene road sections (11.0 km). Again, a significant reduction in DVCs occurred in the short term, by 68% compared with the equivalent period in the year before the trial, or by 67% if assessed against the average number of incidents for the same periods during 2004–2007.

By contrast, a longer duration study undertaken by the German Insurance Association for Accident Research (Voss, 2007) in the Oberbergische region of Germany to establish the effects of five differing roadside measures, including acoustic wildlife reflectors, found no evidence for statistically significant reductions in DVCs from acoustic reflectors or any of the other methods, when compared against a large sample of control sites for which directly comparable before-and-after data were collected. An important aspect of this particular study is that detailed police records for wildlife accidents (of which *c.* 80% related to roe deer) were maintained for all six trial sites (as well as for 37 control road sections) for 3 full years before any of the preventative measures began to be installed (1997–1999), and then a further 3 years (2002 to 2005) after all installations were completed; importantly analyses excluded the intervening period (2000–2002) whilst the differing measures were introduced and thus eliminating any short-term 'novelty' response to the introduction of the deterrents themselves or other associated changes at the roadside.

In order more directly to investigate the true effect of acoustic deterrents on deer behaviour when crossing roads (and thus circumvent the influence of confounding factors such as variation in cull and population numbers on changes in DVCs between years), Langbein (2007b and in preparation) employed periods of remote 24 hr day–night video surveillance at the roadside. Filming was undertaken in two regions of England, one offering high density of wild fallow deer and another with mainly red deer, to assess whether either species notably changed their behaviour when approaching and crossing roads where acoustic deterrents were installed compared with control sections without deterrents; and in particular whether they delay crossing for any longer after traffic where deterrents are present. Video surveillance was spread over an 18-month period following installation during autumn 2005 of (a) acoustic reflectors on two sections of a single carriageway A road in south-west England traversed frequently by red deer, and (b) installation of acoustic reflectors and ultrasound-emitting pillars along two differing sections of a single carriageway B road in south-east England crossed very frequently by fallow deer.

During a total of 93 day/night filming periods undertaken mostly simultaneously at deterrent and control for each road, with similar numbers of filming periods at each in differing seasons, nearly 393 groups of deer (*c.* 1100 individual animals) were filmed at the roadside, including over 273 groups (*c.* 741 individuals) actually seen crossing the roads within the cameras' field of view. Crossing events (defined for analyses as either a single animal or a group crossing together without being separated by traffic) were divided into those occurring in the dark (i.e. between dusk and dawn, when deterrents could be activated by headlights from passing traffic), and those in daylight when deterrents would generally be inactive and traffic flow greater. For fallow deer crossing the road at night, the median delay after the last vehicle passed before the animals entered the roadway lay between 20 and 30 seconds both in the control sections as well as sections fitted with either of the deterrent types. In deterrent and non-deterrent sections a high proportion (>25%) of crossing events commenced within less than 10 seconds of the last traffic. Many fallow were also recorded crossing during daylight with, on average, even shorter delays after traffic. Red deer were only rarely recorded crossing the test road in daylight, and at night the median time interval before crossing was 30–60 seconds after traffic in sections with acoustic reflectors and in control areas. A significant reduction in deer accidents was recorded during the two years post-installation of acoustic reflectors only at the red deer site (Langbein, 2007b), but was of similar magnitude in both control and deterrent sections; the reduction in DVCs here most probably related to a significantly increased cull of red deer known to have been taken locally during winter 2005/6, with lack of any clear evidence that the presence of acoustic deterrents contributed to the reduction.

Similar video observations, using the same type of equipment developed for the above study, were replicated in Slovenia during autumn 2006 on four road sections affected by mostly red deer collisions. In 63 days of surveillance (of which about one third were filmed before and two thirds after deployment of deterrents at the same locations only) 369 individual red deer crossing the road were filmed (Pokorny *et al.*, 2008). Here animals were noted to spend a shorter duration of time on the roadway itself after the installation of deterrents in comparison with the period earlier in autumn when acoustic deterrents were not yet activated or had not yet been implemented (30 seconds versus 35 seconds). The researchers found also that after installation of deterrents red deer in Slovenia appeared to 'escape' somewhat sooner in advance of approaching traffic, but stressed that the trial period was very short and extended only to one species.

Langbein (2007b, 2009) conducted further investigations into the direct response by wild roe, fallow and red deer when exposed to the ultrasonic and low frequency audible signals emitted by Eurocontor Ecopillars. Following placement of the devices at bait stations set up in different areas for each species, video observations were undertaken for several days before and after switching on the pillars (set to emit their various acoustic signals every 15 to 30 seconds throughout the day and night). In case of each of the three species, animals were found to approach and feed calmly close (<1 to 5 m) to the devices within just one day (fallow) to three days (red and roe) after activation, showing little if any overt reaction to the signals, suggesting that habituation to the acoustic signals emitted by this particular device was extremely fast.

Chemical deterrents

Proprietary 'chemical fences' (repellent chemicals encapsulated in slow-release organic foam and applied to roadside posts or trees) have been trialled extensively in Germany, with claims by the manufacturers as well as the German Automobile Association (ADAC) of some efficacy in reducing the frequency of deer–vehicle collisions. From trials on six test sections in Bavaria and northern Westphalia, the manufacturers of a proprietary German scent fence report that 60% of the animals encountering the treated areas withdrew and crossed the road beyond the scent fence at an untreated section; 20% of the animals crossed despite the treatment but crossed very rapidly without delay; the remaining 20% were unaffected. On one section of treated road, reported accidents of roe deer fell within a year from 22 per year to a total of 2 (Kerzel and Kirchberger, 1993). More detailed assessment showed that, although road kills were reduced by 30–80% within some test sections, accidents outside the trial areas actually rose (Lebersorger, 1993), and other, independent, studies have suggested that such scent fences are not in practice as effective as claimed (Lutz, 1994). Scent fences were also included as one of the preventative measures tested in the longer term study by the German Insurance Association (Voss, 2007); as in the case of acoustic and optical reflectors, the numbers of recorded DVCs along the section where a scent fence was installed and regularly maintained and renewed for several years, actually showed an increase overall and failed to provide any evidence of effectiveness of such fences.

Similar results were obtained in a short-lasting trial (year 2005 only) in Slovenia, where efficiency of chemical deterrents was tested on 11 problematic road sections (total length of 13.4 km). Although the number of road-killed deer on these sections decreased by 44% in comparison with comparable periods either in the year before or across the four years before the trial, the number of DVCs at adjacent road sections dramatically increased in the same time, resulting in a total decrease rate of only 15% (which, however, did not differ from the changes on control road sections). Therefore, the positive influence of chemical deterrents as a countermeasure against DVCs was not confirmed (Pokorny *et al.*, 2008).

Car-mounted warning whistles

Various commercial companies offer for sale a device for attachment directly to the front of a motor vehicle, which emits a high frequency whistle claimed to be a deterrent to ungulates or other roadside wildlife. In a study of the response to such whistles, Romin and Dalton (1992) observed no behavioural responses by mule deer to suggest acknowledgement or avoidance of vehicles equipped with such devices, nor could any reduction in the number of DVCs be demonstrated. Unpublished work by scientists from the University of Wisconsin, mentioned in a report by the Insurance Institute for Highways in the USA in 1993, found that neither deer nor humans could actually detect the sound produced by the air-activated whistles in normal operation, and that whistles blown by mouth had no effect on penned deer (see also DVCIC, 2003).

Any effectiveness of 'deer whistles' and horns is dependent on the ability of ungulates to hear and respond to the emitted sound. Manufacturers' claims typically suggest that the air-activated whistles emit sounds at between 16 and 20 kHz at speeds above 30 mph. Overall, there has not been a significant amount of published work on the auditory capabilities of deer; what work there is suggests that the 'range of greatest hearing sensitivity' lies between 1 and 8 kHz – which would be well below the sound range claimed for the various whistle designs in commercial production. In independent experimental trials on six different whistle designs, however, Scheifele *et al.* (1998, 2003) found that the primary operational frequency actually produced by the different whistle designs was 3.3 kHz and 10 kHz. This is closer to the presumed auditory range of (white-tailed) deer; however, it is noted that the 3.3 kHz sound is also within the typical range of normal roadway noise (tyre noise) produced by a vehicle at 45 mph (Scheifele *et al.*, 2003).

An earlier study of the acoustic characteristics of a range of air-activated wildlife warning whistles and also electronic warning horns operated by Schober and Sommer (1984) found that the sound of the latter were at least in principle audible for red and roe deer and hares, but from a moving vehicle no clear positive effects on these animals could be detected for any of the commercial car-mounted devices tested.

Local reductions in ungulate population density

A number of published studies have now demonstrated a relationship between the frequency of deer-related vehicle collisions and local deer densities (e.g. McCaffery, 1973; Schwabe *et al.*, 2002; Rondeau and Conrad, 2003; Wisconsin Department of Natural Resources data summarised on www.deercrash.com), which suggests that more general reduction of ungulate densities, in association with other mitigation techniques, may help to reduce accident frequencies. Despite this, formal studies of the effectiveness of a local reduction in ungulate numbers are few and contradictory.

It is patently clear that there must exist some relationship between ungulate numbers and accident frequency, but this relationship is probably not linear, and other factors may be more important in determining the actual level of accident risk, such that manipulation of population numbers may not achieve the response predicted. Further, to be effective such reductions in density must be maintained against a probable tendency to recolonisation of areas of reduced density through immigration from the surrounding area.

A number of published studies do appear to show evidence of a reduction in frequency of deer collisions with (sustained) local reductions in deer density (e.g. Jones and Witham, 1993; Danielson and Hubbard, 1998; Jenks *et al.*, 2002; Rondeau and Conrad, 2003; Sudharsan *et al.*, 2006); whereas imposed hunting restrictions (Kuser and Wolgast, 1983) or general discontinuation of organised culling (Langbein, 2007a) have been correlated with increases in DVC occurrence. In the latter example the annual number of DVCs attended by the rangers in Ashdown Forest in south-east England rose fivefold from 74 in 2000, to 215 in 2005, and over 315 in both 2006 and 2007.

However convincing such statistics may appear, formal analysis to determine the precise effects of population reduction on the rate of DVCs is extremely difficult. As discussed earlier (see 'Acoustic roadside deterrents') many studies have demonstrated that there is considerable underlying year-to-year variation in the number of such accidents even within control areas where no deliberate action has been taken to prevent DVCs (e.g. Voss, 2007; Langbein, 2007b). Accident frequencies are affected by a number of factors other than simply herd density; and these and other influential factors that may also have altered over the study period (e.g. traffic flow; visitor pressure; forest felling or agricultural cropping; cull level and distribution; climate, etc.) have not been controlled for in what are often rather opportunistic analyses.

Additional, purely stochastic, variation may also be expected between years, not least where annual numbers of accidents are rather low (see above and Voss, 2007). Further, while we might cite a number of other instances where reductions in the density of local deer population have been accompanied by reductions in the frequency of deer–vehicle accidents, there are other published cases where no such relationship has been established (e.g. Waring *et al.*, 1991; Doerr *et al.*, 2001). These differences in outcome indubitably point to the fact that the frequency of ungulate–vehicle collisions is related not to one single or main factor but a multiplicity of causal factors in interaction, amongst which the part played by animal density may be a major factor (e.g. Seiler, 2004; Langbein and Putman, 2006; McShea *et al.*, 2008), or may not. And while it does seem inevitable that there will be some positive association of ungulate densities, traffic flows and ungulate–vehicle collisions at a regional or landscape scale, localised control of animal density alone will not necessarily always lead to a predictable or lasting reduction of accidents (e.g. the converse could occur due to fast recolonisation of favourable habitats by other animals which are less accustomed to live in the vicinity of roads). This once more reinforces our more general view that no single solution to the problem is likely to be effective in isolation, but only when adopted in combination with a suite of other appropriate measures tailored to the local conditions.

8.3.2 Increasing driver awareness and reducing driver speeds

Wildlife warning signs

Warning signs (to increase driver awareness) are the most frequently used approach to reducing ungulate–vehicle accidents. Such cautionary signs warning drivers of the likelihood of wildlife crossing are, however, only likely to be of benefit if erected on approaches to known regular crossing points. In practice, caution signs are relatively rarely precisely targeted but rather tend to warn of increased risk for several kilometres. Further, it is doubtful whether basic signs are in any case very effective in the long term, since drivers readily habituate to them unless the message is reinforced by actual experience of deer crossings (Putman, 1997; Hedlund *et al.*, 2004; Stanley *et al.*, 2006). Furthermore, in many European countries the overabundance of differing road signs increasingly leads to signs being entirely overlooked by drivers. Although basic warning signs have generally been found to be ineffective for reducing the number of ungulate–vehicle collisions, they continue to be widely deployed, since official wildlife warning signs do provide both population managers and road authorities with a degree of judicial safety (absolving their responsibility in the case of collisions with game/wildlife species).

Enhanced or dynamic signs

Various types of enhanced signage, temporary signs, dynamic message boards and animal-activated warning systems have been developed to increase the likelihood that road users will take note of them (Huijser and McGowen, 2003; Huijser *et al.*, 2006; Mastro *et al.*, 2008). Enhanced signs should again be used only to warn of known and regular deer-crossing points along a roadway. Driver habituation might also be reduced if signs were only exposed at particular times or seasons where accidents are known to be more frequent (Iuell *et al.*, 2003). However, Pojar *et al.*

(1972, 1975) found no difference in either vehicle speeds or DVC frequency depending on whether signs were visible or not, or even whether illuminated by permanent or intermittent flashing neon lights; although average speed did decrease by 13 km/h when a deer carcass was present on the highway's emergency lane. Sullivan *et al.* (2004) did report some success with temporary enhanced signs erected only during autumn and spring migration of mule deer (*Odocoilus hemionus*), resulting in a fall in the percentage of speeding vehicles from 19% to 8%, and decrease in DVCs estimated at 50%.

In some cases, dynamic sign systems have been developed further by coupling to sensors capable of detecting animals approaching the roadway. Such signs are thus activated only in direct response to animals present or approaching the carriageway, using a variety of sensors based on heat detection, seismic ground vibrations, or breaking of laser or infra-red beams along the verge. The sensors trigger fibre-optic display enhanced wildlife warning signs and can also be combined with alternative speed detection and display of speed limit signs. Numerous differing versions of such systems have now been installed in Europe and North America. As in the case of other deterrents, firm data on their effectiveness at reducing DVCs remain scarce (Huijser *et al.*, 2006), although Mosler-Berger and Romer (2003) reported a fall in DVCs by near 80% for a series of infra-red activated systems in Switzerland, and several other studies have demonstrated that drivers do slow down in response to activated systems (Gordon *et al.*, 2003; Hammond and Wade, 2004).

Interactive signage, instead of being triggered by animals, may also be triggered by speed of approaching traffic, or to display messages intermittently (e.g. on digital message boards often mounted permanently above major routes or on temporary trailers at the side of the road). Hardy *et al.* (2006) found that temporary messages on such boards warning of wildlife crossings led to lower traffic speeds than either a similar board with traffic information or one not displaying any message; and portable signs seemed more effective than permanent signs.

It does appear that 'dynamic signage' overall has some potential as wildlife mitigation – although it remains unclear what may be the actual cost-effectiveness of animal-sensitive devices versus cheaper and often more reliable devices that are triggered by driver speed, so as to simply enhance warnings against speeding at known accident hot spots. A fuller review of differing types of dynamic or animal-detection and driver-warning signs deployed in Europe and North America is offered by Huijser *et al.* (2006) and Hardy *et al.* (2006).

In-vehicle deer detection systems

Two in-vehicle 'vision systems' have recently been developed designed to enhance driver detection of deer and other ungulates by the roadside, particularly at night. Both use infra-red sensors to offer earlier detection of ungulates or other wildlife either on or beside the carriageway, displaying images continuously on a screen within the dashboard. There are as yet no published studies which evaluate the usefulness or the effectiveness of these technologies. Other future developments being considered are inclusion of wildlife accident hot spot maps for in-car satellite navigations systems, to provide alarms when drivers enter road sections with known high risk of ungulate–vehicle collision, similar to warnings currently used for various other hazards and accident black-spots.

Management of roadside vegetation

Management of vegetation and, specifically, the clearance of woodland or scrub from a margin at the road edge, may have benefits both in increasing driver awareness of ungulates at the roadside, and increasing visibility of oncoming traffic to the ungulates themselves (Waring *et al.*, 1991). Lavsund and Sandegren (1991) report that clearance of a 20-m strip either side of the highway decreased moose collisions by near 20%, and in experimental manipulations to test the effectiveness of vegetation removal along a railway, in reducing the frequency of collisions between trains and moose, Jaren *et al.* (1991) found that removal of vegetation again from a 20–30 m strip on either side of the railway line caused a 56% reduction in the number of recorded accidents. However, for a trial in Germany, Voss (2007) did not find that numbers of roe deer collisions changed significantly between 3 years before/after a more limited strip of around 5 m was cleared of woody scrub and maintained thereafter clear of high vegetation.

While, as noted by Staines *et al.* (2001), one might not advocate a 20-m wide clearance zone more generally alongside all railways or major roads, it seems apparent that vegetation immediately adjacent to such thoroughfares does increase the risk of accident – and a wide strip of vegetation removal in particularly sensitive areas or road sections with poor forward visibility may well be a viable option.

Rea (2003), however, cautions that where vegetation removal is planned, consideration must be given to timing of such clearance and possible increase in subsequent attractiveness of fresh regrowth which in the longer term could potentially actually result in an overall increase in the number of deer utilising the roadside verge. Re-sowing verges with special seed mixtures of mainly grasses and herbs of relatively low nutritional value could help negate this issue.

8.3.3 Safer crossing

Overpasses and underpasses

As noted earlier, highway fencing is at its most effective if it seeks not to prevent animals crossing the road, but rather to direct them to safer crossing points. Such structures may be ones specifically constructed as dedicated overpasses (or 'green bridges') and underpasses beneath main roads, or else adapting crossing structures such as viaduct and agricultural access structures for use by wildlife.

Green bridges and underpasses are becoming increasingly widely used in continental Europe, foremost to help reduce the fragmentation of habitats and associated wildlife populations caused by expanding road and rail infrastructure, as well as to improve road safety. Costs of such structures are inevitably relatively high, though they are no longer always as expensive as commonly assumed (Völk *et al.*, 2001), even when fitted retrospectively to existing roads. Although the majority of major structures to date have been of concrete construction subsequently covered with soil and vegetation, more recently lower-cost green bridges made of wood or metal construction or of reduced dimensions have also proved successful (e.g. Völk *et al.*, 2001; ADAC, 2008).

Specifications and requirements of crossing structures in order to maximise likelihood of use by a wide range and high proportion of wildlife are reviewed in extensive detail in Iuell *et al.* (2003) (and see also Olbrich, 1984; Oord, 1995; Hlavac and Andel, 2002; Georgii *et al.*, 2007; Bissonette and Adair, 2008); thus only a brief overview is attempted here.

An early study of the use of a large sample of farm and forestry accommodation bridges, not necessarily designed primarily or solely for use by wildlife, was undertaken by Olbrich (1984) who assessed the extent to which red, roe and fallow deer used 824 different overpasses and underpasses of differing construction along 823 km of motorways in Germany. Some evidence of use by roe deer was found for 44.7% of all underpasses available but only 22.4% of available overpasses; fallow used 26.3% of underpasses within their distribution and 16.3% of available overpasses; while red deer used only 8.1% of available underpasses and 4.8% of overpasses. Where underpasses are to be provided, Olbrich (1984) suggested minimum height and breadth should be 4 m and stressed that length of underpass should be as short as possible. More specifically, Olbrich found, for all species, that the ratio of aperture size to overall length was critical to use (as (height × breadth)/length). He predicted that red and fallow deer would be least likely to use underpasses where this ratio is less than 1.5; and that for roe deer the ratio should be at least 0.75.

Georgii *et al.* (2007) investigated usage by animals of a range of structures built in Germany more specifically for wildlife during the last two decades, including 20 green bridges, 10 viaducts, 7 underpasses for wild animals, and 6 bridges over waterways and footpaths. Usage was investigated by mapping of tracks and other signs and through filming with infra-red video cameras during March/April on or below a large part of the structures, providing direct evidence of use by animals ranging from red, fallow and roe deer to wild boar, hares, foxes, racoons, polecats and otters. Green bridges and viaducts indicated the most intensive use (85% of all records), while the narrowest wildlife underpasses, bridges over waterways and footpaths for small mammals were only frequented by a few animals. Using multiple regression analysis Georgii *et al.* (2007) showed that use of green bridges was more intensive the wider and older the structures were; while high proportion of wooded vegetation on the structure, noise level peaks, use by humans, and tilling of the land had a negative impact on numbers and species of animals using the crossings.

Pfister *et al.* (1999) also note that while narrower overpasses may indeed be utilised by some individuals, overall use of passages increases with bridge width – and even for roe deer (the smallest species) does not reach an asymptote until widths of 50 m or more.

The wider European COST 341 review (Iuell *et al.*, 2003) recommends width for overpasses of 40–50 m (between the fences) if to be used by red deer and wild boar as well as smaller mammals, suggesting that this width can be lowered to a minimum of 20 m where the topography has a channelling effect leading the animals directly onto the crossing (Iuell *et al.*, 2003). We would note here, however, that many of the recommendations offered by this handbook relate to minimising habitat fragmentation and are aimed at maintaining 'good ecological connectivity' rather than merely reduction of deer/wildlife RTAs. Thus it may be argued that the presence of some smaller passages which those animals determined to cross can use (such as e.g. seasonal movements of male ungulates into female areas during the rut or emigration from natal ranges) may still suffice to bring significant reduction in the frequency of ungulate–vehicle collisions, even if nevertheless still acting as a barrier to free movement of the wider population as a whole.

We would suggest therefore that the above specifications for green bridges should be somewhat less stringently interpreted where improvements for road safety rather than reducing isolation of wildlife populations is the major objective. While specifications presented by COST 341 should be the goal where the extra costs can be justified on the basis of addressing multiple environmental objectives, more modest specifications may nevertheless still be of value in enabling at least occasional use, or use by a small proportion of deer and other ungulates living to either side of major roads. For example, CTGREF (1978) suggested minimum width of overpasses for ungulates of 6 m; Ballon (1985) suggested minimum widths of 8 m and a minimum ratio of width to length of 1 to 10; SETRA (1998) recommended minimal width of overpasses for red deer as 12 m, for roe deer as 7 m; and Langbein (2006) found fallow deer to use both a non-vegetated concrete overpass less than 4 m wide, and another less than 6 m wide with limited soil cover, across the very heavily trafficked M25 motorway orbital route around London.

Whichever form of passage is provided it is clear that it takes a period of time for deer to become used to such corridor structures and to use them freely. Reed *et al.* (1975), Ward (1982), Olbrich (1984), Georgii *et al.* (2007) and Olsson *et al.* (2008) all noted an initial reluctance by deer to use new underpasses until these have 'mellowed' or matured, although use of new bridges by roe and fallow deer have in some cases been observed within a few months of completion (Georgii and Wotschikowski, 2007; authors' own observations).

Overpasses versus underpasses?

Iuell *et al.* (2003) noted that where overpasses and (admittedly rather small) underpasses were available close to each other, moose and deer (*Odocoileus*) preferred to use the overpasses (a conclusion derived from Clevenger *et al.*, 2002; and supported also by Georgii *et al.*, 2007). Note however that in the survey of Olbrich (1984) use of overpasses (i.e. rather than wide green bridges) by red, roe and fallow deer was *lower* than that of underpasses. Small data sets hampered detailed analysis of the factors affecting use, but overall breadth against length seemed the critical consideration. There are few general rules as to when one is more suitable than the other. The choice is partly determined by the topography. In hilly terrain it is often easy to construct both overpasses and underpasses, whereas in flat landscapes underpasses may be easier to construct, if the ground water level is not too high. Overpasses have the advantage that it is easier to provide different microhabitats, because vegetation grows more easily than in underpasses. A wider range of species may therefore use them. However, viaducts, which generally retain quite high and wide original pathways beneath the road, can provide equally good results to landscape bridges (e.g. ECONAT, 1992).

How many passages?

Relatively little work has been undertaken to date on how many overpasses or underpasses may be needed per kilometre of road length to be effective in reducing collisions with vehicles. However, for red deer and roe deer, Hlavac and Andel (2002) recommend that in areas of high importance, passages should be provided at 2–4 km intervals where roe deer are the target species of mitigation; 5–8 km spacing for red deer is suggested by the same authors.

Highway cross-walks

Feasibility and cost of provision of underpasses or overpasses for existing roads will depend largely on the local topography, such that, while retro-fitting of a land bridge may perhaps be most readily achievable in undulating landscapes especially where a road already runs through a cutting, provision of (and landscaping) a similar overpass or tunnel/underpass would be likely to present much greater engineering challenges and higher costs where the existing road runs through a level landscape. Where bridges or accommodation tunnels may not be considered appropriate for existing roadways, an alternative approach is to attempt to provide for safer crossing zones via *cross-walks* across the carriageway surface itself (Lehnert and Bissonette, 1997).

In essence this concept builds upon ideas developed earlier, of using fences or other barriers to guide animals to safer, and well-advertised, crossing places. However, rather than provided as underpasses or overpasses, cross-walks aim to provide relatively safe zones through provision of high-impact signage, speed limits and modifications to the road surface (rumble strips/speed humps) targeted on short sections of road. Road surfaces at the 'cross-walk' itself may also be modified to encourage use by wildlife, with crossings funnelled (and made more predictable) by fencing the roadside in areas where visibility is poor, and permitting crossing of the carriageway only in a limited number of unfenced stretches of roadway where deer and driver visibility has been improved. Lehnert and Bissonette (1997) tested the efficacy of such cross-walks on two-lane and divided four-lane highways in north-eastern Utah. Based on expected kill levels, mortality of mule deer declined by 42.3% and 36.8% along the four-lane and two-lane highways, respectively. Lack of motorist response to warning signs, the tendency for foraging animals to wander from cross-walk boundaries into the carriageway itself, and the ineffectiveness of highway one-way gates in permitting their subsequent escape were considered to contribute most to remaining mortalities within the treatment area, and thus emphasise that careful design, implementation and monitoring of such crossings is critical.

We would propose that consideration might be given to the installation of cattle-grids of appropriate specification to exclude ungulates at such road crossing areas, and also when possible near fence end-runs more generally, to prevent animals entering fenced carriageways. If used with cross-walks such grids would in effect 'link' the fencelines of opposite sides of the carriageway, providing a close-circuit barrier on each side of the cross-walk. Installation of cattle grids on either side of such crossing zones would have the incidental advantage of further reducing traffic speed in these targeted crossing areas.

8.4 Mitigation measures: overall analyses of effectiveness

It would obviously be extremely helpful to our deliberations here if we were able to compare and contrast the absolute efficacy of the various different measures outlined above, or in some way rank them in order of overall effectiveness. In practice this is simply not possible at present and we may offer below only rather general and subjective conclusions.

Firstly, we would note that it is extremely difficult to derive any objective measure of effectiveness for many of the deterrents and other mitigation measures in current use to help reduce collisions with deer and other ungulates, especially where this must be based primarily on analyses of changes in accident frequency before and after installation. This is often the only data available on which to base an assessment of the effectiveness of some measure taken to try and reduce accident frequency, but we should caution:

(i) There is in any case a great deal of variation in accident frequency between years – simple, stochastic year-on-year variation in accident rates (see, for example, Voss, 2007), such that any changes recorded before and after installation of some mitigation measure cannot necessarily be attributed unequivocally to the deterrent measure installed; therefore, adequate numbers of control (unprotected) road sections should be selected, and ideally those where deterrents are installed should be chosen at random from among all road sections considered.

(ii) Accident frequencies on given (monitored) stretches of road are in any case likely to be relatively few in number (often between 0 and 5 or perhaps at best 0 to 10 in most studies), unless trials monitor sections of road over many miles or consider numerous directly comparable replicates. This very restricted range of candidate values also contributes to further difficulty in determining any statistically valid difference between periods before and after the installation of any attempt at mitigation of accident frequency, simply because such differences will be proportionally extremely low (see also Lehnert and Bissonette, 1997; Danielson and Hubbard, 1998).

(iii) Numbers of ungulates killed along the same section of road in consecutive years cannot be assumed to be statistically independent, with, for example, years of relatively high collision frequency (and resulting mortality) potentially reducing overall collision risk in the next if all other factors (level of cull) remain identical.

Records of ungulate–vehicle collisions are also often obtained retrospectively for past years once a trial commences, which can lead to differing levels of accuracy between control sites and sites already identified as 'problem' areas requiring mitigation. Furthermore, few studies continue to be monitored for a sufficient number of years after installation due to financial constraints; trials monitored for just one or two years make it difficult to be certain whether any reduction in collision rates encountered truly relates to the specific type of deterrent or other measure installed, or simply 'novelty avoidance' of any form of change introduced along the trial sections. Even when longer monitoring is possible, there is often an unintentional tendency to discontinue maintenance and monitoring first on those roads where least success has been reported (partly due to difficulty in justifying costs to local authorities in absence of positive results), thus potentially skewing results towards those where more reductions were noted (even if not necessarily resulting from the deterrents).

Roedenbeck *et al.* (2007) also suggest that such practical and financial limitations on study designs have often led to high levels of uncertainty and low strength of inference in studies of the effectiveness of mitigation measures (Roedenbeck *et al.*, 2007; see also Lehnert and Bissonette, 1997; Danielson and Hubbard, 1998), and Mastro *et al.* (2008) conclude that as a consequence, substantial amounts of time and money have been expended on dubious methods of reducing rates of ungulate–vehicle collisions, such as car-mounted whistles, reflectors, as well as permanent signage which may warn of ungulate crossings over distances of many miles.

To overcome some of the above problems associated with studies based solely on changes in recorded collision rates, a few studies have employed video surveillance to assess effects of roadside deterrents on deer behaviour more directly (D'Angelo *et al.*, 2006; Langbein, 2007b; Pokorny *et al.*, 2008). While such studies can help to determine quite quickly whether deer do respond at all to a given deterrent and thus dismiss those devices with little potential early on, where positive results are found, studies need to be replicated in future years to establish the extent to which animals habituate to that deterrent.

In addition to such reservations about objective assessment of the actual effectiveness of different deterrents or mitigation measures, it is equally clear that (even if it were possible to offer objective appraisal of their relative value for reducing the frequency of deer-related accidents), the different methods available to reduce ungulate–vehicle collisions do not necessarily have an *absolute* effectiveness, but many of them will have different utility and effectiveness in different contexts. For example: as already noted, optical warning reflectors, if effective at all, can only be effective in reducing accidents at night, and for these as well as acoustic or chemical deterrents, any analysis of the comparative effectiveness of different measures is thus necessarily context related. Indeed, while the long-term effectiveness of most types of deterrents remains unproven, the potential effectiveness of, for example, acoustic deterrents would be expected to be quite different, from the outset, between countries or regions of very high traffic flow and which often have high levels of other sound pollution (e.g. England: Langbein, 2007b), and more remote areas with relatively low level of traffic noise and thus a lower likelihood that animals may habituate as quickly to new sound or light signals (e.g. Slovenia: Pokorny and Poličnik, 2008).

Further, final decisions about which form of mitigation should be deployed will clearly also be significantly influenced by relative costs (see Putman *et al.*, 2004 for fuller review). Good evidence now exists for the positive effects of well-designed green bridges to reduce the barrier effects posed by major roads for wildlife populations as well as also reducing animal–vehicle collisions (e.g. Pfister *et al.*, 1997; Hatt, 2000; Clevenger and Waltho, 2000; Iuell *et al.*, 2003; Georgii *et al.*, 2007); however, the high cost of such structures will tend to be possible to justify only in limited numbers of areas where a range of nature conservation objectives as well as animal welfare and road safety issues are likely to benefit.

In view of the relatively high costs associated also with animal-activated signage, as well as the difficulty of fitting crossing structures such as overpasses and underpasses retrospectively to existing road structures, we would anticipate that mitigation on major national strategic routes in most countries for the foreseeable future will remain primarily dependent on fencing high-risk sections, where possible directing ungulates towards joint-use crossing structures. The increased installation on major national highways of overhead variable message signs to forewarn drivers of congestion or other hazards has also begun to be used to forewarn of heightened collision risk at particular times of the year in some areas, and such intermittent use for this purpose may increase in future as such ‘road furniture’ becomes more commonly available.

For lesser county and local roads in areas of high DVC risk we would suggest the answer will usually lie in using a combination of approaches

Table 8.2 *Which, if any, individual measures designed to reduce deer vehicle collisions are worth considering?*

Note – in general best results are achieved by selection of a range of complementary measures, rather than reliance on any one of the individual approaches listed

Mitigation measures	Suitable situations and supporting measures	Potential effectiveness/Advantages	Disadvantages
Fencing	Major high-risk roads of high traffic flow; most effective when leads to safer crossing point, and contains escape ramps/leaps.	Well-proven effectiveness where of appropriate mesh size and height, and sufficient length to prevent 'end-runs'. [1, 2, 3, 4, 5]	High maintenance cost; barrier effect also to other wildlife. [6]
Overpasses and green bridges	Major high-risk roads; most effective with lead-in fencing, and natural ground cover.	Well-proven effectiveness; ungulate usage increases with width; but smaller structures can also help alleviate wildlife collisions. [7, 8, 9]	High cost; feasibility dependent on landscape. More readily installed on new-build than for existing roads. [8]
Underpasses and viaducts	Major high-risk roads; most effective with lead-in fencing, and natural ground cover.	Good – where of adequate specification. Mostly lower cost than overpasses of similar size. [7, 9, 10]	High cost; feasibility dependent on landscape. Often longer delay before used by ungulates than in case of overpasses. [7, 9]
Highway cross-walks	Low to medium speed routes; needs to be supported by fencing, signage, speed restriction, and ideally deer-grids.	Good – if well signed. [11]	Not likely to be acceptable on major routes where traffic has to be kept flowing.
Optical wildlife warning reflectors	Roads of low traffic volume providing some traffic-free periods. Vegetation around reflectors needs to be kept clear.	Limited convincing evidence of success. Relatively low cost; do not prevent normal range use. [12, 13]	Rapid habituation where lit up by frequent traffic. Can at best only function during night. Many trials indicate ineffective. [14, 15, 16, 17, 18]
Acoustic wildlife warning devices	Roads of low traffic volume, where habituation is least likely, and providing safe crossing periods.	Variable evidence. Lasting effects likely to depend on type and variability of signals. [19, 20]	General effectiveness remains unproven. Limited potential on roads of high traffic volume. Much higher (× 10) cost than optical reflectors. [17, 21]

Chemical/olfactory deterrents	Roads of low to moderate traffic flow.	Limited convincing evidence of success. Most intend to raise level of alertness, rather than prevent animals crossing. [22]	Limited independent evidence of effectiveness. Requires renewal at regular intervals. Likely habituation. [17, 19, 23, 24]
Vehicle-mounted ultrasound whistles and electronic horns		Poor effectiveness. [25] Some types very cheap to install.	No convincing evidence of effectiveness. Signals mostly drowned out by traffic noise. [26, 27, 28]
Standard wildlife warning signage	Any road type, but should be targeted to forewarn of short, well-defined sections of high risk.	Can help absolve legal responsibility of road authorities or population managers. Moderate cost.	Over-abundance of wildlife and other signage lead to reduced effect on driver behaviour. Low effectiveness (if any) at reducing collisions. [29, 30, 31]
Interactive speed-activated wildlife + speed signage	Any road type, but should be targeted to forewarn of short, well-defined sections of high risk.	Some potential, but yet unproven for DVC reduction. Increased driver perception. [32, 33]	Driver habituation over time, if not reinforced by seeing animals near the crossing point, and as digital signage in general becomes more common. [34, 35]
Interactive animal-activated signage	Major well-defined animal crossing points on roads of moderate traffic flow.	Promising effects on driver awareness and local speed reduction. [36, 37, 38]	High cost compared with standard or speed-activated signage. Variable reliability of differing sensor types. [35]
Speed limits	Low to moderate traffic flow routes. Speed sign at same site as wildlife sign preferable.	Good – provided well enforced. Reduces severity of accidents if not necessarily frequency. [for refs. see 4]	Feasibility/acceptability for major roads is limited.
Reduction of local deer density	Prevention of increase, if not reduction, of deer numbers required in order for most other measures (including fencing) to remain effective.	Good – provided undertaken over wide area, and as one part of overall DVC reduction strategy. [39, 40, 41, 42]	Localised culling may shift rather than reduce collisions, and destabilise population. Public understanding of need to control wildlife is limited. [14, 43]
Immuno-contraception	Isolated, self-contained populations.	Non-lethal; higher public acceptability in some countries/situations than culling. Limited/short-term effectiveness. [44]	Requires high proportion of herd inoculated. Ethically questionable. Very high cost. [5]

Table 8.2 (*cont.*)

Mitigation measures	Suitable situations and supporting measures	Potential effectiveness/Advantages	Disadvantages
Reducing animal disturbance	Forests with high human/dog disturbance.	High potential – where dog walking and human activity often panics deer to cross roads. Low cost if achieved through restrictions on activity in specific high-risk areas.	Difficulty to achieve compliance; e.g. keeping dogs on leads. May be contrary to other policies to increase public use of forests and countryside.
Verge clearance and maintenance	All roads. Ideally verges re-sown with grass mixtures of low digestibility. Clear verges also a prerequisite if reflectors in use.	Promising. Improved forward visibility for drivers and animals; dependant on width possible to clear. [45, 46, 47]	Effect on collisions reduction not fully proven. Increased forage production on verge may attract animals if not timed carefully. [17, 48]
Public awareness raising and driver education	Increasing importance as traffic and collision risk escalates. Animal hazard awareness should be built into national driver syllabuses.	High potential – relatively low cost if based on leaflets and printed media. Can be integrated with other road safety campaigns.	Effects unclear; may be short lived unless replicated. Responsiveness of driving public questionable.

1, Reed *et al.* (1982); 2, Ward (1982); 3, Ballon (1985); 4, Putman *et al.* (2004); 5, Mastro *et al.* (2008); 6, Feldhamer *et al.* (1986); 7, Olbrich (1984); 8, Iuell *et al.* (2003); 9, Georgii *et al.* (2007); 10, ECONAT (1992); 11, Lehnert and Bissonette (1997); 12, Schafer and Penland (1985); 13, Gladfelter (1982); 14, Waring *et al.* (1991); 15, Reeve and Anderson (1993); 16, Woodard *et al.* (1973); 17, Voss (2007); 18, D'Angelo *et al.* (2006); 19, Pokorny *et al.* (2008); 20, Pokorny and Poličnik (2008); 21, Langbein (2007b); 22, Kerzel and Kirchberger (1993); 23, Lebersorger (1993); 24, Lutz (1994); 25, Tracy (2003; in DVCIC; 2003); 26, Romin and Dalton (1992) 27, Schober and Sommer (1984); 28, Scheifele *et al.* (2003); 29, Putman (1997); 30, Hedlund *et al.* (2004); 31, Stanley *et al.* (2006); 32, Sullivan *et al.* (2004); 33, Hardy *et al.* (2006); 34, Pojar *et al.* (1975); 35, Huijser *et al.* (2006); 36, Gordon *et al.* (2003); 37, Hammond and Wade (2004); 38, Mosler-Berger and Romer (2003); 39, McCaffery (1973); 40, Schwabe *et al.* (2002); 41, Rondeau and Conrad (2003); 42, Sudharsan *et al.* (2006); 43, Doerr *et al.* (2001); 44, Rutberg and Naugle (2008); 45, Jaren *et al.* (1991); 46, Staines *et al.* (2001); 47, Lavsund and Sandegren (1991); 48, Rea (2003).

rather than focusing on any one individual measure aimed at prevention of accidents. Raising general public awareness via the media, provision of local information sheets giving information on problematic road sections, and inclusion of more animal hazard awareness during instruction for driver education may all have a part to play, together with enhanced signage targeted on known common crossing locations for ungulates, sensitive management of roadside verges to improve forward visibility without increasing their attractiveness for grazing, culling or other measures to prevent escalation of local population numbers, and continued search for more efficient roadside deterrents as well as fencing. Best results overall are likely to be achieved where road authorities work in close partnership with ungulate ecologists, population managers, and forest and land managers to integrate those roadside measures most suited to the local situation with action to raise public awareness and management of the wildlife population.

References

ADAC (2008) *Grünbrücken: Empfehlungen für die Praxis*. [Advice note on green-bridges] (In German) München, Germany: ADAC.

Allen, R.E. and McCullough, D.R. (1976) Deer-car accidents in Southern Michigan. *Journal of Wildlife Management* **40**, 317–325.

Andersen, N.P. and Madsen, A.B. (2007) *Trafi kdræbte større dyr i Danmark – kortlægning og analyse af påkørselsforhold*. (In Danish with English summary) Faglig rapport fra DMU nr. 626. Aarhus, Denmark: Danmarks Miljøundersøgelser, Aarhus Universitet. Online: www2.dmu.dk/Pub/FR626a.pdf

Andersen, R., Lund, E., Solberg, E. and Sæther, B.-E. (2010) Ungulates and their management in Norway. In M. Apollonio, R. Andersen and R. Putman (eds.) *European Ungulates and their Management in the 21st Century*. Cambridge, UK: Cambridge University Press, pp. 14–36.

Apollonio, M., Andersen, R. and Putman, R. (eds.) (2010) *European Ungulates and their Management in the 21st Century*. Cambridge, UK: Cambridge University Press, 604 pp.

Armstrong, J.J. (1992) *An Evaluation of the Effectiveness of Swareflex Deer Reflectors*. Toronto, Canada: Research and Development Branch, Ministry of Transportation.

Ballon, P. (1985) Bilan technique des aménagements réalisés en France pour réduire les impacts des grandes infrastructures linéaires sur les ongulés gibiers. In *Actes du XVII ème Congrès de l'Union Internationale des Biologistes du Gibier*, pp. 679–689.

Bashore, T.L., Tzilkowski, W.M. and Bellis, E.D. (1985) Analysis of deer-vehicle collision sites in Pennsylvania. *Journal of Wildlife Management* **49**, 769–774.

Bissonette, J.A. and Adair, W. (2008) Restoring habitat permeability to roaded landscapes with isometrically-scaled wildlife crossings. *Biological Conservation* **141**, 482–488.

Bissonette, J.A. and Hammer, M. (2000) Effectiveness of earthen ramps in reducing big game highway mortality in Utah. *Utah Cooperative Fish and Wildlife Research Unit Report Series* **1**, 1–29.

Bissonette, J.A., Kassar, C. and Cook, L.J. (2008) An assessment of costs associated with deer-vehicle collisions: human death and injury, vehicle damage, and deer loss. *Human–Wildlife Conflicts* **2**, 17–27.

Carranza, J. (2010) Ungulates and their management in Spain. In M. Apollonio, R. Andersen, and R. Putman (eds.) *European Ungulates and their Management in the 21st Century*. Cambridge, UK: Cambridge University Press, pp. 419–440.

Carsignol, J. (1989) Dix années d'observations des collisions vehicules-grand mammifères gibiers sur l'autoroute A4 en Alsace-Lorraine et sur le réseau routier du département de la Moselle. *Office National de la Chasse, Bulletin Mensuelle* **135**, 32–37.

Clevenger, A.P. and Waltho, N. (2000) Factors influencing the effectiveness of wildlife underpasses in Banff National Park, Alberta, Canada. *Conservation Biology* **14**, 47–56.

Clevenger, A.P., Chruszcz, B. and Gunson, K.E. (2001) Highway mitigation fencing reduces wildlife-vehicle collisions. *Wildlife Society Bulletin* **29**, 646–653.

Clevenger, A.P., Chruszcz, B., Gunson, K.E. and Wierzchowski, J. (2002) *Roads and Wildlife in the Canadian Rocky Mountain Parks: Movement, Mortality and Mitigation*. Final report to Parks Canada, Banff, Alberta, Canada, 432 pp.

Conover, M.R., Pitt, W.C., Kessler, K.K., DuBow, T.J. and Sanborn, W.A. (1995) Review of human injuries, illnesses, and economic losses caused by wildlife in the United States. *Wildlife Society Bulletin* **23**, 407–414.

CTGREF (1978) *Autoroute et Grand Gibier*. Note technique 42. Paris: Ministère de l'Agriculture, Groupement Technique Forestier.

D'Angelo, G.J., D'Angelo, J.G., Gallagher, G.R., *et al.* (2006) Evaluation of wildlife warning reflectors for altering white-tailed deer behavior along roadways. *Wildlife Society Bulletin* **34**, 1175–1183.

Danielson, B.J. and Hubbard, M.W. (1998) A literature review for assessing the status of current methods of reducing deer-vehicle collisions. Unpublished report to Iowa Department of Transportation and Iowa Department of Natural Resources.

Decker, D.J., Loconti Lee, K.M. and Connelly, N.A. (1990) *Incidences and Costs of Deer-related Vehicular Accidents in Tomkins County, New York*. HDRU Series 89–7, revised February 1990. New York: Department of Natural Resources, New York State Agricultural College and Cornell University, 22 pp.

Deer-Vehicle Crash Information Clearing House (DVCIC) (2003) On-line manual of mitigation measures and perceived effectiveness. Online www.deercrash.com/toolbox

Desire, G. and Recorbet, G. (1990) Resultats de l'enquete réalisée de 1984 à 1986 sur les collisions entre les vehicules et les grands mammifères sauvages. *Office National de la Chasse, Bulletin Mensuelle* **143**, 38–47.

Doerr, M.L., McAninch, J.B. and Wiggers, E.P. (2001) Comparison of 4 methods to reduce white-tailed deer abundance in an urban community. *Wildlife Society Bulletin* **29**, 1105–1113.

ECONAT (1992) Protection de la faune dans les projets de nouveau tracés ferroviaires. Yverdon-les-Bains, Switzerland: ECONAT, 51 pp.

Etter, D.R., Van Deelen, T.R., Ludwig, D.R., Kobal, S.N. and Warner, R.E. (2002) Survival and movements of white-tailed deer in suburban Chicago, Illinois. *Journal of Wildlife Management* **66**, 500–510.

Fehlberg, U. (1994) Ökologische Barrierewirkung von Strassen auf wild-lebende Säugetiere. *Deutsche Tierärzliche Wochenschrift* **101**, 81–132.

Feldhamer, G.A., Gates, J.E., Harman, D.M., Loranger, A.J. and Dixon, K.R. (1986) Effects of interstate highway fencing on white-tailed deer activity. *Journal of Wildlife Management* **50**, 497–503.

Finder, R.A., Roseberry, J.L. and Woolf, A. (1999) Site and landscape conditions at white-tailed deer collision locations in Illinois. *Landscape and Urban Planning* **44**, 77–85.

Ford, S.G. and Villa, S.L. (1993) *Reflector Use and the Effect they have on the Number of Mule Deer Killed on California Highways*. Report FHWA/CA/PD94/01. Sacramento, CA: California Department of Transport.

Forman, R.T.T. and Alexander, L.E. (1998) Roads and their major ecological effects. *Annual Review of Ecology and Systematics* **29**, 207–231.

Forman, R.T.T., Friedman, D.S., Fitzhenry, D., *et al.* (1997) Ecological effects of roads: toward three summary indices and an overview for North America. In K. Canters, A. Piepers and D. Hendriks-Heersma (eds.) *Habitat Fragmentation and Infrastructure*. Proceedings of the International Conference on Habitat Fragmentation, Infrastructure and the role of Ecological Engineering, 17–21 September 1995. Delft, the Netherlands: Ministry of Transport Public Works and Water Management, pp. 40–54.

Georgii, B. and Wotschikowski, U. (2007) *Strassen und Wildtiere*. [In German] Bonn, Germany: Bundesministerium für Verkehr, Bau und Stadtentwicklung.

Georgii, B., Peters-Ostenberg, E., Henneberg, M., *et al.* (2007) *Nutzung von Grünbrücken und anderen Querungsbauwerken durch Säugetiere. Gesamtbericht zum Forschungs- und Entwicklungsvorhaben*. [In German] Forschung Strassenbau und Strassenverkehrstechnik 971. Bonn, Germany: Bundesministerium für Verkehr.

Gilbert, J.R. (1982) *Evaluation of Deer Mirrors for Reducing Deer-Vehicle Collisions*. United States Federal Highway Administration Report FHWA/RD/82/061. Washington, DC: United States Federal Highway Administration, 16 pp.

Gill, R.M.A. (1990) *Monitoring the Status of European and North American Cervids*. GEMS Information Series, 8. Nairobi, Kenya: Global Environment Monitoring Systems, United Nations Environment Programme, Nairobi. 277 pp.

Gladfelter, J.R. (1982) *Effect of Wildlife Highway Warning Reflectors on Deer-Vehicle Accidents*. Iowa Highway Research Board Project HR-210. Iowa Department of Transportation, 11 pp.

Gleason, J.S. and Jenks, J.A. (1993) Factors influencing deer–vehicle mortality in east central South Dakota. *Prairie Naturalist* **25**, 281–288.

Gordon, K.M., Anderson, S., Gribble, B. and Johnson, M. (2003) *Evaluation of the FLASH (flashing light animal sensing host) system in Nugget Canyon, Wyoming*. Report FHWAWYL-01/03F. Laramie, WY: Wyoming Cooperative Fish and Wildlife Research Unit.

Groot Bruinderink, G.W.T.A. and Hazebroek, E. (1996) Ungulate traffic collisions in Europe. *Conservation Biology* **10**, 1059–1067.

Grovenburg, T.W., Jenks, J.A., Klaver, R.W., *et al.* (2008) Factors affecting road mortality of white-tailed deer in eastern South Dakota. *Human–Wildlife Conflicts* **2**, 48–59.

Haikonen, H. and Summala, H. (2001) Deer-vehicle crashes – extensive peak at 1 hour after sunset. *American Journal of Preventive Medicine* **21**, 209–213.

Hammond, C. and Wade, M.G. (2004) *Deer Avoidance: the Assignment of Real World Enhanced Deer Signage in a Virtual Environment*. Saint Paul, MN: Minnesota Department of Transportation.

Hansen, C.S. (1983) Costs of deer-vehicle accidents in Michigan. *Wildlife Society Bulletin* **11**, 161–164.

Hardy, A., Lee, S. and Al-Kaisy, A.F. (2006) Effectiveness of animal advisory messages on dynamic message signs as a speed reduction tool. *Transportation Research Record* **1973**, 64–72.

Hartwig, D. (1991) Erfassung der Verkehrsunfälle mit Wild im Jahre 1989 in Nordrhein-Westfalen im Bereich der Polizeibehörden. *Zeitschrift für Jagdwissenschaft* **37**, 55–62.

Hartwig, D. (1993) Auswertung der durch Wild Verursachten Verkehrsunfälle nach der Statistik für Nordrhein-Westfalen. *Zeitschrift für Jagdwissenschaft* **39**, 22–33.

Hatt, S. (2000) Grünbrücke Lotterbuck A 4.2.9: Eine Erfolgskontrolle nach drei Jahren. *Schweitzerische Zeitschrift für Forstwessen* **151**, 290–297.

Hedlund, J.H., Curtis, P.D. and Williams, A.F. (2004) Methods to reduce traffic crashes involving deer: what works and what does not. *Traffic Injury Prevention* **5**, 122–131.

Hlavac, V. and Andel, P. (2002) *On the Permeability of Roads for Wildlife: A Handbook*. Prague: Agency for Nature Conservation and Landscape Protection of the Czech Republic.

Hubbard, M.W., Danielson, B.J. and Schmitz, R.A. (2000) Factors influencing the location of deer-vehicle accidents in Iowa. *Journal of Wildlife Management* **64**, 707–713.

Huijser, M.P. and McGowen, P.T. (2003) Overview of animal detection and animal warning systems in North America and Europe. In C.L. Irwin, P. Garrett and K. P. McDermott (eds.) *Proceedings of the 2003 International Conference on Ecology and Transportation*, held Lake Placid, New York. Raleigh, NC: Center for Transportation and the Environment, North Carolina State University.

Huijser, M.P., McGowen, P.T., Camel, W., *et al.* (2006) *Animal Vehicle Crash Mitigation Using Advanced Technology Phase 1: Review, Design and Implementation*. Oregon Department of Transportation Research Unit, Salem, Oregon, and Federal Highway Administration, Washington, DC.

Hussain, A., Armstrong, J.B., Brown, D.B. and Hogland, J. (2007) Land-use pattern, urbanization, and deer–vehicle collisions in Alabama. *Human–Wildlife Conflicts* **1**, 89–96.

Imesch-Bebié, N., Gander, H. and Schnidrig-Petrig, R. (2010) Ungulates and their management in Switzerland. In M. Apollonio, R. Andersen and R. Putman (eds.) *European Ungulates and their Management in the 21st Century*. Cambridge, UK: Cambridge University Press, pp. 357–391.

Insurance Institute for Highway Safety (IIHS) (1993) Deer, moose collisions with motor vehicles peak in spring and fall. *Status Report* **28**(4).

Insurance Institute for Highway Safety (IIHS) (2004) Lots of approaches are under way to reduce deer collisions, but few have proven effective. *Status Report* **39**(1), 5–7.

Insurance Institute for Highway Safety (IIHS) (2008) Collisions with deer and other animals spike in November; fatal crashes up 50% since 2000. *IIHS Press Release*, 30 October 2008. Online www.iihs.org/news/2008/iihs_news_103008.pdf

Iuell, B., Bekker, G.J., Cuperus, R. *et al.* (2003) *Wildlife and Traffic: A European Handbook for Identifying Conflicts and Designing Solutions*. Action 341 on 'Habitat Fragmentation due to Transportation Infrastructure'. Brussels: European Commission.

Jared, D. (1992) *Evaluation of Wild Animal Highway Warning Reflectors.* Report 98003. Atlanta, GA: Georgia Department of Transportation.

Jaren, V., Andersen, R., Ulleberg, M., Pedersen, P.H. and Wiseth, B. (1991) Moose-train collisions: the effects of vegetation removal with a cost-benefit analysis. *Alces* **27**, 93–99.

Jenks, J.A., Smith, W.P. and DePerno, C.S. (2002) Maximum sustained yield harvest versus trophy management. *Journal of Wildlife Management* **66**, 528–535.

Jones, J.M. and Witham, J.H. (1993). Urban deer 'problem-solving' in northeast Illinois: an overview. In J.B. McAninch (ed.) *Proceedings of the 55th Midwest Fish and Wildlife Conference: Urban Deer: A Manageable Resource.* St. Louis, MO: The Wildlife Society, pp. 58–65.

Kerzel, H. (2005) Wildunfälle: Zur Notwendigkeit von Verkehrsschutzzäunen und Grünbrücken. In *Grünbrücken für den Biotopverbund.* Schriftenreihe des Landesjagdverbandes Bayern e.V. 14, pp. 77–86.

Kerzel, H. and Kirchberger, U. (1993) Erfolge im Kampf gegen Wildunfälle. *Die Pirsch* **18**, 3–5.

Kuser, J.E. and Wolgast, L.J. (1983) Deer road-kill increases with no-firearms-discharge law. *Bulletin of the New Jersey Academy of Science* **28**, 71–72.

Langbein, J. (1985) *North Staffordshire Deer Survey 1983–1985. I. Research and Development.* Fordingbridge, UK: British Deer Society.

Langbein, J. (2006) *Conservation and Management of Deer in Epping Forest and its Buffer Land Estate.* Report to the Corporation of the City of London.

Langbein, J. (2007a) *National Deer-Vehicle Collisions Project: England 2003–2005.* Final Report to the Highways Agency. Wrexham, UK: The Deer Initiative.

Langbein, J. (2007b) Use of remote video surveillance to investigate deer behaviour in relation to wildlife deterrents, roads and vehicles. Presentation at 'Deer on our Roads Seminar', Ashridge, UK, October 2007. Online www.deercollisions.co.uk/pages/workshop2.html]

Langbein, J. (2009) *Ecopillar Acoustic Wildlife Warning Devices: Acoustic Characteristics and Effectiveness at Alerting Deer and Deterring them from Crossing in Front of Traffic.* Research Report 09/04 to the Highways Agency. Wrexham, UK: The Deer Initiative.

Langbein, J. and Putman, R.J. (2006) *National Deer-Vehicle Collisions Project; Scotland, 2003–2005.* Report to the Scottish Executive, June 2006.

Lavsund, S. and Sandegren, F. (1991) Moose-vehicle relations in Sweden: a review. *Alces* **27**, 118–126.

Lebersorger, P. (1993) Verkehrspartner Wild. *Weidwerk* **11**, 47–48.

Leblond, M., Dussault, C., Ouellet, J., *et al.* (2007) Electric fencing as a measure to reduce moose-vehicle collisions. *Journal of Wildlife Management* **71**, 1695–1703.

Lehnert, M.E. and Bissonette, J.A. (1997) Effectiveness of highways crosswalk structures in reducing deer-vehicle collisions. *Wildlife Society Bulletin* **25**, 809–818.

Lode, T. (2000) Effect of a motorway on mortality and isolation of wildlife populations. *Ambio* **29**, 163–166.

Lutz, W. (1994) Ergebnisse der Anwendung eines sogenannten Duftzaunes zur Vermeidung von Wildverlusten durch den Strassenverkehr nach Gehege- und Freilandorientierungen. *Zeitschrift für Jagdwissenschaft* **40**, 91–108.

Madsen, A.B., Strandgaard, H. and Prang, A. (2002) Factors causing traffic killings of roe deer, *Capreolus capreolus* in Denmark. *Wildlife Biology* **8**, 55–61.

Maillard, D., Gaillard, J.-M., Hewison, M., *et al.* (2010) Ungulates and their management in France. In M. Apollonio, R. Andersen and R. Putman (eds.) *European Ungulates and their Management in the 21st Century*. Cambridge, UK: Cambridge University Press, pp. 441–474.

Malo, J.E, Suarez, F. and Diaz, A. (2004) Can we mitigate animal-vehicle accidents using predictive models? *Journal of Applied Ecology* **41**, 701–710.

Mastro, L.L., Conover, M.R. and Frey, S.N. (2008) Deer-vehicle collision prevention techniques. *Human–Wildlife Conflicts* **2**, 80–92.

McCaffery, K.R. (1973) Road-kills show trends in Wisconsin deer populations. *Journal of Wildlife Management* **37**, 212–216.

McShea, W.J., Stewart, C.M., Kearns, L.J., Liccioli, S. and Kocka, D. (2008) Factors affecting autumn deer–vehicle collisions in a rural Virginia county. *Human–Wildlife Conflicts* **2**, 110–121.

Mladenhoff, D.J., Sickely T.A. and Wydeve A.P. (1999) Predicting gray wolf landscape recolonization: logistic regression models vs field data. *Ecological Applications* **9**, 37–44.

Mosler-Berger, C. and Romer, J. (2003) Wildwarnsystem CALSTROM. [In German] *Wildbiologie* **3**, 1–2.

Nielsen, C.K., Anderson, R.G. and Grund, M.D. (2003) Landscape influences on deer-vehicle accident areas in an urban environment. *Journal of Wildlife Management* **67**, 46–51.

Ng, J.W., Nielsen, C. and St. Clair, C.C. (2008) Landscape factors influencing deer–vehicle collisions in an urban environment. *Human–Wildlife Conflicts* **2**, 34–47.

Olbrich, P. (1984) Untersuchung der Wirksamkeit von Wildwarnreflektoren und der Eignung von Wilddurchlässen. *Zeitschrift für Jagdwissenschaft* **30**, 101–116.

Olsson, M.P.O, Widen, P. and Larkin, J.L. (2008) Effectiveness of a highway overpass to promote landscape connectivity and movement of moose and roe deer in Sweden. *Landscape and Urban Planning* **85**, 133–139.

Oord, J.G. (1995) *Handreiking maatregelen voor de fauna langs weg en water, Rijkswaterstaat*. Delft, Netherlands: Dienst Weg- en Waterbouwkunde & Dienst Landinrichting en Beheer Landbouwgronden, 278 pp.

Pafko, F. and Kovach, B. (1996) *Experience with Deer Reflectors: Trends in Assessing Transportation Related Wildlife Mortality*. St Paul, MN: Minnesota Department of Transportation.

Pepper, H.W. (1999) Road traffic accidents and deer reflectors: a comparative trial of the efficacy of standard red and new blue/green roadside reflectors at preventing motor vehicle and wild deer collisions. Internal report on Project 257. London: Forestry Commission, Forest Research.

Pepper, H.W., Chadwick, A.H. and Packer, J.J. (1998) *Deer Reflectors and Road Traffic Accidents through Forestry Commission Forests: A Review of Traffic Accident Records for Roads where Deer Warning Reflectors have been Installed.* Appendix to contract report VC 0317. London: Ministry of Agriculture Fisheries and Foods.

Pfister, H.P., Keller, V., Reck H. and Georgii, B. (1997) *Bio-ökologische Wirksamkeit von Grünbrücken über Verkehrswege*. Forschung Straßenbau und Straßenverkehrstechnik 756. Bonn, Germany: Bundesministerium für Verkehr.

Pfister, H.P., Heynen, D., Georgii, B., Keller, V. and von Lerber, F. (1999) *Häufigkeit und Verhalten ausgewählter Wildsäuger auf unterschiedlich breiten Wildtierbrücken (Grünbrücken)*. Sempach, Switzerland: Schweizerische Vogelwarte, 49 pp.

Pojar, T.M., Reseigh, T.C. and Reed, D.F. (1972) Deer crossing signs may prove valuable in reducing accidents and animal deaths. *Highway Research News* **46**, 20–23.

Pojar, T.M., Prosence, R.A., Reed, D.F. and Woodard, T.N. (1975) Effectiveness of a lighted, animated deer crossing sign. *Journal of Wildlife Management* **39**, 87–91.

Pokorny, B. (2006) Roe deer-vehicle collisions in Slovenia: situation, mitigation strategy and countermeasures. *Veterinarski Arhiv* **76**, 177–187.

Pokorny, B. and Poličnik, H. (2008) *Monitoring učinkovitosti izvedenih ukrepov za preprečevanje trkov vozil z divjadjo.* [Monitoring of effectiveness of countermeasures implemented for reducing the number of game-vehicle collisions] (In Slovene) Final report for Slovene Directorate for Roads, Contract no. 2415–07–000721/0. Velenje, Slovenia: ERICo, 82 pp.

Pokorny, B., Marolt, J. and Poličnik, H. (2008) *Ocena učinkovitosti in vplivov zvočnih odvračalnih naprav kot sredstva za zmanjšanje števila trkov vozil z veliko divjadjo.* [Assessment of the effectiveness and impacts of acoustic deterrents as a countermeasure for reducing the number of big game-vehicle collisions] (In Slovene) Final report for Slovene Hunters Association, Contract no. LZS-04/1298. Velenje, Slovenia: ERICo, 107 pp.

Puglisi, M.J., Lindzey, J.S. and Bellis, E.D. (1974) Factors associated with highway mortality of white-tailed deer. *Journal of Wildlife Management* **38**, 799–807.

Pürstl, A. (2006) Tierärztlichles Gutachten zum Farbsehvermögen von Rot und Rehwild. Vienna: Tierambulanz Türkenschanzplatz. (unpublished report)

Putman, R.J. (1997) Deer and road traffic accidents: options for management. *Journal of Environmental Management* **51**, 43–57.

Putman, R.J., Langbein, J. and Staines, B.W. (2004) *Deer and Road Traffic Accidents; A Review of Mitigation Measures: Costs and Cost-Effectiveness.* Report to the Deer Commission for Scotland, Contract RP 23A.

Rea, V. (2003) Modifying roadside vegetation management practices to reduce vehicular collisions with moose, *Alces alces. Wildlife Biology* **9**, 81–91.

Reed, D.F., Pojar, T.M. and Woodard, T.N. (1974) Use of one-way gates by mule deer. *Journal of Wildlife Management* **38**, 9–15.

Reed, D.F., Woodard, T.N. and Pojar, T.M. (1975) Behavioral response of mule deer to a highway underpass. *Journal of Wildlife Management* **39**, 361–367.

Reed, D.F., Beck, T.D.I. and Woodard, T.N. (1982) Methods of reducing deer-vehicle accidents: benefit-cost analysis. *Wildlife Society Bulletin* **10**, 349–54.

Reeve, A.F. and Anderson, S.H. (1993) Ineffectiveness of Swareflex reflectors at reducing deer-vehicle collisions. *Wildlife Society Bulletin* **21**, 127–132.

Roedenbeck, I.A., Fahrig, L., Findlay, C.S., *et al.* (2007) The Rauischholzhausen agenda for road ecology. *Ecology and Society* **12**, 11. Online www.ecologyandsociety.org/vol12/iss1/art11/

Romin, L.A. and Bissonette, J.A. (1996) Deer-vehicle collisions: status of state monitoring activities and mitigation efforts. *Wildlife Society Bulletin* **24**, 276–283.

Romin, L.A. and Dalton, L.B. (1992) Lack of response by mule deer to wildlife warning whistles. *Wildlife Society Bulletin* **20**, 382–384.

Rondeau, D. and Conrad, J.M. (2003) Managing urban deer. *American Journal of Agricultural Economics* **85**, 266–281.

Rutberg, A.T. and Naugle, R.E. (2008) Deer-vehicle collision trends at a suburban immunocontraception site. *Human–Wildlife Conflicts* **2**, 60–67.

Ruusila, V. and Kojola, I. (2010) Ungulates and their management in Finland. In M. Apollonio, R. Andersen and R. Putman (eds.) *European Ungulates and their Management in the 21st Century*. Cambridge, UK: Cambridge University Press, pp. 86–102.

Sanders, W. (1985) Fallow deer: a study of the possible factors influencing the annual cycle of road casualties. HND dissertation, Seal Hayne College, UK.

Schafer, J.A. and Penland, S.T. (1985) Effectiveness of Swareflex reflectors in reducing deer-vehicle accidents. *Journal of Wildlife Management* **49**, 774–776.

Scheifele, P.M., Browning, D.G. and Scheifele, L.M. (1998) Measurement of several types of 'deer whistles' for motor vehicles: frequencies, levels, and animal threshold responses. *Journal of the Acoustical Society of America* **104**(3), 1811.

Scheifele, P.M., Browning, D.G. and Scheifele, L.M. (2003) Analysis and effectiveness of 'deer whistles' for motor vehicles: frequencies, levels, and animal threshold responses. *Journal of the Acoustical Society of America ARLO* **4**, 13.

Schober, F, and Sommer, F. (1984) Untersuchungen akustischer Wildwarngeräten für Kraftfahrzeuge. *Zeitschrift für Jagdwissenschaft* **30**, 164–176.

Schwabe, K.A., Schuhmann, P.W. and Tonkovich, M. (2002) A dynamic exercise in reducing deer-vehicle collisions: management through vehicle mitigation techniques and hunting. *Journal of Agricultural and Resource Economics* **27**, 261–280.

Seiler, A. (2004) Trends and spatial pattern in ungulate-vehicle collisions in Sweden. *Wildlife Biology* **10**, 301–313.

SETRA (1998) *Collisions véhicules-grands mammifères sauvages: évolutions des inventaires de 1984–1986 et 1993–1994*. Note d'information 60. Bagneux, France: Service d'Etudes Techniques des Routes et Autoroutes/Ministère de l'Environnement.

SGS Environment (1998) *The Prevention of Wildlife Casualties on Roads through the Use of Deterrents: Prevention of Casualties among Deer Populations*. Report to UK Highways Agency SW335/V3/11–98. Ellesmere Port, UK: SGS UK Ltd.

Sheets, R. and Cason, T. (2005) What deer see and hear. *Research Magazine*. Athens. GA: University of Georgia.

Staines, B.W., Langbein, J. and Putman, R.J. (2001) *Road Traffic Accidents and Deer in Scotland*. Report to the Deer Commission, Scotland.

Stanley, L., Hardy, A. and Lassacher, S. (2006) Responses to enhanced wildlife advisories in a simulated environment. *Transportation Research Record* **1980**, 126–133.

Sudharsan, K., Riley, S.J. and Winerstein, S.R. (2006) Relationships of autumn hunting season to the frequency of deer–vehicle collisions in Michigan. *Journal of Wildlife Management* **70**, 1161–1164.

Sullivan, T.A., Williams, A.F., Messmer, T.A., Hellinga, L.A. and Kyrychenko, S.Y. (2004) Effectiveness of temporary warning signs in reducing deer–vehicle collisions during mule deer migrations. *Wildlife Society Bulletin* **32**, 907–915.

Ückermann, E. (1964) Erhebung uber die Wildverluste durch den Strassenverkehr und die Verkehrsunfälle durch Wild. *Zeitschrift für Jagdwissenschaft* **10**, 142–168.

Ückermann, E. (1983) Die Auswirkung der Wildverluste durch den Stassenverkehr auf die Nutzungsfähigkeit der Reviere. *Zeitschrift für Jagdwissenschaft*, **29**, 264–265.

Ujvari, M., Baagoe, H.J. and Madsen, A.B. (1998) Effectiveness of wildlife warning reflectors in reducing deer-vehicle collisions: a behavioral study. *Journal of Wildlife Management* **62**, 1094–1099.

van Wieren, S.E. and Groot Bruinderink, G.W.T.A. (2010) Ungulates and their management in the Netherlands. In M. Apollonio, R. Andersen and R. Putman (eds.) *European Ungulates and their Management in the 21st Century*. Cambridge, UK: Cambridge University Press, pp. 165–183.

VerCauteren, K.C. and Pipas, M.J. (2003) A review of the color vision in white tailed deer. *Wildlife Society Bulletin* **31**(3), 684–691.

VerCauteren, K.C., Gilsdorf, J.M., Hygnstrom, S.E., *et al.* (2006) Green and blue lasers are ineffective for dispersing deer at night. *Wildlife Society Bulletin* **34**, 371–374.

Völk, F, Glitzner, I. and Wöss, M. (2001) Kostenreduktion bei Grünbrücken durch deren rationellen Einsatz. Kriterien – Indikatoren – Mindesstandards. *Schriftenreihe des Bundesministerium für Verkehr, Innovation, Technologie u. Stassenforschung, Vienna* **513**, 112 pp.

Voss, H. (2007) *Unfallhäufungen mit Wildunfällen – Modellversuche im Oberbergischen Kreis*. [In German] Gesamtverband der Deutschen Versicherungswirtschaft e.V. Berlin: Unfallforschung der Versicherer.

Ward, A.L. (1982) Mule deer behavior in relation to fencing and underpasses on Interstate 80 in Wyoming. *Transportation Research Record* **859**, 8–13.

Waring, G.H., Griffis, J.L. and Vaughn, M.E. (1991) White-tailed deer roadside behavior, wildlife warning reflectors and highway mortality. *Applied Animal Behaviour Science* **29**, 215–223.

Witmer, G.W. and de Calesta, D.S. (1992) The need and difficulty of bringing the Pennsylvania deer herd under control. *Proceedings of the Eastern Wildlife Damage Control Conference*, **5**, 130–137.

Woodard, T.N., Reed, D.F. and Pojar, T.M. (1973) Effectiveness of Swareflex wildlife warning reflectors in reducing deer-vehicle accidents. Internal Report. Colorado Division of Wildlife, Fort Collins, Colorado.

Wu, E. (1998) Economic analysis of deer-vehicle collisions in Ohio. In *Proceedings of International Conference on Wildlife Ecology and Transportation*, held Fort Myers, Florida. Washington, DC: Federal Highway Administration, pp. 43–52.

Zacks, J.L. (1986) *Do White-tailed Deer Avoid Red: An Evaluation of the Premise Underlying the Design of Swareflex Wildlife Reflectors*. Transportation Research Record 1025. Washington, DC: Transportation Research Board, National Research Council.

9

Large herbivores as 'environmental engineers'

CHRIS SMIT AND RORY PUTMAN

9.1 Large herbivores as 'keystone species' in ecosystems

It is clear from many chapters in this book that large herbivores may have a very significant impact in the ecological systems of which they are a part – with a major shaping effect on the structure and dynamics of vegetational systems – even at a landscape scale. These effects include consumption of vegetation, redistribution of nutrients via deposition of urine and dung, soil compaction and erosion via trampling, dispersal of seeds via fur or dung, and secondarily, through these effects on the vegetation, alteration and creation of habitats or microenvironments for other plants and animals.

Some of these effects are direct and straightforward: consumption of vegetation by large herbivores obviously leads to profound changes of the morphology of individual plants (height, stature, structure), but also to changes of the physical three-dimensional architecture of the community. The stunted and 'hedged' shrubs in the understorey of grazed woodlands or the distinct browseline on the underside of the canopy trees form nice examples. Consumption also leads to changes in species composition, with elimination from the community of grazing-intolerant species, and increase of species which are tolerant to defoliation, or have specific defences against attack: spines, thorns, or chemical defences rendering them less palatable (Callaway *et al.*, 2000). Together with dunging and trampling, consumption further creates gaps within closed swards for the establishment of ephemerals.

More indirectly, consumption of dominant vigorous plant species reduces competition for light and nutrients for weaker competitors and thus generally enhances the overall plant species richness through the phenomenon of 'competitive release' (Grant, 1972; Milchunas *et al.*, 1988; Olff and Ritchie,

Ungulate Management in Europe: Problems and Practices, eds. Rory Putman, Marco Apollonio and Reidar Andersen. Published by Cambridge University Press.

1998; Rook *et al.*, 2004). Too heavy a grazing pressure may have reverse effects, driving the community towards a species-poor assemblage of a few hardy and resistant species; intermediate grazing pressure is likely to lead to maximum diversity (see also Connell, 1978).

While it is generally well known that selective grazing and trampling by large herbivores may increase diversity in this way (by suppression of dominant plant species in favour of others), it is perhaps less well appreciated that facilitation (i.e. positive interaction between species) is also strongly affected by large herbivores. Particularly unpalatable species – physically (spines and thorns) or chemically (alkaloids or high lignin/cellulose content) defended species – may act as grazing refuges in the landscape with positive effects for the heterogeneity in the vegetation structure, flowering, seed set and establishment of several plant species, and overall plant species richness (Olff *et al.*, 1999; Rousset and Lepart, 1999, 2000; Milchunas and Noy-Meir, 2002; Bakker *et al.*, 2004; Smit *et al.*, 2005, 2006, 2007). The importance of these grazing refuges for conservation of species richness increases with increased herbivore densities (Callaway *et al.*, 2000, 2005); however, when herbivore densities are too high, the protective effects of unpalatable plants disappear because they get more and more damaged by trampling and browsing (Smit *et al.*, 2007, Vandenberghe *et al.*, 2008).

By feeding in one place and dunging in another, large herbivores create discontinuities in nutrient flows through the system – and the fact that many of the system's nutrients are taken out of circulation for a period (retained in the body tissues of the herbivore itself until it dies) imposes further heterogeneity in nutrient availability. Herbivores may also affect the productivity of the vegetation grazed: while heavy levels of grazing or browsing may suppress growth rates, lighter levels of off-take commonly result in an actual increase in productivity, stimulating production of side shoots, unfurling of new leaves (potentially leading to 'grazing lawns', McNaughton, 1984). All these changes in turn change the plant species composition and relative abundance at all levels; the importance of grazing in maintaining and restoring small and large scale heterogeneity of structure and species composition in plant communities has been thoroughly described by Bakker (1998).

However, the effects of grazing on the community do not stop at the level of the vegetation. The altered vegetation composition and structure generally also affects the abundance and diversity of invertebrates and small mammals, and thus in turn affects the diversity of raptors or mammalian predators (see Section 9.3, and also Hirons, 1984; Hill, 1985; Tubbs and Tubbs, 1985; Putman, 1986, 1996; Putman *et al.*, 1989; Petty and Avery, 1990; Feber *et al.*, 2001; Flowerdew and Ellwood, 2001; Fuller, 2001; Stewart, 2001).

Responses to grazing of this kind – with changes in species composition and relative abundance of invertebrates, small mammals and birds – illustrate only too clearly that changes in grazing intensity in any ecological system will have implications far beyond the immediate consequences for the vegetation itself or upon direct competitors, which is why they are considered as 'keystone species' (*sensu* Paine, 1969) in natural ecosystems.

In a European context, for example, Vera has argued convincingly that now-extinct large herbivores (aurochs, tarpan) may have had a major impact in the development of both structure and composition of woodlands by keeping open for longer the clearings within temperate forests initially created by death and tree-fall of mature individual canopy trees – thus facilitating the persistence within the understorey of shade-intolerant species such as, for example, oak (*Quercus robur* and *Q. petraea*) and hazel (*Corylus avellana*) and contributing in this way to maintenance of diversity in the regenerating canopy (Vera, 2000; Vera *et al.*, 2006). Vera considers that the shifting pattern of herbivory, in association with senescence and tree-fall, also contributes to a horizontal diversity within such forests, with a shifting pattern over time of new glades, old clearings now abandoned and thus available as establishment areas and areas of fully established mature canopy woodland. Although Vera's theory that Europe's primeval lowlands were not closed forests but half-open park-like landscapes is under debate (Svenning, 2002; Birks, 2005; Mitchell, 2005; Szabo, 2009), its influence on present-day grazing management in Europe is undoubtedly large.

In fairness, many of the more extreme effects of herbivores on vegetation or on the structure and dynamics of the wider community have been recorded in what are, perhaps, rather artificial situations (where for example, herbivores are maintained at rather higher densities than might have occurred in truly 'natural' systems with a full complement of predators and competitors; see, for example, Putman, 1986, 1996) – and evidence for a wider effect in truly natural systems is surprisingly scant. Whether or not under wholly natural conditions large herbivores may truly be claimed as 'keystone species' in any natural system is thus still a matter of some debate (and see, for example, Bond, 2001).

9.2 Deliberate manipulation of herbivore populations to effect a change in the dynamics and structure of the managed ecosystems

Whatever may be their potential role in fully natural systems, the fact remains that in many human-managed or human-modified systems large ungulates can have a major impact on ecosystem structure or dynamics. If large herbivores can have such a significant and far-reaching effect in shaping the structure and composition of the vegetational systems, it follows that a change in the density of one or another species, removal of a species altogether, or alternatively the addition to the community of a completely new species, may itself have a profound effect on the subsequent vegetational dynamics of that system – offering managers a potentially very powerful tool for deliberate manipulation of vegetational systems to some preferred, alternative steady state.

Over the last 25 years the popularity of exploiting the impacts of large herbivores as a management tool in this way in (semi)natural and cultural ecosystems has increased considerably in Europe. The ambition is generally:

- to try to suppress dominance of particularly aggressive plant species (such as *Molinia caerulea* or *Deschampsia* sp.) within the field layer (van Wieren, 1991; Bokdam and Gleichman, 2000; Cresswell, 2008)
- to engineer a greater heterogeneity of structure (heterogeneity of sward height and in creation of greater patchiness within the overall community mosaic) in order to support a greater overall biodiversity within the community itself (chiefly of plants and invertebrates) (Bakker *et al.*, 1983, 1984; Bakker, 1989, 1998; Olff and Ritchie, 1998; Olff *et al.*, 1999; Rook *et al.*, 2004; Evans *et al.*, 2006; Dennis *et al.*, 2008).
- to try and develop a particular species composition or structure within the community to favour particular target species, for example of lepidoptera (Feber *et al.*, 2001; Öckinger *et al.*, 2006; Schtickzelle *et al.*, 2007) or birds (e.g. Vulink, 2001; Batáry *et al.*, 2007).

In some instances, introductions of large herbivores may also be directed towards the active conservation of threatened breeds or races of the animals themselves.

Despite the current popularity of the approach, however, we still know relatively little about the processes and mechanisms that are involved, or how these may individually be controlled or enhanced in management terms. Managers often manipulate herbivore impacts in one way or another – but because they are managers, not scientists, they often fail to monitor the effects in detail, simply modifying management by trial and error until they achieve what they want.

We can distinguish two types of herbivory in ecosystems: 'natural' and 'cultural' grazing. While both types use large herbivores as 'keystones' or 'environmental engineers' to try to reach various management goals, these types do differ in essence from each other.

Natural grazing focuses on the enhancement of natural processes and associated heterogeneity and biodiversity in the landscape, while keeping human impact and interference to a minimum. Large herbivores – wild ungulates or (re)introduced semi-wild ungulates such as Konik horses and Highland cattle – are present year-round at densities that are generally based on the availability of food during winter months (if applicable). This type of (mostly low intensity) grazing is often applied in nature areas, forests and 'new wilderness' nature development projects.

Cultural grazing aims at restoring and conserving particular habitats and species that once evolved under centuries of extensive human land use. These species, habitats and landscapes are presently disappearing rapidly due to intensification or abandonment of management. In order to preserve or restore these habitats and adapted species it is necessary to continue or restore the traditional management and land use. It includes (seasonal) grazing by domestic cattle, horses, pigs, sheep and goats, and generally occurs at higher intensities than at natural grazing. Here, human interference is highly necessary and an intrinsic part of the system.

We present below a number of examples of such 'manipulations' to offer illustration and to try and explore the underlying mechanisms.

9.2.1 Modification of density of existing herbivore species

In the Highlands of Scotland, red deer (*Cervus elaphus*) are in large part influential in maintaining the 'open moorland' mosaic of heathland, acid grassland and blanket bog (although grazing impacts on these habitats are imposed also by free-ranging flocks of domestic sheep). Where densities of deer are artificially reduced below around 4 or 5 deer per 100 hectares (or where grazers are completely excluded from fenced enclosures), the reduction of browsing impact is apparent in an explosive regeneration of woodland or scrub – at least within those parts of the open hill range with suitable edaphic and microclimatic conditions and at altitudes below the effective treeline. Clearly, therefore, there remains this natural potential for succession towards woodland or scrub, which is arrested, or suppressed by unmanaged grazing levels (e.g. Scott *et al.*, 1996; Stewart, 1996; Putman, 2003, 2008). At higher grazing pressure still, we may observe a decline in both the size and integrity of heather patches (*Calluna vulgaris*) and an expansion of *Molinia caerulea* and other acid grasses within the overall moorland mosaic (Hester and Baillie, 1998; Hester *et al.*, 1999).

There have been a number of instances where landowners or land managers have engineered dramatic reductions in impacts of red deer or sheep, either by total exclusion of larger herbivores from an area, through fencing, or by actual reductions of herbivore population density (removal of sheep, increased cull levels of red deer). The effects on the vegetation of fenced exclosure, in particular, are often immediately apparent to the naked eye – but in practice, the response of the vegetation has been objectively monitored in surprisingly few of these 'natural experiments'.[1]

However, in one particular instance, the effects on the vegetation, both in recording the regeneration of broadleaved woodland tree species and in changes apparent within the composition and structure of the vegetation of areas of open moorland, have been monitored on a regular basis, following a deliberate reduction in populations of red deer (Putman, 2003, 2008; Putman *et al.*, 2006). Annual recording of vegetation structure and seedling establishment along fixed transects and within fixed quadrats has been carried out within the Creag Meagaidh National Nature Reserve in the Highlands of Scotland since 1987, over a period when deer densities were systematically reduced from densities of the order of 17.5 deer per km^2 in 1986. (Deer densities reached their lowest levels, 1.27 deer per km^2, in 1990/1991, and although numbers have risen somewhat since, deer populations are still kept at minimum presence through sustained and heavy culling; Putman, 2003; Putman *et al.*, 2005.)

In this case, reductions in herbivore impacts were deliberately undertaken in order to engineer a change in the vegetation – specifically to encourage woodland regeneration in those parts of the reserve capable of sustaining native woodland cover; and during the period of reduced deer presence, there was indeed observed a strong pulse of recruitment of saplings of birch, rowan and willow (Putman, 2003). To a large extent, this pulse consisted of the 'escape' of an existing cohort of recruits already present within the vegetational matrix, but simply suppressed by repeated browsing, so that they had remained below the surrounding heather canopy and had never previously had the opportunity to grow on beyond this level. In addition to this 'escape' of existing but suppressed saplings, there was also apparent a small annual additional recruitment of new seedlings, but this was relatively small, and rates of recruitment declined over the years as (through lack of disturbance from hooves and mouthparts) litter layers became more and more densely packed (Stewart, 1996; Lamont, 1998).

Changes recorded in open hill sites where tree regeneration would not be expected were less marked. Of the permanent quadrats established and recorded from 1988 to 2002, there would appear to have been a general increase in the percentage cover of heather (as *Calluna vulgaris*); approximately equal

[1] It should be stressed that these reductions in grazing impact have been undertaken for a variety of reasons, and not always with the main intention of achieving a change in the vegetation.

numbers of sample plots showed an increase or a decline, or no net change in dominance of the grass *Molinia caerulea* and no consistent trends could be detected among other plant groups (Putman, 2008).

The effects on vegetation of altering red deer densities by culling and exclosure treatments and subsequent impact on plant species composition, structure and flowering of plant communities have also been studied on the Isle of Rum in north-west Scotland (Virtanen *et al.*, 2002). In this study, effects were compared between different densities and also between productive and unproductive vegetation types. While the impact of the culling and exclosure treatments was pronounced on more productive grasslands, with lower deer densities leading to a decline in species richness and loss of rosette-forming low-growing species such as *Carex flacca*, *Hypochaeris radicata*, *Pilosella officinarum* and *Plantago lanceolata*, the impact on unproductive grasslands was negligible. Despite some limitations of this observational study, with potential confounding factors of soil, microclimate and herbivore assemblage (cattle, ponies and goats where also present), it appears that, particularly in productive plant communities, exclusion or reduction of deer densities can lead to a decrease of plant species richness and dominance of only a few species.

9.2.2 Addition of new species to the community

In order to break up the dense mat of litter and reduce the canopy-dominating effect of coarse grasses such as *Molinia caerulea* and *Deschampsia flexuosa*, populations of free-ranging cattle are often introduced to these same upland systems. Once again, the effects of such introductions on the vegetation are relatively rarely quantified. However, regular monitoring of vegetational changes following the introduction of cattle to an upland grazing system at Cruach on Rannoch Moor has been carried out on behalf of the Rannoch Trust by Cresswell (2008). Cattle were introduced in 1989 to a moorland grazing system (heathland, acid grassland, blanket bog) already supporting populations of native red deer.

We choose to cite this particular example since it deals with exactly the same vegetational system as we have just discussed above – but in this case subject to a different manipulation: additional of an extra species of grazer rather than reduction of existing herbivore densities. In this case, annual monitoring of fixed point quadrats showed no significant increase overall in heather cover (although increases were apparent in some parts of the site). However, significant declines were noted in both dominance and percentage cover of *Molinia* in all plots (Cresswell, 2008). Once again, no consistent trends were apparent in any other taxa.

There are also several instances in Continental Europe where large herbivores have been reintroduced *de novo* in order to restore typical vegetation types, plant species and species richness of 'cultural systems' (heathlands, moorlands, wood pastures), which were themselves perhaps initially shaped by a tradition of extensive livestock grazing, but which have subsequently deteriorated through reduction or complete removal of former grazing schedules.

In the Netherlands, for example, it is common to restore or rejuvenate *Calluna* heather in grass-rich heaths by (re)introducing traditional sheep grazing, commonly combined by sod cutting; year-round grazing by free-ranging cattle is also becoming increasingly common. In one instance effects of introduced free-ranging cattle on vegetation development, tree recruitment and plant species richness was particularly well monitored via permanent plots and exclosures in grass-rich heathlands during a 10-year period (Bokdam and Gleichman, 2000). Grazing with *c.* 0.2 animals/ha led to an increase of plant species richness in the first five years and stabilised in the last five years. Pine and birch encroachment occurred regularly in the open heathlands, which was removed by the site manager. Without this additional cutting, cattle grazing alone leads to a more complex, dynamic, tree–grass–heather mosaic.

Since the outbreak of the rabbit haemorrhagic disease virus (RHDV) in the late 1980s and 1990s in Europe, which led to dramatic declines of rabbit populations by 60–90% (Drees *et al.*, 2007; Forrester *et al.*, 2006), managers also try to replace rabbit grazing by introducing large grazers to avoid extensive shrub and grass encroachment and potential loss of specific open habitats or plant species. In several coastal dune areas in the Netherlands, for example, Highland cattle, often combined with Konik and Exmoor horses, were introduced or existing densities were increased for this purpose (Aptroot *et al.*, 2007; van Wijk *et al.*, 2006). In some cases, the succession from species-rich herb towards species-poor shrub vegetation in the absence of rabbits was indeed retarded by inclusion of large herbivores, but negative effects (local species loss due to nutrient enrichment or trampling) are also reported. Fortunately, rabbit populations are now recovering in most areas in the Netherlands as they become more resistant to this virus (data from Network Ecological monitoring, VZZ Dutch Mammal Society).

Low intensity grazing (~0.4 animal/ha) has also been commonly introduced in various Dutch 'nature development projects' – where attempts are made not so much to 'restore' old semi-natural habitats but to 'create' more naturalistic habitats on abandoned agricultural land or in floodplains, with the aim being to create half-open heterogeneous and species-rich landscapes with mosaics of short grassland, tall swards, shrubs and trees – similar to the ancient grazed woodlands.

Often, the aim is also to reintegrate the abandoned agricultural fields into adjacent larger nature areas or forests via introduced grazers. This approach appears quite successful: sharp boundaries between agricultural fields and nature reserves disappear and vegetation mosaics consisting of short grassland, tall swards, shrub thickets and associated tree recruitment may develop quite rapidly on the former homogeneous agricultural fields and floodplains (Kuiters and Slim, 2003; van Uytvanck *et al.*, 2008a, 2008b). However, the time span at which such half-open heterogeneous landscapes develop depends largely on the soil types. For example, a lush mosaic with patchy shrubs and associated trees occurred within 10 years at the richer riverine floodplains, but it took 20–30 years before any shrub appeared at the poorer Pleistocene sandy soils (Kuiters, 2004).

Another typical example of reintroduction of (cultural) grazing as a restoration tool for plant communities comes from the characteristic dry alvar grasslands of the island Öland of south-east Sweden. These alvar grasslands contain species-rich limestone plant communities on dry, shallow, nutrient-poor soils. Traditional alvar management consisted of livestock grazing (sheep and cattle) in combination with removal of juniper scrub for firewood, while no further significant agricultural activities or fertiliser application ever took place. Long-term abandonment of the traditional grazing and shrub removal (over some 80 years) led to encroachment of *Juniperus communis* and the subsequent disappearance of the typical alvar species diversity, also from the mostly transient or short-lived seed bank (Bakker *et al.*, 1996). Reintroduction of cattle and sheep grazing, in combination with shrub removal, resulted in the recovery of species richness (Rosen and Bakker, 2005).

Year-round or seasonal grazing by cattle or sheep within woodlands is also becoming an increasingly well-established management practice for much the same purpose – in restoration of wood–pasture systems or more specifically targeted for the effect of their selective grazing impact on *Molinia* or *Deschampsia* or other aggressive grasses which might otherwise tend to outcompete shade-sensitive plant species; and in addition for the impact of trampling in breaking up dense mats of leaf litter and creating regeneration niches for tree seedlings (see, for example, Mitchell and Kirby, 1990; Armstrong, 2008).

In the Veluwe area of the central Netherlands, Highland cattle (1 animal/15 ha) were introduced 25 years ago with the main objective of tackling the large build-up of organic material of *Deschampsia flexuosa* (up to 40 tonnes of dry material per hectare). Since their introduction, the litter layer reduced and spaces appeared where regeneration of heather was apparent, while in the pine forest, conditions for regeneration of Scots pine and birch seemed to improve. Over large areas across the whole system, *Deschampsia* grassland gradually transformed into shrub vegetation dominated by either *Calluna* or *Vaccinium* (Natuurmonumenten, pers. comm.). However, while it is likely that these reported vegetation changes over the years can be attributed to the introduction of cattle grazing, strong reductions in atmospheric nitrification since the 1980s also clearly played a role here. Permanent quadrats established in grazed areas and exclosures are currently being revisited in order to separate out these effects (C. Smit, unpublished).

In much the same way, domestic pigs are increasingly being used in various woodland restoration schemes in the UK and elsewhere, to mimic the effect of wild boar in removal of litter, breaking up of the vegetation mat and rooting into bare earth, to increase the number of potential regeneration sites for tree seeds and enhance the potential of a site for woodland regeneration (e.g. Bruinderink and Hazebroek, 1996). Domestic and free-ranging herbivores are increasingly seen to be beneficial for dispersal of plants: species with fur-assisted dispersal are over-represented among the declining plant species in north-west Europe (Ozinga *et al.*, 2009).

More recently, European bison have been introduced from the last remaining wild populations in Eastern Europe (Poland and Estonia) into natural areas in France, Germany and the Netherlands, and more plans are being made for introductions in other areas (Veluwe, the Netherlands; Ardennes, Belgium). The primary aim of these introductions is to contribute to the protection of this still highly endangered European bison by enlarging its area of distribution in Europe. Risk of extinction is still currently relatively high as the remaining population is highly concentrated in Eastern Europe; the expansion of European bison towards Western European countries would help to reduce this risk. In addition, however, these introductions offer excellent opportunities to study the ecology of the European bison and its impact on the environment, since the available literature and knowledge on this once nearly extinct animal is very limited.

Between 2007 and 2008 a total of six European bison were introduced in the Kraansvlak, a dune area of approximately 200 ha situated in the west of the Netherlands (Cromsigt *et al.*, 2007). It is a unique study in the sense that the behaviour, ecology and impact of these introduced animals on their environment has been carefully studied for several years. The expectation is that the bison, as Europe's largest grazer, has stronger 'environmental engineering' impact than introduced cattle and horses, as bison are expected to consume higher proportions of (bark of) woody species and to create more gaps in forests by rubbing and pushing over trees (Smit *et al.*, 2008). Preliminary results of this Kraansvlak study already seem to confirm this idea: consumption of bark of the woody *Euonymus europeus* by bison was much higher than consumption by cattle (J. Cromsigt, pers. comm.). Also, bison dung appears to contain more germinating seeds than cattle dung, and also the species that germinate from these seeds seem to differ from those from cattle dung. Ongoing monitoring and analyses will soon reveal more about the impact of this large herbivore as an environmental engineer.

9.2.3 Multi-species grazing systems may be more effective than single species systems

From the above-given examples, it becomes clear that introducing or maintaining herbivores at low densities generally has more positive impact on heterogeneity of vegetation structure, and subsequent floral and faunal diversity, than does high-density herbivory which can even have detrimental effects.

Most of these examples, however, come from single species grazing systems, while little empirical knowledge is available for multi-species grazing systems which, in fact, may exhibit more positive effects than single species grazing alone. Indeed, one example from the uplands in the UK showed that low intensity mixed cattle–sheep grazing resulted in more breeding territories of meadow pipits (*Anthus pratensis*) than low-intensity sheep grazing alone (Evans *et al.*, 2006). Although it is not yet fully understood why adding cattle led to these positive effects, it is likely to be caused by the different grazing and trampling impacts of the two herbivore species that together create higher heterogeneity in the vegetation structure than sheep grazing alone (Dennis *et al.*, 2008). Furthermore, the diversity and abundance of arthropods associated with (cattle) dung may increase, which may improve the food quality for several bird species.

Hence, introduction of a more complex, multi-species assemblage may be more effective in achieving management objectives than imposition of grazing by a single herbivore. In various parts of the Netherlands, horses have been introduced in a number of coastal sand-dune area reserves where they generally graze together with cattle or sheep (most Dutch islands and a dozen coastal dune areas in the provinces of North Holland, South Holland and Zeeland) – once again with the general aim of reducing the dominance of coarse grasses. In a number of cases the introduction of ponies has taken place after initial grazing regimes with cattle alone failed to cause sufficient impact on coarser grass species such as *Carex arenaria* and *Calamagrostis epigejos* – and it is found that the addition of horses to the herbivore assembly better achieves the desired response. Horses also prove more effective at opening up dense thickets of *Crateagus monogyna* and *Prunus spinosa* (H. Piek, pers. comm.).

The Oostvaardersplassen nature reserve (56 km^2) is a wetland reserve, located in the young polder Zuidelijk Flevoland in the centre of the Netherlands (52°26′ N, 5°19′ E). It is designated as an SAC (European Special Area of Conservation) and RAMSAR site. To maintain open shallow pools and grass lawns as feeding ground for wetland birds, such as greylag geese and barnacle geese (*Anser anser*; *Branta leucopsis*), spoonbills (*Platalea leucorodia*), herons (*Ardea cinerea*), great white egrets (*Casmarodius albus*) and others, free-ranging populations of primitive breeds of both cattle and horses were introduced in 1983 and 1984 (Heck cattle, a phenotypic reconstruction from a number of early breed types of the extinct aurochs *Bos primigenius*, and Konik horses, a breed believed to be little modified from the forest tarpan (Cornelissen and Vulink, 2001)). Populations are unmanaged (except for humane destruction of animals *in extremis*) and entirely self-sustaining.

However, when grazed only by cattle and ponies, it was noted that elder (*Sambucus nigra*) showed significant expansion in abundance and distribution within open areas (Vulink *et al.*, 2000; P. Cornelissen, pers. comm.). In response to this, red deer (*Cervus elaphus*) were also introduced to the system (in 1992) deliberately to target this encroachment of open areas by elder, which by 1990 covered large parts of the border zone at the edge of the reserve (Cornelissen and Vulink, 2001; Vulink, 2001).

Frequency, density, canopy cover and height of *Sambucus* were assessed in 1996 and 2002. While *Sambucus* had expanded rapidly under a grazing regime imposed only by cattle and horses, both density and overall cover of *Sambucus* within the reserve showed a significant decline over the six-year study period following the introduction of red deer, with debarking by deer appearing to be the main factor responsible for the observed mortality. The extent to which this was due to the specific impact of the addition of a different species (red deer) to the existing system or due simply to increased competition among herbivores as a result of increased herbivore numbers overall is not clear, but direct observations of behaviour of the herbivores and the observed bite marks on *Sambucus* stems during field work suggested that certainly the majority of this bark stripping was carried out by red deer (P. Cornelissen *et al.*, in preparation).

As populations have increased in size over the years (570 Heck cattle, 1140 horses and 2320 deer in winter 2008) significant impacts have been recorded on open vegetational systems (dry grassland, wet grasslands, *Phragmites* reedbeds). Thus there has been a gross change in the relative proportion of the entire area occupied by the different habitat types (an overall increase in the proportional area of dry grassland, a decrease in the percentage area of reedbeds and comparative stasis in wet grassland or tall herb communities: Vulink and van Eerden, 1998; Cornelissen and Vulink, 2001; Cornelissen, 2006) as well as a change, within each community type, in the actual physical structure of the sward (with an increase in each case of the proportional area of short grass or 'grazed reedbed'; Vulink *et al.*, 2000; Vulink 2001).

In general, increasing the number of species of grazers in managed reserves is mainly through the introduction of large grazing herbivores in systems where either no other ungulates are present or where only smaller species of herbivores may occur. In many areas in central Netherlands, however, cattle (and sometimes horses) have been introduced where roe deer, red deer, fallow deer and wild boar already occur. Dune areas to which grazing by cattle and/or horses has been added were also commonly already grazed by roe deer or fallow deer. Similarly, the European bison has recently been introduced in an area where fallow deer, roe deer and rabbits were already present. The reason for these additions of larger domestic or wild herbivores is essentially due to a recognition of the specific effects that come through the grazing (and trampling) effects of true large grazers, effects that are not apparent with the presence of the smaller herbivores only. Furthermore, the presence of different herbivore species in a system may favour 'niche differentiation': each species takes its own niche in the landscape due to the competition for resources between the different herbivore species. This process is not occurring when only a single herbivore species is present in a system. Thus, it is possible that the clearest effects of grazing are reached – and declared management goals achieved – when a more complete herbivore assemblage is present.

We might note, however, that introductions of additional herbivores may equally have negative effects on population density and performance of species already present within an area. In one instance, the introduction of red deer, mouflon and wild boar in two new forest areas at the Hoge Veluwe National Park in the Netherlands, until then exclusively inhabited by roe deer, negatively affected the density and performance of roe deer populations (Smit, 2002). Direct competition between roe deer and red deer for woody forage seems to be responsible for these changes, but the reduction of habitat quality through reduced woody understorey may also have played an important role. Thus any such introductions must be undertaken with some caution. While managers mostly aim at changing heterogeneity of vegetation composition or species diversity, such alterations of herbivore interactions may have unforeseen consequences.

9.3 Effects on other animals

As our last example already illustrates, manipulating herbivore composition or density not only has consequences for the vegetation but also for many associated animal groups. In many instances, these higher-order effects on the fauna may be incidental (although managers should of course be aware of the wider influences of grazing regimes manipulated to achieve some other purpose); in some cases however, grazing regimes may be introduced or manipulated specifically for the effects that this may have on animal diversity. Effects may be indirect, through alterations of the vegetation composition and structure, or more direct, for example in enhancement of dung-feeding invertebrates (e.g. Putman *et al.*, 1989; Stewart, 2001). Below we give examples of the potential power of such relatively simple manipulations of large herbivore populations for the associated animal community (invertebrates, smaller mammals, birds – and their predators).

In the extreme, very heavy grazing generally has a rather depressive effect on diversity overall. In the New Forest of southern England, free-ranging cattle and ponies, together with populations of red, fallow, sika and roe deer have imposed a regime of extremely heavy grazing over many centuries (see, for example, Putman, 1986, 1996). As a result the New Forest grasslands and heathlands clearly lack many of the possible structural vegetation layers. The woodlands lack virtually *any* ground, field or shrub layer; and indeed the whole structural 'layer' between ground level and the clear browse horizon at 1.8 m is almost completely missing.

Work undertaken in the early 1980s in comparison of the species diversity and population sizes of small mammals within the New Forest with those recorded in equivalent vegetation types in areas outside the forest boundary (grazed by deer but not by domestic livestock), reveals striking and consistent differences (Hill, 1985; Putman, 1986; Putman *et al.*, 1989). All ungrazed woodland areas studied supported substantial populations of woodmice *Apodemus sylvaticus* and bank voles *Clethrionomys glareolus*, with lower densities recorded of yellow-necked mice *Apodemus flavicollis*, and both common and pygmy shrew (*Sorex araneus*, *S. minutus*). Rodent communities of grazed woodlands within the New Forest were characterised by healthy populations of woodmice, but all other species were rare or absent (Hill, 1985). The New Forest's heathlands and grasslands are equally profoundly affected by reduced structural 'depth', providing scant cover from predation. While heathland plots beyond the forest boundary supported large, permanent populations of woodmice and harvest mice (*Micromys minutus*), and grasslands in turn supported strong populations of woodmice and field voles

(*Microtus agrestis*) small mammals were almost completely absent from open communities within the New Forest itself.

The reduced diversity and overall abundance of small mammals in the heavily grazed woodlands and open 'wastes' have been shown in their turn to have had an effect on the species composition and foraging behaviour of predators. Two independent studies of the diets of foxes (*Vulpes vulpes*) within the New Forest (Senior, unpublished data, quoted in Putman, 1986; Farley, 1986) reveal that while New Forest animals did consume small rodents when available, the frequency and relative proportion in the diet was lower than that recorded in other areas: few birds were taken and the foxes clearly rely heavily on invertebrate material (particularly earthworms and beetles), carrion and autumn fruit. Perhaps in response to scarcity of prey the overall density of foxes within the forest is also unusually low: estimated as only 2 per km^2, with an adult density of 0.75 per km^2 (Insley, 1977).

Avian predators are also affected by the low rodent abundance. The breeding attempts and breeding success of buzzards (*Buteo buteo*) in the New Forest was negatively correlated with the population density of grazing cattle and ponies (Tubbs and Tubbs, 1985). Similar effects have been demonstrated for other raptors within the New Forest whose diets would normally be expected to contain high numbers of rodents. Tawny owls (*Strix aluco*) compensated for low availability of rodents with increased intake of invertebrate prey, particularly beetles (*Geotrupes* and *Typhoeus* species), but the overall density of individuals and number of breeding pairs of owls is significantly lower within the New Forest as compared with areas outside (Hirons, 1984).

Other studies too have recently emphasised that heavy grazing by wild deer, within woodlands in particular, may seriously impact upon the abundance and breeding success of a number of woodland bird species (especially shrub-breeding and insectivorous species such as nightingales *Luscinia megarhynchos*; Fuller, 2001), although there are equally species which may derive positive advantage from such heavy grazing. Wood warblers (*Phylloscopus sibilatrix*), pied flycatchers (*Ficedula hypoleuca*) and redstarts (*Phoenicurus phoenicurus*) all depend on the park-like conditions of traditional wood pastures (Stowe, 1987; Mitchell and Kirby, 1990).

For example, research undertaken by the British Forestry Commission (Petty and Avery, 1990) has shown that immediately after the cessation of grazing in upland areas about to be afforested, populations of field voles (*M. agrestis*) show an initial surge in population density. This increased food supply is exploited by predators such as short-eared owls *Asio flammeus*, long-eared owls *Asio otus*, kestrels *Falco tinnunculus*, and barn owls *Tyto alba*, whose breeding success and population sizes also increase until canopy closure within the new plantations (see Evans *et al.*, 2006; Wheeler, 2008 who also demonstrate an increase in densities of *M. agrestis* in direct response to reduction of sheep densities in upland areas of the UK).

In general, reduction in grazing intensity has positive effects on many species groups, but not on all. Abundance and species richness of butterflies and grasshoppers, as well as abundance of some taxa of ground-dwelling arthropods, increased after a reduction in grazing intensity in semi-natural grasslands in the UK, Germany, France and Italy (Wallis de Vries *et al.*, 2007). Also, the species richness of nectar-seeking butterflies and bumble bees declined with increasing grazing intensity in Sweden (Soderstrom *et al.*, 2001). Similar positive results were found for reduced sheep grazing in upland UK on spiders, bugs and beetles, while no such effects were found for brachyceran flies, caterpillars and craneflies (Dennis *et al.*, 2008). In particular, foliar arthropods, an important food source for many bird species, profit from reduced stocking densities.

But while availability of nectar for flower-feeding species of invertebrates often declines with increasing grazing pressure, in some contexts grazing is positively beneficial to a number of species of woodland butterflies in maintaining open areas for basking and/or maintaining a suitable ground flora of larval food plants (see, for example, Petley-Jones, 1995; Feber *et al.*, 2001). Altering game densities of mouflon and red deer had clear impact on butterfly populations in the Czech Milovicky woods. 'Overstocking' at densities of 1.02 animals per ha in the 1980s led to disappearance of some specialists and common species from the forest meadows, but at the same time these high densities also maintained a degree of openness in the forests which supported some open woodland specialists (Benes *et al.*, 2006). In general, opening up of shady high forests by coppicing or increasing numbers of large herbivores would appear to have positive effects for woodland butterflies.

9.4 The future potential of grazing manipulations

Grazing manipulations are increasingly being utilised in practical conservation management in many countries. Denmark and Sweden already have a long tradition of grazing in their nature reserves and as noted, grazing management by sheep, cattle and pigs is becoming more widespread within the UK. In Belgium many reserves are grazed in a similar way as in the Netherlands, while in Germany plans are being made to develop large areas with 'naturalistic' grazing in Brandenburg. The European bison has been introduced in France (Sainte-Eulalie), Germany (Hardehousen and Damerower Werden) and Sweden (Eriksberg) and the possibility of reintroductions in large nature reserves in the Netherlands (Veluwe and Oostvaardersplassen) and Belgium (Ardennes) is being considered.

It is important, however, to recognise that use of large herbivores in this way is no panacea . . . and imposition of an appropriate grazing regime does not solve all conservation problems! In effect, while grazing regimes may be aimed at delivering a high structural diversity within vegetation and thus a high intrinsic biodiversity, this in itself does not guarantee that all the management goals for a given site will be met. The main reason for this is that, next to a more general 'biodiversity' goal, there is almost always also a 'species' goal. External policy documents such as the Habitats Directive and the Bird Directive of the EU (respectively Directive 92/43/EEC on the Conservation of Natural Habitats and Directive 79/409/EEC on the Conservation of Wild Birds), the RAMSAR convention or, more recently, the Natura 2000 requirements often set targets not only for overall biodiversity or species richness but also for the maintenance of populations of specific target species within broad population limits.

In many instances these objectives are in direct conflict, or at the very least may not be deliverable under a simple, single, grazing regime. In the Zwanewater, a 600-ha dune area in the Netherlands, management must seek to maintain existing populations, or engineer an increase in population for 10 species of raptors, 9 species of colonial breeding birds, 1 amphibian species and 2 mammal species, while also numerous specified vegetation types have to be maintained or developed. A simple grazing regime with large herbivores (currently cattle and horses) cannot ensure that all these goals can be met, with the result that some species are decreasing and others are increasing.

In a similar way, at the Junner Koeland, a 100-ha floodplain nature reserve in the Netherlands, present management tries to combine a 'species approach' (maintenance of river dune vegetation with *Dianthus deltoides*, *Thymus pulegioides* and *Carex caryophyllea*) with a 'system approach' (maintenance of shifting mosaics of grasslands, shrubs and trees in space and time). Although these rare shifting mosaics typically occur at low grazing intensities ($<$ 0.4 cows/ha), grazing intensity was recently much increased and 15 Iceland ponies were added to the system with the aim of enlarging populations of some river dune plants. While it is uncertain whether this last goal can be met by increased grazing intensity (yearly flooding seems to be the most important factor for dynamics of river dune vegetations), it is obvious that this increase will not do much for the development of shifting mosaics and associated species.

In most cases when cultural grazing is applied, therefore, additional management may be required (such as mowing, removal of woody regeneration, cutting sods and burning; see again, for example, Bakker, 1989; Bokdam and Gleichman, 2000). In effect, both science and practical experience make it clear that when the management goals are many and specific, grazing can only contribute a partial solution to management and should be considered as one element only of a wider management package.

The less specified the goals and the more 'system-oriented' the management aim may be – typically when natural or unmanaged grazing systems are applied – the more easily the grazing effects (or the lack thereof) can be accepted. A system approach is applied in the Oostvaardersplassen (see above) and the Veluwezoom National Park, a 5000-ha area of heathland and forests in central Netherlands. In the Oostvaardersplassen the evaluation of the management is positive because the system is always right!

Despite the increased popularity of the idea of exploiting grazing as a powerful management tool in manipulating vegetational systems, there are as yet relatively few accepted general rules as to which species (or species combination), at which density, and in what seasonal period best deliver a given outcome in any particular set of circumstances. As we have already hinted, relatively few manipulations have been subjected to any formal or objective programme of monitoring and for the most part the development of a programme of grazing by large herbivores in any given situation remains a process of trial and error (or perhaps an exercise in rediscovery of the wheel! See also Armstrong, 2008). Indeed, from the grazing research that *has* been carried out in the past 25 years (mainly focused on grassland ecosystems and woodlands), only two general insights emerge. The most important of these is that there seems to be a humpshape relationship between grazing intensity and heterogeneity of vegetation structure with subsequent overall biodiversity (frequently plant species richness) (Milchunas *et al.*, 1988, van Wieren and Bakker, 2008), while the second is that the effects of grazing are much more pronounced when soil nutrient levels are high (Bakker *et al.*, 2006).

We feel that given the potential of such manipulations of grazing impact, and given the already widespread use of this as a conservation tool, there is an urgent need for a more formal, research-driven set of protocols which are more truly science based . . . and by the same token, an urgent need for a more formal programme of scientific study to underpin such application. Management actions based on scientific principles can be much better justified than trial and error alone.

We ourselves are aware (see also Armstrong, 2008) that despite the growing popularity of the technique, and the increasing number of examples of use of herbivores to deliver specific management aims,

- much of this is not properly carried out
- much of it is achieved through trial and error rather than real understanding of the processes going on (whereas if we had better understanding, this might lead more quickly to effective solutions, and avoid repeating the same errors!)
- many grazing systems are continually altered and changed from season to season, without waiting to see what might be the long-term effects of the original prescription
- in many cases the actual effects of a given grazing regime are not adequately monitored or recorded.

We believe that a more scientific approach is required if we are ever to develop proper, science-based approaches to achieving the objectives sought. Managers themselves can help a great deal to meet this need and it often requires relatively simple action. First of all, managers should be encouraged fully to document their management actions undertaken over time, including basic information such as number and type of animals, length of grazing season, size of areas and dominant vegetation type (preferably via a dozen permanent plots). Recordings of vegetational structure and diversity should be recorded both at the outset and after a change in the management. Given the number of 'experiments' which are being undertaken as managers increasingly exploit grazers to achieve particular vegetational ends, such improved recording would potentially offer an excellent information base – but could be made more factual and objective.

References

Aptroot, A., Van Dobben, H.F., Slim, P.A. and Olff, H. (2007) The role of cattle in maintaining plant species diversity in wet dune valleys. *Biodiversity and Conservation*, **16**, 1541–1550.

Armstrong, H.E. (2008) The impacts of cattle in woodlands. In W. Cresswell *et al.* (eds.) *Grazing Management of Upland Habitats for Nature Conservation*. Stroud, UK: Cresswell Associates/The Rannoch Trust.

Bakker, E.S., Olff, H., Vandenberghe, C., *et al.* (2004) Ecological anachronisms in the recruitment of temperate light-demanding tree species in wooded pastures. *Journal of Applied Ecology* **41**, 571–582.

Bakker, E.S., Ritchie, M.E., Olff, H., Milchunas, D.G. and Knops, J.M.H. (2006) Herbivore impact on grassland plant diversity depends on habitat productivity and herbivore size. *Ecology Letters* **9**, 780–788.

Bakker, J.P. (1989) *Nature Management by Grazing and Cutting*. Dordrecht, Netherlands: Kluwer Academic Publishers.

Bakker, J.P. (1998) The impact of grazing on plant communities. In M.F. Wallis de Vries, J.P. Bakker and S.E. van Wieren (eds.) *Grazing and Conservation Management*. Dordrecht, Netherlands: Kluwer Academic Publishers, pp. 137–184.

Bakker, J.P., de Bie, S., Dallinga, J.H., Tjaden, P. and de Vries, Y. (1983) Sheep-grazing as a management tool for heathland conservation and regeneration in the Netherlands. *Journal of Applied Ecology* **20**, 541–560.

Bakker, J.P., de Leeuw, J. and van Wieren, S.E. (1984) Micro-patterns in grassland vegetation created and sustained by sheep grazing. *Vegetatio* **55**, 153–161.

Bakker, J.P., Bakker, E.S., Rosen, E., Verweij, G.L. and Bekker, R.M. (1996) Soil seed bank composition along a gradient from dry alvar grassland to *Juniperus* shrubland. *Journal of Vegetation Science* **7**, 165–176.

Batáry, P., Báldi, A. and Erdos, S. (2007) Grassland versus non-grassland bird abundance and diversity in managed grasslands: Local, landscape and regional scale effects. *Biodiversity and Conservation* **16**, 871–881.

Benes, J., Cizek, O., Dovala, J. and Konvicka, M. (2006) Intensive game keeping, coppicing and butterflies: the story of Milovicky Wood, Czech Republic. *Forest Ecology and Management* **237**, 353–365.

Birks, H.J.B. (2005) Mind the gap: how open were European primeval forests? *Trends in Ecology and Evolution* **20**(4), 154–156.

Bokdam, J. and Gleichman, M. (2000) Effects of grazing by free-ranging cattle on vegetation dynamics in continental north-west European heathland. *Journal of Applied Ecology* **37**, 415–431.

Bond, W. (2001) Keystone species: hunting the snark? *Science* **292**, 63–64.

Bruinderink, G.W.T.A. and Hazebroek, E. (1996) Wild boar (*Sus scrofa scrofa* L.) rooting and forest regeneration on podzolic soils in the Netherlands. *Forest Ecology and Management* **88**, 71–80.

Callaway, R.M., Kikvidzc, Z. and Kikvidzc, D. (2000) Facilitation by unpalatable weeds may conserve plant diversity in overgrazed meadows in the Caucasus Mountains. *Oikos* **89**, 275–282.

Callaway, R.M., Kikodze, D., Chiboshvili, M. and Khetsuriani, L. (2005) Unpalatable plants protect neighbors from grazing and increase plant community diversity. *Ecology* **86**, 1856–1862.

Connell, J.H. (1978) Diversity in tropical rain forests and coral reefs. *Science* **199**, 1302–1310.

Cornelissen, P. (2006) *Vegetatie, grote herbivoren en ganzen in de Oostvaardersplassen. Evaluatie 1996–2005*. RIZA werkdocument nr. 2006.044X. Lelystad, Netherlands: Rijksinstituut voor Integraal Zoetwaterbeheer en Afvalwaterbehandeling, RIZA.

Cornelissen, P. and Vulink, J.T. (2001) Effects of cattle and horses on vegetation structure. Are cattle and horses browsers enough to stop development of shrubs and trees? In B. Gerken and M. Görner (eds.). *Neue Modelle zu Masznahmen der landschaftsentwicklung mit groszen Pflanzenfressern und praktische Erfahrungen bei der Umsetzung*. Höxter/Jena, Germany: Natur- und Kulturlandschaft **4**, pp. 189–197.

Cresswell, W. (2008) A 15-year experiment to assess the impact of cattle grazing on the floral composition of Rannoch Moor. In W. Cresswell *et al.* (eds.) *Grazing Management of Upland Habitats for Nature Conservation*. Stroud, UK: Cresswell Associates/The Rannoch Trust.

Cromsigt, J., Linnartz, L., Verdonschot, R., Esselink, H. and Olff, H. (2007) De Kraansvlak pilot: wilde Wisenten in een Nederlandse context. *De Levende Natuur* **108**, 206–208.

Dennis, P., Skartveit, J., McCracken, D.I., *et al.* (2008) The effects of livestock grazing on foliar arthropods associated with bird diet in upland grasslands of Scotland. *Journal of Applied Ecology* **45**, 279–287.

Drees, J.M., Dekker, J.J.A., Lavazza, A. and Cappucci, L. (2007) *Voorkomen en verspreiding van Rabbit Haemorrhagic Disease en Myxamatose in Nederlandse konijnenpopulaties*. VZZ rapport 2007.17. Arnhem, Netherlands: Zoogdiervereniging VZZ.

Evans, D.M., Redpath, S.M., Evans, S.A., *et al.* (2006) Low intensity, mixed livestock grazing improves the breeding abundance of a common insectivorous passerine. *Biology Letters* **2**, 636–638.

Farley, A.E. (1986) Diets of foxes in the south of England. BSc Honours thesis (Biology), University of Southampton, UK.

Feber, R.E., Brereton, T.M., Warren, M.S. and Oates, M. (2001) The impacts of deer on woodland butterflies: the good, the bad and the complex. *Forestry* **74**, 271–276.

Flowerdew, J.R. and Ellwood, S.A. (2001) Impacts of woodland deer on small mammal ecology. *Forestry* **74**, 277–288.

Forrester, N.L., Trout, R.C., Turner, S.L., *et al.* (2006) Unravelling the paradox of rabbit haemorrhagic disease virus emergence, using phylogenetic analysis; possible implications for rabbit conservation strategies. *Biological Conservation* **131**, 296–306.

Fuller, R.J. (2001) Responses of woodland birds to increasing numbers of deer: a review of evidence and mechanisms. *Forestry* **74**, 289–298.

Grant, P.R. (1972) Convergent and divergent character displacement. *Biological Journal of the Linnean Society* **4**, 39–68.

Hester, A.J. and Baillie, G.J. (1998) Spatial and temporal patterns of heather utilisation by sheep and red deer within natural heather:grass mosaics. *Journal of Applied Ecology* **35**, 772–784.

Hester, A.J., Gordon, I.J, Baillie, G.J. and Tappin, E. (1999) Foraging behaviour of sheep and red deer within natural heather:grass mosaics. *Journal of Applied Ecology* **36**, 133–146.

Hill, S.D. (1985) Influences of large herbivores on small rodents in the New Forest. PhD thesis, University of Southampton, UK.

Hirons, G.J.M. (1984) The diet of tawny owls (*Strix aluco*) and kestrels (*Falco tinnunculus*) in the New Forest, Hampshire. *Proceedings of the Hampshire Field Club and Archaeological Society* **40**, 21–26.

Insley, H. (1977) An estimate of the population of the red fox (*Vulpes vulpes*) in the New Forest, Hampshire. *Journal of Zoology (London)* **183**, 549–553.

Kuiters, A.T. (2004) *Ontwikkeling van mozaieklandschappen onder invloed van begrazing; Een drietal casestudies*. Alterra-rapport 1105. Wageningen, Netherlands: Alterra.

Kuiters, A.T. and Slim, P.A. (2003) Tree colonisation of abandoned arable land after 27 years of horse-grazing: the role of bramble as a facilitator of oak wood regeneration. *Forest Ecology and Management* **181**, 239–251.

Lamont, L. (1998) Birch regeneration and deer grazing in the Scottish Highlands. MSc thesis, University of Edinburgh, UK.

McNaughton, S.J. (1984) Grazing lawns: animals in herds, plant form, and coevolution. *American Naturalist* **124**, 863–886.

Milchunas, D.G. and Noy-Meir, I. (2002) Grazing refuges, external avoidance of herbivory and plant diversity. *Oikos* **99**, 113–130.

Milchunas, D.G., Lauenroth, W.K and Sala, O.E. (1988) A generalized model of the effects of grazing by large herbivores on grassland community structure. *American Naturalist* **132**, 87–106.

Mitchell, F.J.G. (2005) How open were European primeval forests? Hypothesis testing using palaeoecological data. *Journal of Ecology* **93**, 168–177.

Mitchell, F.J.G. and Kirby, K. (1990) The impact of large herbivores on the conservation of semi-natural woods in the British uplands. *Forestry* **63**, 333–353.

Öckinger, E., Eriksson, A.K. and Smith, H.G. (2006) Effects of grassland abandonment, restoration and management on butterflies and vascular plants. *Biological Conservation* **133**, 291–300.

Olff, H. and Ritchie, M.E. (1998) Effects of herbivores on grassland plant diversity. *Trends in Ecology and Evolution* **13**, 261–265.

Olff, H., Vera, F.W.M., Bokdam, J., *et al.* (1999) Shifting mosaics in grazed woodlands driven by the alternation of plant facilitation and competition. *Plant Biology* **1**, 127–137.

Ozinga, W.A., Romermann, C., Bekker, R.M., *et al.* (2009) Dispersal failure contributes to plant losses in NW Europe. *Ecology Letters* **12**, 66–74.

Paine, R.T. (1969) A note on trophic complexity and community stability. *American Naturalist* **103**, 91–93.

Petley-Jones, R. (1995) Deer or butterflies? A woodland dilemma. *Enact* **3**(3), 8–10.

Petty, S.J. and Avery, M.I. (1990) *Forest Bird Communities*. Occasional Paper 26. Edinburgh, UK: Forestry Commission.

Putman, R.J. (1986) *Grazing in Temperate Ecosystems; Large Herbivores and their Effects on the Ecology of the New Forest*. London: Croom Helm/Chapman and Hall, 200 pp.

Putman, R.J. (1996) *Competition and Resource Partitioning in Temperate Ungulate Assemblies*. London: Chapman and Hall, 200 pp.

Putman, R.J. (2003) *An Analysis of Trends in Deer Populations and the Vegetation of Creag Meagaidh National Nature Reserve 1986–2001: Review and Update 2003*. Research Reports. Edinburgh, UK: Scottish Natural Heritage.

Putman, R.J. (2008) The effects of reducing grazing pressure from red deer and sheep on vegetation within the Creag Meagaidh National Nature Reserve. In W. Cresswell *et al.* (eds.) *Grazing Management of Upland Habitats for Nature Conservation*. Stroud, UK: Cresswell Associates/The Rannoch Trust.

Putman, R.J., Edwards, P.J., Mann, J.C.E., How, R.C. and Hill, S.D. (1989) Vegetational and faunal changes in an area of heavily grazed woodland following relief of grazing. *Biological Conservation* **47**, 13–32.

Putman, R.J., Duncan, P. and Scott, R. (2005) Demographic changes in a Scottish red deer population (*Cervus elaphus*. L.) in response to sustained and heavy culling: an analysis of trends in deer populations of Creag Meagaidh National Nature Reserve 1986–2001. *Forest Ecology and Management* **206**, 263–281.

Putman, R.J., Duncan, P. and Scott, R. (2006) Tree regeneration without fences? An analysis of vegetational trends within the Creag Meagaidh National Nature Reserve, 1988–2001 in response to significant and sustained reductions in grazing pressure. *Practical Ecology and Conservation* **6**, 52–65.

Rook, A.J., Dumont, B., Isselstein, J., *et al.* (2004) Matching type of livestock to desired biodiversity outcomes in pastures: a review. *Biological Conservation* **119**, 137–50.

Rosen, E. and Bakker, J.P. (2005) Effects of agri-environment schemes on scrub clearance, livestock grazing and plant diversity in a low-intensity farming system on Oland, Sweden. *Basic and Applied Ecology* **6**, 195–204.

Rousset, O. and Lepart, J. (1999) Shrub facilitation of *Quercus humilis* regeneration in succession on calcareous grasslands. *Journal of Vegetation Science* **10**, 493–502.

Rousset, O. and Lepart, J. (2000) Positive and negative interactions at different life stages of a colonizing species (*Quercus humilis*). *Journal of Ecology* **88**, 401–412.

Schtickzelle, N., Turlure, C. and Baguette, M. (2007) Grazing management impacts on the viability of the threatened bog fritillary butterfly *Proclossiana eunomia*. *Biological Conservation* **136**, 651–660.

Scott, D., Hirst, D., Staines, B.W. and Elston, D. (1996) *Red Deer and Tree Regeneration in the Cairngorms and Monadhliaths*. Report on contract RASD/019/95. Edinburgh, UK: Scottish Natural Heritage.

Smit, C., Béguin, D., Buttler, A. and Mueller-Schaerer, H. (2005) Safe sites for tree regeneration in wooded pastures: a case of associational resistance? *Journal of Vegetation Science* **16**, 209–214.

Smit, C., Den Ouden, J. and Muller-Scharer, H. (2006) Unpalatable plants facilitate tree sapling survival in wooded pastures. *Journal of Applied Ecology* **43**, 305–312.

Smit, C., Vandenberghe, C., Den Ouden, J. and Mueller-Schaerer, H. (2007) Nurse plants, tree saplings and grazing pressure: changes in facilitation along a biotic environmental gradient. *Oecologia* **152**, 265–273.

Smit, C., Dekker, J. and Cromsigt, J. (2008) Ruimte voor de wisent in de lage landen. *De Levende Natuur* **109**, 32–33.

Smit, R. (2002) The secret life of woody species. A study on woody species establishment, interactions with herbivores and vegetation succession. PhD thesis, Wageningen University, The Netherlands.

Soderstrom, B., Svensson, B., Vessby, K. and Glimskar, A. (2001) Plants, insects and birds in semi-natural pastures in relation to local habitat and landscape factors. *Biodiversity and Conservation* **10**, 1839–1863.

Stewart, A.J.A. (2001) The impact of deer on lowland woodland invertebrates: a review of the evidence and priorities for future research. *Forestry* **74**, 259–270.

Stewart, F. (1996) The effects of red deer (*Cervus elaphus*) on the regeneration of birch (*Betula pubescens*) woodland in the Scottish Highlands. PhD thesis, University of Aberdeen, UK.

Stowe, T.J. (1987) The management of sessile oakwoods for pied flycatchers. *RSPB Conservation Review* **1**, 78–83.

Svenning, J.C. (2002) A review of natural vegetation openness in north-western Europe. *Biological Conservation* **104**, 133–148.

Szabo, P. (2009) Open woodland in Europe in the Mesolithic and in the Middle Ages: can there be a connection? *Forest Ecology and Management* **257**, 2327–2330.

Tubbs, C.R. and Tubbs, J.M. (1985) Buzzards (*Buteo buteo*) and land use in the New Forest, Hampshire, England. *Biological Conservation* **31**, 46–65.

van Uytvanck, J., Decleer, K. and Hoffmann, M. (2008a) Establishment patterns of woody species in low intensity-grazed pastures after the cessation of intensive agricultural use. *Forest Ecology and Management* **256**, 106–113.

van Uytvanck, J., Maes, D., Vandenhaute, D. and Hoffmann, M. (2008b) Restoration of woodpasture on former agricultural land: the importance of safe sites and time gaps before grazing for tree seedlings. *Biological Conservation* **141**, 78–88.

van Wieren, S.E. (1991) The management of populations of large mammals. In I.F. Spellerberg, F.B. Goldsmith and M.G. Morris (eds.) *The Scientific*

Management of Temperate Communities for Conservation. Oxford, UK: Blackwell, pp.103–127.
van Wieren, S.E. and Bakker, J.P. (2008) The impact of browsing and grazing herbivores on biodiversity. In I.J. Gordon and H.H.T. Prins (eds.) *The Ecology of Browsing and Grazing*. Ecological Studies 195. Berlin: Springer, pp.263–292.
van Wijk, M.N., Sanders, M.E., de Jong, J.J. and van Veen, M.P. (2006) Natuurbeheer in de duinen. In *Achtergronden bij de natuurbalans 2005*. Rapport 408763012/2006. Bilthoven, Netherlands: Milieu en Natuurplanbureau.
Vandenberghe, C., Smit, C., Frelechoux, F., Pohl, M. and Buttler, A. (2008) The influence of grazing intensity on deciduous and coniferous tree sapling responses to facilitation by a thorny shrub. *Basic and Applied Ecology*, **10**, 427–436.
Vera, F.W.M. (2000) *Grazing Ecology and Forest History*. Wallingford, UK: CABI Publishing.
Vera, F.W.M, Bakker, E.S. and Olff, H. (2006) Large herbivores: missing partners of western European light-demanding tree and shrub species? In K. Danell, P. Duncan, R. Bergström and J. Pastor (eds.) *Large Herbivore Ecology, Ecosystem Dynamics and Conservation*. Cambridge, UK: Cambridge University Press, pp.203–231.
Virtanen, R., Edwards, G.R. and Crawley, M.J. (2002) Red deer management and vegetation on the Isle of Rum. *Journal of Applied Ecology* **39**, 572–583.
Vulink, J.T. (2001) Hungry herds: management of temperate lowland wetlands by grazing. PhD thesis, University of Groningen, the Netherlands.
Vulink, J.T. and van Eerden, M.R. (1998) Hydrological conditions and herbivory as key operators for ecosystem development in Dutch artificial wetlands. In M.F. Wallis de Vries, J.P. Bakker and S.E. van Wieren (eds.) *Grazing and Conservation Management*. Dordrecht, Netherlands: Kluwer Academic Publishers, pp.217–252.
Vulink, J.T., Drost, H. and Jans, L. (2000) The influence of different grazing regimes on *Phragmites*-shrub vegetation in the well-drained zone of a eutrophic wetland. *Applied Vegetation Science* **3**, 73–80.
Wallis de Vries, M.F., Parkinson, A.E., Dulphy, J.P., Sayer, M. and Diana, E. (2007) Effects of livestock breed and grazing intensity on biodiversity and production in grazing systems. 4. Effects on animal diversity. *Grass and Forage Science* **62**, 185–197.
Wheeler, P. (2008) Effects of sheep grazing on abundance and predators of field vole (*Microtus agrestis*) in upland Britain. *Agriculture Ecosystems and Environment* **123**, 49.

10

Ungulate–large carnivore relationships in Europe

WŁODZIMIERZ JĘDRZEJEWSKI, MARCO APOLLONIO,
BOGUMIŁA JĘDRZEJEWSKA AND ILPO KOJOLA

10.1 Introduction

Wherever carnivores are present in a hunting area, they are blamed for reducing ungulate numbers. In the past, predator control was often carried out to promote game species (Filonov, 1989; Jędrzejewska *et al.*, 1996; Jędrzejewski *et al.*, 1996). By contrast, in regions where large predators have long been exterminated, ungulate numbers show pronounced fluctuations between years and their high densities are accused of damaging forest regeneration and farmland crops or reducing biodiversity (Chapter 6 this volume, and references therein). Recently, protection or restitution of top predators has been viewed as a necessary step towards conservation of ecosystem biodiversity (Ray, 2005). In this chapter, we review the results of European studies on large carnivore predation on ungulates. We attempt to point out what practical implications for game management have emerged from those studies. We also briefly present the theoretical predictions regarding predator–prey relationships.

Ungulate populations are influenced by many factors. Food supply (connected with habitat productivity) determines how many individuals can live and reproduce in a given area (Sinclair, 2003). Severe winter conditions may directly cause deaths (Okarma *et al.*, 1995; Del Giudice *et al.*, 2002; Jonas *et al.*, 2008), but climate also has indirect effects through altering food availability (Jonas *et al.*, 2008). Predators, diseases, hunters and road traffic are important factors of mortality (Gazzola *et al.*, 2005; Chapters 8 and 11 this volume). Population density itself, alone or in conjunction with weather or other factors, may affect reproduction and mortality rates through competition for food following a density-dependent mechanism (Bonenfant *et al.*, 2002). One of the essential questions asked by game managers is whether the mortality caused by

Ungulate Management in Europe: Problems and Practices, eds. Rory Putman, Marco Apollonio and Reidar Andersen. Published by Cambridge University Press.

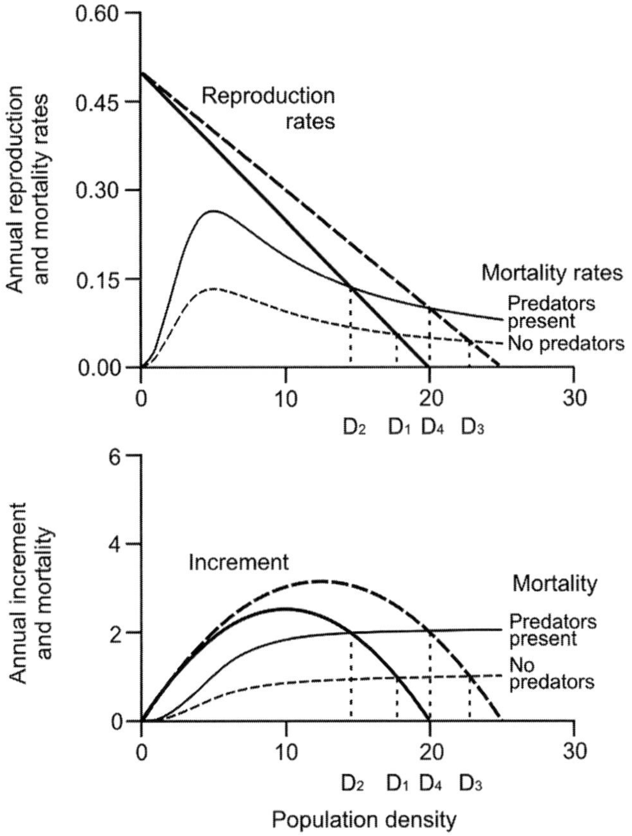

Figure 10.1 Schematic model of predator-caused mortality compared with other mortality, annual increment and population density. (Upper graph) Reproduction rate (young born as percentage of population number) declines with growing population density (solid line, normal productivity of habitat; broken line, improved productivity of habitat). Mortality rates (annual losses as percentage of population density) are shown in situations with (solid line) and without predators (broken line). If mortality caused by predation is additive to other deaths, total mortality rate will increase in the presence of predators. (Lower graph) Reproduction (increment) and mortality, shown in absolute values, in relation to population density. Increment is highest at intermediate densities. Intersections of reproduction and mortality rates as well as increment and mortality show population densities (D_1 to D_4) at various levels of habitat productivity and predation impact.

predators brings ungulate numbers below the carrying capacity of habitat and below the level observed in the absence of predators (Figure 10.1). More specific questions are: how many prey individuals do predators kill annually? How big is the toll taken by them compared with the annual production of young? Do they effectively limit the densities of their prey?

The studies which attempt to answer these questions have to quantify two components of predation rates: (1) the numbers of predators and their response to changes in prey density (called numerical response), and (2) the rates at which predators kill their prey and its relationships with changing ungulate numbers (called functional response: Holling, 1959; Messier, 1995). In Europe the studies on these subjects are quite complex as both ungulate and predator populations are strongly influenced by human activities, which confound the direct cause and effect relationship between predator and prey fluctuations in number.

Yet another important issue from a management point of view is the question of whether predation on ungulates is additive or compensatory to other mortality factors, including any hunting harvest. If predation is additive then introduction of a new predator will increase the total mortality of an ungulate species and lead to declining density. If predation is compensatory, predators would kill only those individuals that would otherwise die from other causes (such as disease or harsh winter conditions) and would bring no serious consequences for population density. However, if hunting pressure is intense, i.e. more than 80–85% of total ungulate mortality as reported from many European countries (Putman, 2008), there are very limited possibilities for compensatory mortality as hunting itself removes a vast proportion of individuals that might otherwise have died from other causes. Work to date by Kramer (1990), Lindström *et al.* (1994), Aanes and Andersen (1996), and Melis *et al.* (2010) all suggest that mortality through human hunting is largely additional to any mortality resulting from natural predation.

10.2 Large carnivores and ungulate populations in Europe

In Europe, two species of large predators rely on ungulates as their main prey: the wolf *Canis lupus* and the Eurasian lynx *Lynx lynx*. In the past their ranges covered most of Europe, but in historical times they shrank to the eastern (wolf) or north-eastern (lynx) part of the continent and a few isolated, usually mountainous, regions in western and southern Europe (Figure 10.2).

Two other species, the brown bear *Ursus arctos* and the wolverine *Gulo gulo*, are also capable of regularly killing adult ungulates. In many regions, the diet of bears is dominated by vegetal matter, invertebrates and carrion (Elgmork and Kaasa, 1992; Frąckowiak and Gula, 1992; Grosse *et al.*, 2003) but at northern latitudes bears can also successfully hunt large ungulates, particularly moose (Filonov, 1989). The wolverine is an important predator of reindeer *Rangifer tarandus* wherever these two species coexist (Landa *et al.*, 1997). However, the geographic ranges of these species are even more

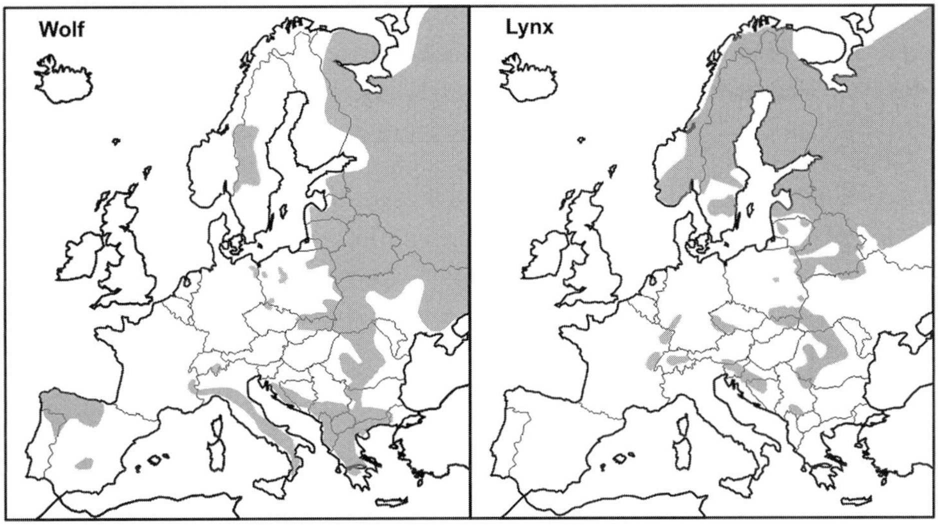

Figure 10.2 Schematic ranges of wolf (*Canis lupus*) and lynx (*Lynx lynx*) occurrence in Europe. Sources: Boitani (2000), Breitenmoser *et al.* (2000), W. Jędrzejewski, unpublished data, V.E. Sidorovich, unpublished data.

restricted than those of wolves and lynx, due to either natural reasons (wolverine, a typical boreal species) or past persecution by humans (brown bear) and thus any impacts on ungulate populations tend to be highly localised. Finally, ungulate remains – interpreted as evidence of scavenging – were reported in scats of golden jackals *Canis aureus*, but it remains unknown whether the jackals are actually capable of killing young ungulates (Lanszki and Heltai, 2002), for instance wild boar piglets (Lanszki *et al.*, 2006). In this chapter, therefore, we will focus largely on wolves and lynx, as predation by these obligate ungulate hunters has the most serious consequences for populations of large game.

The red fox *Vulpes vulpes* is another, smaller carnivore which may occasionally take ungulate prey. This is an opportunistic predator whose diet may include a wide range of items (e.g. Lanszki *et al.*, 2006; Sidorovich *et al.*, 2006; Webbon *et al.*, 2006; Dell'Arte *et al.*, 2007; Rosalino and Santos-Reis, 2009). However, in some contexts red fox may take both adult ungulates (of smaller species such as roe deer) and neonatal juvenileds (e.g. Linnell *et al.*, 1995; Aanes and Andersen, 1996; Kjellander and Nordström, 2003; Jarnemo *et al.*, 2004; Panzacchi, 2007).

The nature of fox predation is, however, rather distinct from that of the larger carnivores. As noted, this is essentially an opportunistic predator (taking ungulates simply because they offer another potential food source, rather than being an ungulate specialist). In addition, while red foxes can kill adult roe deer in winter when snow conditions give them an advantage (Borg, 1962; Cederlund and Lindström, 1993; Melis *et al.*, 2009) they usually prey primarily on juveniles in their first summer – or indeed within the first 60 days of life (Linnell *et al.*, 1995; Aanes and Andersen, 1996; Kjellander and Nordström, 2003; Jarnemo *et al.*, 2004; Panzacchi, 2007; Melis *et al.*, 2009) with the result that any impacts of predation are highly seasonal. This contrasts strongly with lynx or wolf, which are primarily ungulate specialists and which prey upon ungulates year round. Because of the very different nature of predation by foxes, they are not included within the wider treatment of this chapter. However, we recognise that in certain contexts, predation on neonatal ungulates by red foxes may have a significant effect on some ungulate populations in some areas and some situations. We therefore offer a separate treatment on this rather different form of predation in Box 10.1.

As regards ungulate species taken as prey, of 12 species indigenous to Europe (Mitchell-Jones *et al.*, 1999), four have the broadest geographic distribution (Figure 10.3) and are most important for both hunting economy and ecological relations with large carnivores. They are: the moose *Alces alces*, the red deer *Cervus elaphus*, the roe deer *Capreolus capreolus*, and the wild boar *Sus scrofa*. These species will therefore also be given special consideration in this chapter.

10.3 Large carnivore densities and numerical relationships with ungulate prey

In Europe, the territories held by wolf packs (minimum convex polygons with 100% of radio-locations) increase with latitude by an order of magnitude: from 100–200 km^2 in the Mediterranean countries to over 1000 km^2 in Fennoscandia (Figure 10.4). A similar increase is shown in home ranges of the Eurasian lynx, with female ranges consistently smaller than those of adult males (Figure 10.4). Thus, population densities of large carnivores become lower at high latitudes. Densities of wolves range from 4.7 wolves/100 km^2 in Italy, to 2.4 in Poland and 0.3 in Finland (Okarma *et al.*, 1998; Apollonio *et al.*, 2004; I. Kojola *et al.* unpublished). In Switzerland and eastern Poland, 1–3 lynx/100 km^2 (kittens excluded) were recorded compared with 0.3 lynx/100 km^2 in Norway and Sweden (Jędrzejewski *et al.*, 1996; Linnell *et al.*, 2001; Molinari-Jobin *et al.*, 2002).

The most feasible explanation for such a great biogeographical variation in abundance of large carnivores is the parallel south–north gradient in the densities of their most important prey such as the roe deer (Melis *et al.*, 2009), red deer (M. Apollonio *et al.*, in prep.), and wild boar (Melis *et al.*,

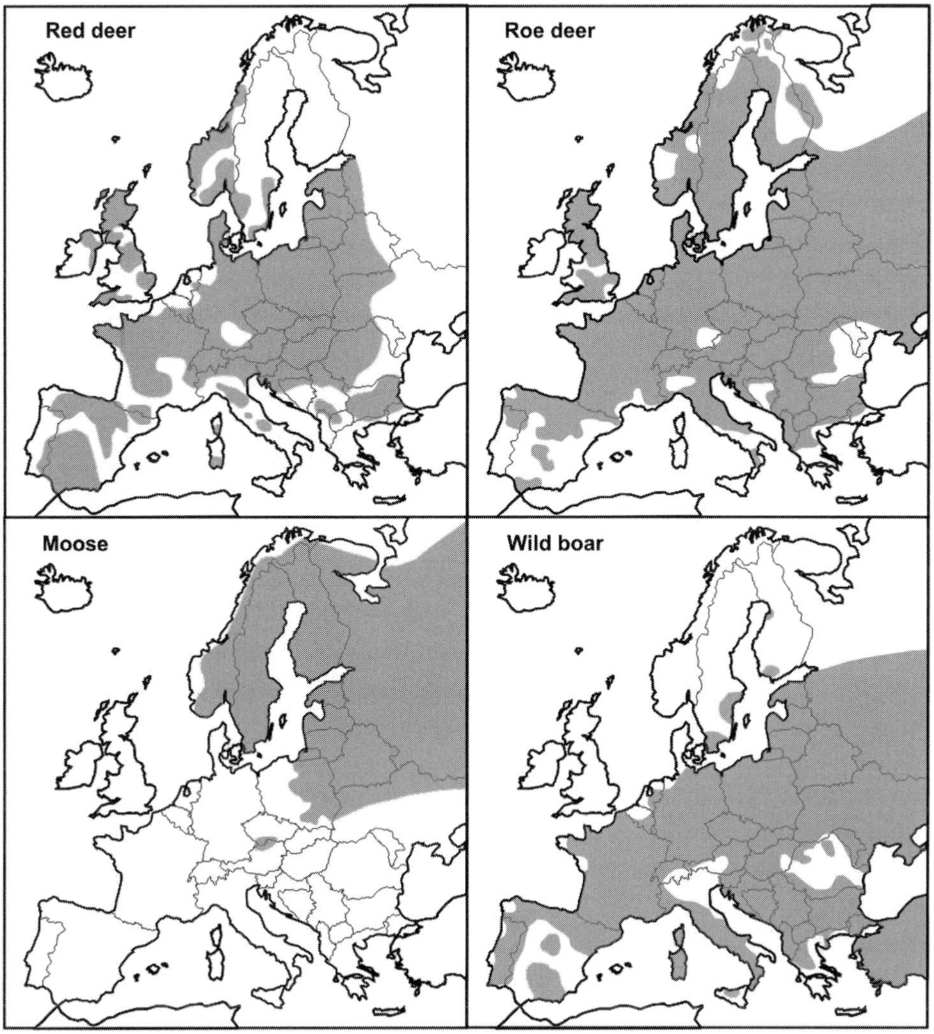

Figure 10.3 Schematic ranges of red deer (*Cervus elaphus*), roe deer (*Capreolus capreolus*), moose (*Alces alces*), and wild boar (*Sus scrofa*) occurring in Europe. Sources: Bobek *et al.* (1992), Fruziński (1992), Dzięciołowski and Pielowski (1993), Andersen *et al.* (1998), Mitchell-Jones *et al.* (1999).

2006), all of which decline from south to north. Also, the number of sympatric ungulate species decreases at higher latitudes (Okarma, 1995). However, it is still unknown whether prey biomass per predator (i.e. predator:prey ratio) is stable or varies over geographic scale.

At the local scale, the numerical response of large predators to changes in prey abundance (i.e. increase in predator densities with growing abundance of prey, and decline of predators with declining prey numbers) has not been

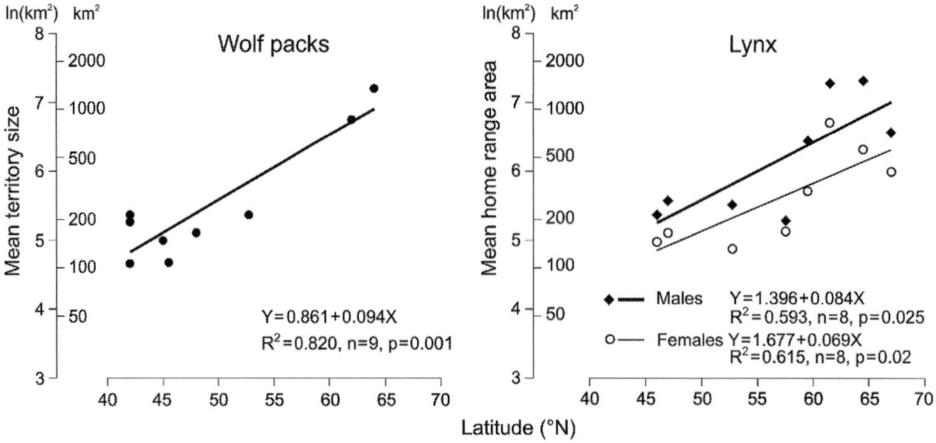

Figure 10.4 Increase of territory size of wolf packs and home ranges of adult female and male lynx with latitude in Europe. Based on radio-telemetry studies on wolves in Portugal (Pereira *et al.*, 1985), Spain (Vila *et al.*, 1990), Italy (Ciucci *et al.*, 1997), Croatia (Kusak and Huber, 2000), Romania (Promberger *et al.*, 1998), Slovakia (Findo and Chovancova, 2004), Poland (Jędrzejewski *et al.*, 2007), Sweden and Norway (Sand *et al.*, 2000), Finland (Kojola *et al.*, 2004); and on lynxes in: France (Stahl *et al.*, 2002), Switzerland (Breitenmoser *et al.*, 1993), Poland (Schmidt *et al.*, 1997), Latvia (Ornicans *et al.*, 2005), Sweden and Norway (Linnell *et al.*, 2001).

documented in Europe so far. The main reason for lack of any reported relationship may be the fact that wolf and lynx numbers have usually been shaped by human control or persecution rather than by natural factors. As demonstrated by over 100-year long data from Białowieża Primeval Forest (Poland and Belarus), not only was the positive numerical response of wolves and lynx absent but the predator numbers were *negatively* correlated to the abundance of their prey (Jędrzejewska and Jędrzejewski, 1998). Such a strong negative relationship was driven by extermination of wolves and lynx, and the fact that in periods when predators were absent, their main prey (red and roe deer) rapidly increased in numbers (Jędrzejewska *et al.*, 1997). In north-eastern Belarus, wolf numbers in 1990–2000 were not correlated with the wild ungulate densities, but were shaped by hunting harvest in the two preceding years (Sidorovich *et al.*, 2003).

However, that European predators may normally show a positive numerical response to prey numbers under natural conditions was suggested by more recent data on roe deer and lynx numbers and reproductive success in Białowieża Forest (Poland). In 1992–1998, hunters reduced the densities of roe deer from over 4 to 1 individual/100 km^2. The lynx responded to the decline of their main prey by markedly lower breeding success (from 2 to

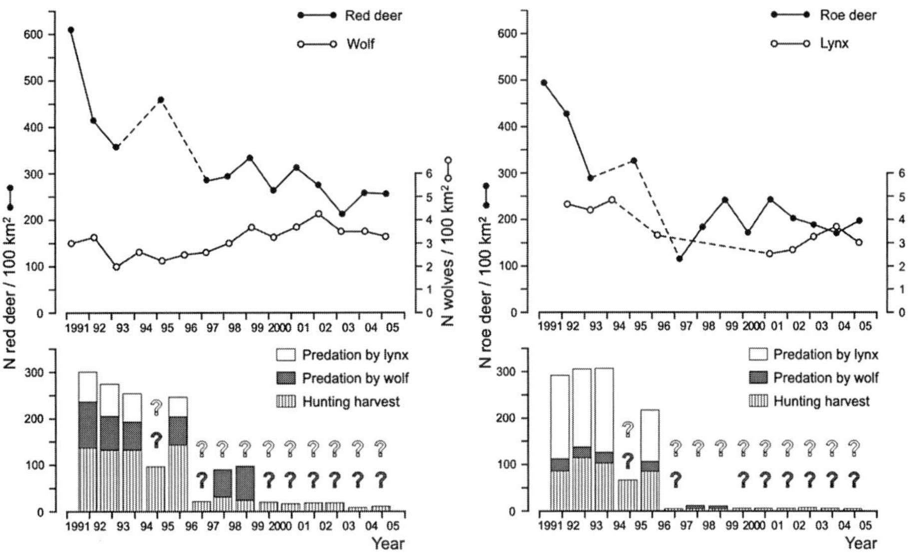

Figure 10.5 Winter densities of red deer and roe deer in Białowieża Primeval Forest (east Poland) in 1991–2005 in relation to densities of their main predators (wolf and lynx), annual hunting harvest, and annual predation impact (in the latter case data were available for some years; question marks signify lack of data). Sources: Okarma *et al.* (1997), Jędrzejewska and Jędrzejewski (1998), Jędrzejewski *et al.* (2000, 2002), and data from Regional Directorate of State Forests in Białystok.

<1 kitten/100 km^2), which eventually led to lower densities of the lynx population (Figure 10.5). Also, home range size of lynx increased markedly, which led to a consequent decline in overall population density (Schmidt, 2008).

Interestingly, the parallel reduction of red deer by hunters in Białowieża Forest in 1991–1998 had no effect on the number of wolves, which even continued to increase and exerted heavy pressure on the declining deer population. This may be explained by the fact that wolf is a more opportunistic predator that, unlike the specialist lynx, could rely on a wide array of potential prey. After the increased level of hunting ceased, red deer could not recover but remained at low density (Figure 10.5).

At a continental scale, however, densities of large carnivores seem to be shaped predominantly by various human actions, such as legal control in Eastern Europe (e.g. Sidorovich *et al.*, 2003), poaching, which occurs even in protected populations (e.g. Jędrzejewska *et al.*, 1996; Jędrzejewski *et al.*, 1996; Capitani *et al.*, 2006), a combination of the two (Andrén *et al.*, 2006), and reintroductions (Breitenmoser *et al.*, 2001). Nonetheless, in recent decades, the changing attitude to large carnivores in many countries (Boitani, 2000; Breitenmoser *et al.*, 2000), protection, and several successful reintroduction programmes have resulted in a generally decreasing role of direct human control over their populations (cf. Linnell *et al.*, 2005). All that has contributed to the recovery of wolves, increase of their geographic range and local densities. No such spectacular revival of lynx has been observed yet.

An additional factor that comes into play and, in the near future, may appear the most important for wolf and lynx populations is the large-scale alteration of the environment, especially the fragmentation of forests and other suitable habitats by a dense network of human settlements (towns, villages) and transportation infrastructure (Jędrzejewski *et al.*, 2004, 2005; Kramer-Schadt *et al.*, 2004; Niedziałkowska *et al.*, 2006).

10.4 Ungulates in diets of large predators

Feeding ecology of both lynx and wolf shows great spatial and temporal variation. Wild ungulates and – in situations of their scarcity, also domestic ones – are the dominant prey of European wolves (Okarma, 1995; Meriggi and Lovari, 1996). Throughout the continent, these predators coexist with one to five species of hoofed animals (Okarma, 1995). Wolves are capable of killing all extant wild ungulates ranging in size from roe deer to European bison (body mass of adult females below 20 kg and over 400 kg, respectively).

The proportion of a given species in the wolf diet depends on several factors. Based on results of 26 studies on wolf prey composition in Europe, we present, in Figures 10.6 and 10.7, the relative proportions of moose, red deer, roe deer and wild boar among wolf kills compared with their respective shares in local communities of ungulates. The broken line in each graph denotes killing of prey proportional to its occurrence.

The role of moose as wolf prey seems to increase in direct proportion to its availability (simply increasing as the percentage of that species increased within the overall ungulate community; Figure 10.6). Despite marked variation among localities, no clear positive or negative selection of moose by wolves occurs at a landscape or global scale. By contrast, in a number of local studies, red deer has been reported as a clearly preferred prey of wolves (e.g. Jędrzejewski *et al.*, 2000; Gazzola *et al.*, 2005). Indeed, throughout Europe, the percentage of red deer in wolf diet was generally bigger than expected from its relative frequency within the ungulate community. Thus, the red deer was strongly positively selected from the local community of ungulates (Figure 10.6). Wolves show no dietary response to changes in roe deer relative abundance, as long as roe deer made 20–80% of all wild ungulate

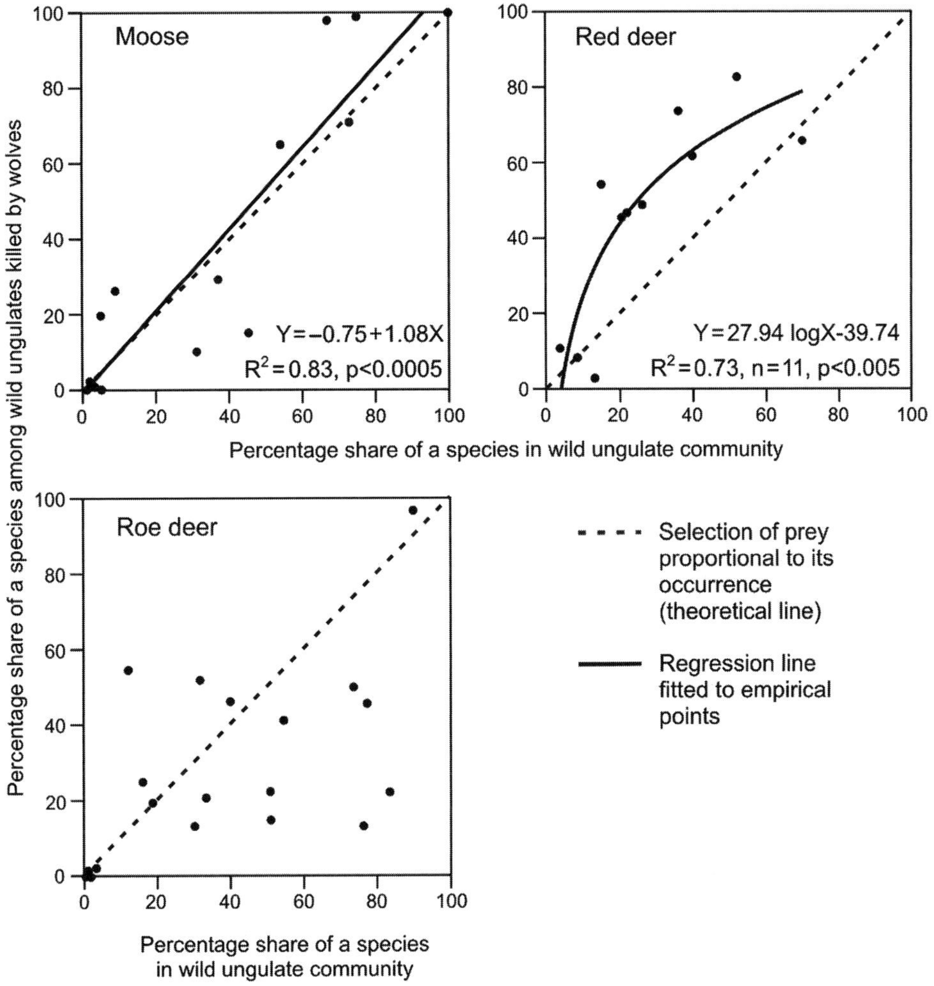

Figure 10.6 Wolf dietary responses to varying relative abundance of ungulate prey in Europe. The percentage contributions by moose, red deer, and roe deer to total number of ungulates killed by wolves plotted against percentage shares of those species in the local communities of wild ungulates. Each point denotes one study site. Localities and sources: Kavkazskii Reserve (Russia) after Kudaktin (1978) and Filonov (1989), eastern Carpathians (Ukraine) after Kerechun (1979), Bryansk region (Russia) after Vatolin (1979), Berezinskii and Pripyatskii Reserves (Belarus), Darvinskii, Ilmenskii, Laplandskii, Mordovskii, Okskii, and Pechoro-Ilychskii Reserves (Russia) after Filonov (1989), Białowieża Primeval Forest (Polish and Belarusian parts) after Jędrzejewska and Jędrzejewski (1998), Bieszczady Mountains (Poland) after Śmietana (1998) and Śmietana *et al.* (2000), north Karelia (Finland) after Gade-Jorgensen and Stagegaard (2000), Vitebsk region (Belarus) after Sidorovich *et al.* (2003), Arezzo region (Italy) after Mattioli *et al.* (2004), Susa Valley (Italy) after Gazzola *et al.* (2005), western Beskidy Mountains (Poland) after Nowak *et al.* (2005).

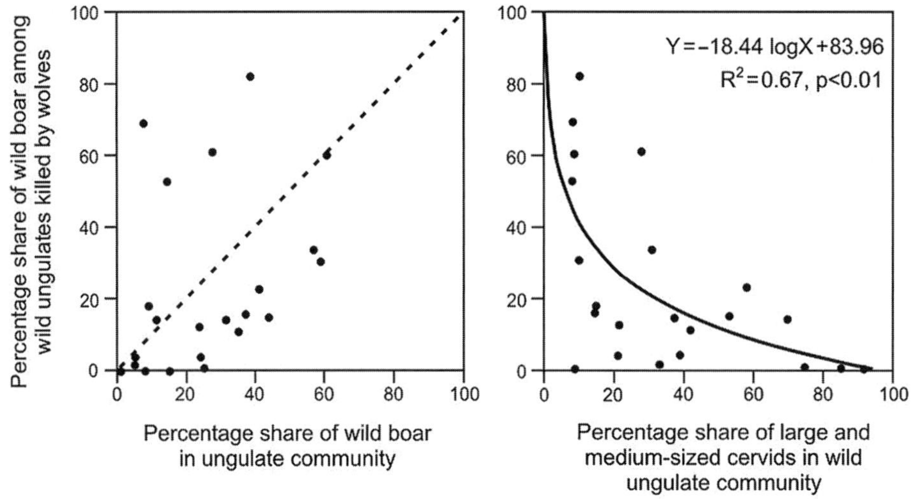

Figure 10.7 Percentage contributions by wild boar to all ungulates killed by wolves in relation to the relative abundance of wild boar in local communities of ungulates (left graph) and relative abundance of large and medium-sized cervids (moose and red deer) in the community of ungulates (right graph). Each point denotes one study site in Europe. Localities and sources: Kavkazskii Reserve (Russia) after Kudaktin (1978) and Filonov (1989), eastern Carpathians (Ukraine) after Kerechun (1979), Bryansk region (Russia) after Vatolin (1979), Berezinskii and Pripyatskii Reserves (Belarus), Darvinskii, Ilmenskii, Mordovskii, and Okskii Reserves (Russia) after Filonov (1989), Białowieża Primeval Forest (Polish and Belarusian parts) after Jędrzejewska and Jędrzejewski (1998), Bieszczady Mountains (Poland) after Śmietana (1998) and Śmietana *et al.* (2000), Vitebsk region (Belarus) after Sidorovich *et al.* (2003), Susa Valley (Italy) after Capitani *et al.* (2004), Arezzo region (Italy) after Mattioli *et al.* (2004), western Beskidy Mountains (Poland) after Nowak *et al.* (2005).

numbers (Figure 10.6). Only in the Ilmenskii Reserve (Russia), where roe deer formed 90% of ungulate numbers, was this species a staple prey of wolves.

Perhaps more complicated is the role of wild boar in wolf diet. Some local studies, especially those from Central and Eastern Europe, reported a significant avoidance of wild boar by wolves (e.g. in Ukraine, Belarus, Poland; Kerechun, 1979; Filonov, 1989; Jędrzejewski *et al.*, 2000), whereas data from southern Europe (e.g. Italy; Mattioli *et al.*, 2004) have documented a dominating role of boar in wolf diet and even its selection by wolves from the community of potential prey. Our review of studies throughout Europe revealed that eagerness of wolves to hunt wild boar is determined not only by the relative abundance of that species, but also by the presence or absence (and density) of wolves' primary prey (red deer and moose) (Figure 10.7). At similar proportions of wild boar among wild ungulates, say 10–40%, wolves avoid this species in central and Eastern Europe, where red deer and/or moose are frequent, but are shown to prefer it actively in southern Europe, where the only alternative prey is roe deer.

Due to their smaller size and solitary hunting, lynx usually prey on small-sized ungulates: roe deer, chamois, reindeer, and the females and calves of red deer. Only one study in Europe (Berezinskii Reserve, Belarus) did document a significant share of moose among ungulates killed by lynx (Filonov, 1989). The degree to which Eurasian lynx hunt on ungulates is also strongly dependent on latitude. In northern and north-eastern Europe, lynx diet is dominated by hare, mainly the blue hare *Lepus timidus*, and, where present, roe deer (Odden *et al.*, 2006; Sidorovich, 2006); predation on semi-domesticated reindeer where present may also be substantial (Danell *et al.*, 2006). Further to the south, the proportion of ungulates in lynx diet increases (review in Jędrzejewski *et al.*, 1993). In contrast to wolves, lynx have a far lower impact on domestic animals, but in some regions they may cause significant problems as predators of sheep (e.g. in central Norway, Sunde *et al.*, 2000; French Jura Mountains, Stahl *et al.*, 2002).

At the European scale, dietary responses by lynx to varying proportions of ungulates in a community can be explored for three widespread species: roe deer, red deer, and wild boar (Figure 10.8). Roe deer was always strongly positively selected by the lynx and taken more frequently than expected from their relative share in the ungulate community. The proportion of animals that died from lynx predation in one long-term study in Norway (1995–2005; Melis *et al.*, 2010) was significantly higher at low density (0.01–0.25 individuals harvested/km^2), with respect to medium to high density (0.26–2.50 individuals harvested/km^2), so that in practice at low density of roe deer, predation by lynx removed annually about 22% of the population, compared with 9% at medium and high density. This inverse density dependence of lynx predation on roe deer has been previously described by Jędrzejewska and Jędrzejewski (1998) in the multi-predator system of Białowieża Primeval Forest (Poland).

Lynx did not respond in any clear way to varying abundance of red deer (Figure 10.8). Wild boar were markedly avoided by that predator though their incidence in lynx diet increased with growing percentage share of boar in the ungulate community.

10.5 Kill rates by wolves and lynx

A few studies have attempted to quantify the kill rates or total predation by wolves based on their metabolic rates or daily food consumption (Głowaciński and Profus, 1997; Gazzola *et al.*, 2007). These are, however, underestimates of the true kill rates. In Europe, so far two field studies – in

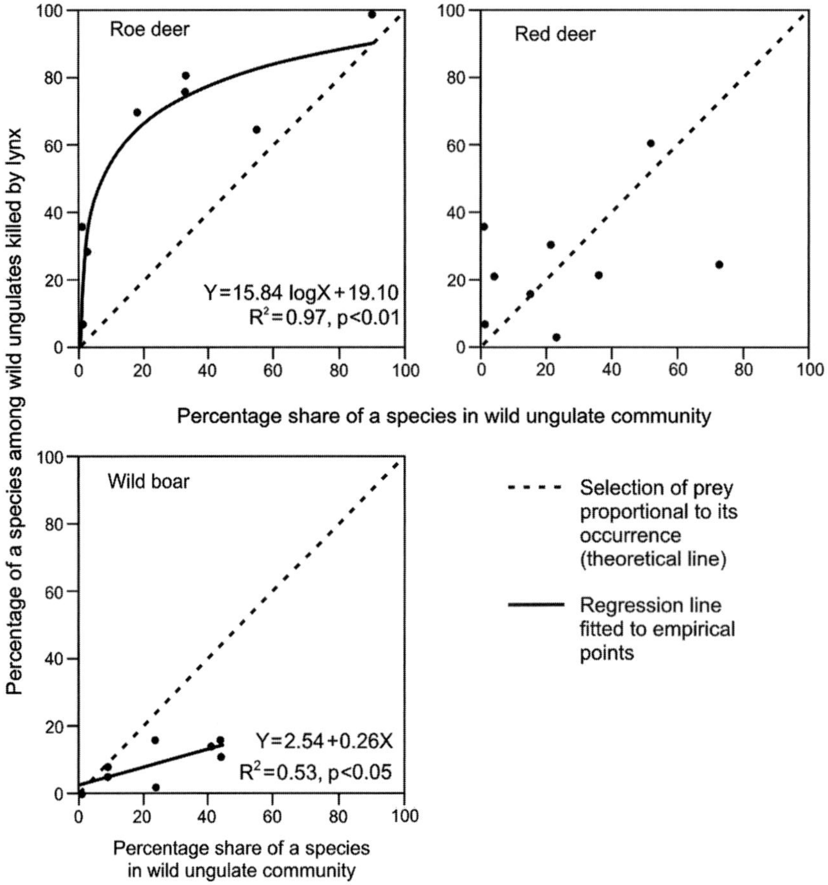

Figure 10.8 Lynx dietary responses to varying relative abundance of ungulate prey in Europe. The percentage contributions by roe deer, red deer, and wild boar to total number of ungulates killed by lynx plotted against percentage shares of those species in the local communities of wild ungulates. Each point denotes one study site. Localities and sources: eastern Carpathians (Ukraine) after Kerechun (1979), Berezinskii Reserve (Belarus), Ilmenskii and Kavkazskii Reserves (Russia) after Filonov (1989), Białowieża Primeval Forest (Polish and Belarusian parts) after Jędrzejewska and Jędrzejewski (1998), Bieszczady Mountains (Poland) after Śmietana *et al.* (2000).

Scandinavia and eastern Poland – endeavoured to estimate the actual kill rates by wolves. Scandinavian wolves hunted predominantly moose. One predator killed, on average, 29 moose and 1.5 roe deer per year (Sand *et al.*, 2005; Table 10.1). Polish wolves, which coexisted with a multi-species community of ungulates, killed, on average, 42 large prey (mainly red deer) per year per capita and their mean daily consumption under natural conditions was 5.6 kg of food per wolf (Jędrzejewski *et al.*, 2002; Table 10.1).

Table 10.1 *Kill rates and consumption rates by wolves in eastern Poland (Białowieża Primeval Forest, after Jędrzejewski* et al.*, 2002) and Scandinavia (Tyngsjo, south-central Sweden and Grafjell, south-eastern Norway, after Sand* et al.*, 2005)*

Prey	Scandinavia	Eastern Poland	
	Kill rate (*N* prey per wolf per year)	Kill rate (*N* prey per wolf per year)	Consumption rate (kg per wolf per day)
Moose	29.1	0.2	0.06
Red deer	–	27.4	4.31
Roe deer	1.5	2.2	0.10
Wild boar	–	12.4	0.54
Small prey[a]	?	2.9	0.07
Carrion, own old prey	?	–	0.48
Total	30.6	45.3	5.58

[a]Beaver, brown hare, red fox, domestic dog.

Table 10.2 *Kill rates of Eurasian lynx (mean number of ungulate prey killed by one lynx annually) in Switzerland (calculated from: Molinari-Jobin* et al.*, 2002), Poland (Okarma* et al.*, 1997) and Norway (Karlsen, 1997)*

Prey species	Number of prey killed annually by a lynx		
	Switzerland, Jura Mountains	Poland, Białowieża Forest	Norway, Hedmark
Red deer	–	18	–
Roe deer	50	48	40
Alpine chamois	12	–	–
Total	62	66	40

Interestingly, kill rates did not depend on pack size, but increased substantially with snow cover. Deep snow made it easier for wolves to hunt successfully for red deer (Jędrzejewski *et al.*, 2002).

In the case of lynx, kill rates estimated in the Polish and Swiss populations yielded very similar results: annually 62–66 ungulate prey, mainly roe deer, per lynx (Table 10.2). There were great sex- and age-related differences among lynx. Females rearing kittens showed the highest kill rates (female with 3 kittens: 116 deer per year), whereas subadults, which had recently began to live on their own, had the lowest rates of successful kills (43 deer/year) (Okarma *et al.*, 1997). The Norwegian study (Karlsen, 1997) yielded markedly smaller estimates of the lynx kill rates than the two former ones (Table 10.2).

10.6 Impact of predation on ungulate densities and population dynamics

Regular records of ungulate carcasses in an area can provide an indirect, relative estimate of the role of predation among various mortality factors. Such long-term records on natural causes of ungulate mortality are available from several areas in Europe and Western Siberia (Figure 10.9). They suggest that – wherever ungulates coexist with large predators (brown bear, wolf and lynx) – predation is the most important cause of natural mortality in moose, red deer and roe deer (on average, 60–80% of natural deaths are due to predators). Predation is markedly less important for the wild boar (Figure 10.9). These records show also which species of predators matter for a given ungulate species: moose suffer mostly from wolves and brown bear, red deer from wolves, roe deer is preyed upon by lynxes and wolves, and wild boar by wolves.

The magnitude of wolf and/or lynx predation on ungulates was estimated in four localities in Europe, ranging from Italy to Finland (Table 10.3). In eastern Poland (1991–2000), wolves annually took 57–105 red deer, 4–25 roe deer, 19–38 wild boar and 0–2 moose per 100 km^2 (Jędrzejewski *et al.*, 2000, 2002). In the Italian Western Alps (2000–2003), wolves killed 20–72 red deer, 21–124 roe deer, and 7–30 chamois per year per 100 km^2 (Gazzola *et al.*, 2007). In these two studies, predation by wolves amounted to 2–17% of the spring–summer (seasonally highest) densities of ungulates, and 7–69% of their total (including hunting harvest) annual mortality (Table 10.3).

In eastern Poland (1991–1996), lynx killed annually 110–181 roe deer and 42–70 red deer (fawns and females, only) per 100 km^2 (Okarma *et al.*, 1997). In Switzerland (1988–1997), lynx took an average annual toll of 50 roe deer and 88 chamois per 100 km^2 (Molinari-Jobin *et al.*, 2002). Compared with population density of their main prey species, the lynx appeared to harvest 9–26% of spring–summer numbers of deer or chamois, and they contributed 17–56% to the total annual mortality of prey (Table 10.3).

Furthermore, the results of the Polish study (Białowieża Primeval Forest, where the two large carnivores coexist) showed that the effects of wolf and lynx predation were additive. The total predation impact was stronger there than in sites where only one species of predator lived (see also Aanes and Andersen, 1996; Melis *et al.*, 2009).

10.7 What is the effect of predation on ungulate populations?

One of the important questions in modern ecology is the extent to which predation may be density dependent. If predation rates increase with densities of prey, predators are believed to regulate prey numbers. But if,

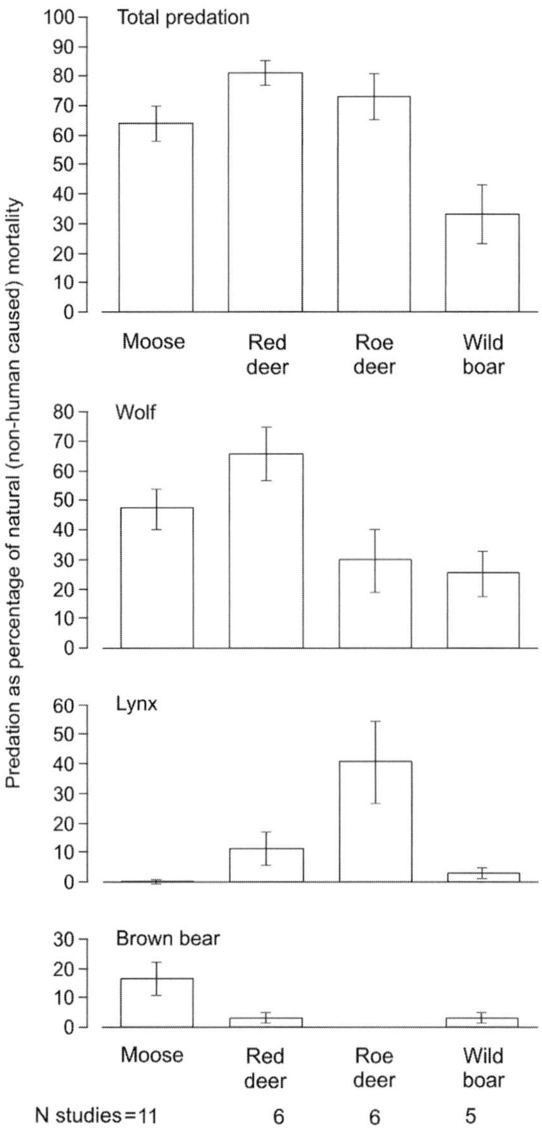

Figure 10.9 Predation as percentage of natural (non-anthropogenic) mortality of moose, red deer, roe deer, and wild boar. Mean (± SE) estimates, based on records of ungulate carcasses, in several studies in Europe and western Siberia. Regions and sources: moose – Arkhangelsk and Petersburg regions, Bashkirskii, Darvinskii, Laplandskii, Mordovskii, Okskii and Pechero-Ilychskii Reserves, north-west regions (Russia) after Timofeeva (1974), Rusakov (1979), Filonov (1989), Yevtikov (1991), Berezinskii Reserve (Belarus) after Filonov (1989), Białowieża Forest (Poland) after Okarma *et al.* (1995); red deer – Bashkirskii and Altaiskii Reserves (Russia) and Berezinskii Reserve (Belarus) after Filonov (1989), eastern Carpathian Mountains (Ukraine) after Kerechun (1979), Białowieża Forest (Poland) after Jędrzejewska

in turn, predators act in a density-independent or inversely density-dependent way, predation can only limit (lower) prey numbers (Sinclair, 1989; Messier, 1991).

The study on European wolves in the Białowieża Forest (eastern Poland) documented that, in 1991–2000, percentage predation on red deer and wild boar (number of deer or boar killed by wolves annually as percentage of their respective densities) showed a tendency to be negatively related to densities of those prey. Wolf predation did not vary with changing densities of roe deer (Jędrzejewski *et al.*, 2002). Thus, wolves are not capable of regulating the populations of their prey, but do have the potential to limit their numbers. Lynx predation on roe deer also shows inverse density dependence (Jędrzejewska and Jędrzejewski, 1998; Melis *et al.*, 2009) and kill rates once again do not vary with varying roe deer density; thus predation by lynx will also have a potentially limiting, but not a regulating effect, on prey populations. (By contrast, red fox predation on roe is density dependent (see for example, Aanes *et al.*, 1998; Panzacchi *et al.*, 2008a); in consequence fox predation may genuinely be expected to have a regulating effect on prey populations; see Box 10.1.)

The characteristics of predation by both wolf and lynx show that these species at least have the potential to limit numbers of their prey. Does predation by larger carnivores ever bring ungulate density below the carrying capacity of the habitat and below the level observed in the absence of predators in actual practice? The ability of European large carnivores to limit ungulate numbers below the habitat carrying capacity has been shown in both local studies and biogeographic reviews. In Białowieża Primeval Forest (Poland and Belarus), wolves and lynx were exterminated twice during the last 100 years. During the whole long-term data series, the combined abundance of five species of ungulates (European bison, moose, red and roe deer, and wild boar, expressed as crude biomass of all species

Caption for Figure 10.9 (cont.)
and Jędrzejewski (1998), western Alps (Italy) after Gazzola *et al.* (2005); roe deer – north-west Russia, Ilmenskii Reserve (Russia) after Rusakov (1979) and Filonov (1989), Carpathian Mountains (Ukraine) after Kerechun (1979), Białowieża Forest (Poland) after Jędrzejewska and Jędrzejewski (1998), western Alps (Italy) after Gazzola *et al.* (2005), Jura Mountains (Switzerland) after Molinari-Jobin *et al.* (2002); wild boar – Petersburg and Pskov regions (Russia) after Rusakov and Timofeeva (1984), Berezinkii Reserve (Belarus) after Filonov (1989), Carpathian Mountains (Ukraine) after Kerechun (1979), Białowieża Forest (Poland) after Okarma *et al.* (1995), northern Apennines (Italy) after A. Gazzola *et al.*, unpublished.

Table 10.3 *Impact of wolves and lynx on ungulate populations in Europe: annual predation expressed as percentage of summer (seasonally highest) density of prey and percentage of total (including human-caused) mortality*

Locality and prey	Annual predation by wolves as percentage of:		Annual predation by lynx as percentage of:	
	Summer numbers	Total mortality	Summer numbers	Total mortality
Italy, Alta Valle di Susa[a]				
Red deer	8 (17)	32 (69)	–	–
Roe deer	8 (17)	16 (35)	–	–
Chamois	2 (5)	7 (16)	–	–
Italy, Northern Apennine				
Wild boar	8	8.5	–	–
Roe deer	3.5	19	–	–
Spain, Asturias				
Wild boar	?	12	–	–
Switzerland, Jura Mountains				
Roe deer	–	–	9	31
Chamois	–	–	11	56
Poland, Białowieża Primeval Forest				
Moose	10	20	0	0
Red deer	12	40	10	17
Roe deer	3	7	26	47
Wild boar	6	24	+	+
Finland, Kainuu				
Reindeer	?	34	–	–

Minus (–) means absence of a given predator from the study area; ? no data available, + denotes negligible impact of a predator.

[a]In the Italian study, the lower numbers are estimates based on metabolic rates of wolves, higher numbers (in parentheses) are calculated based on daily food consumption in the wild.

Sources: Italy – Gazzola *et al.* (2007), A. Gazzola *et al*, unpublished; Spain – Nores *et al.* (2008); Switzerland – Molinari-Jobin *et al.* (2002); Poland – Okarma *et al.* (1997), Jędrzejewska and Jędrzejewski (1998), Jędrzejewski *et al.* (2000, 2002); Finland – Kojola *et al.* (2004).

per unit area) was strongly positively correlated with annual temperature. Biomass of ungulates increased as the climate warmed up (Figure 10.10). Strikingly, the positive correlation between temperature and ungulate biomass held in both situations: when predators were scarce or absent, and when they were numerous (two parallel lines in Figure 10.10). The relative roles of temperatures and predation differed greatly among species of ungulates. Climate was crucial for the bison and wild boar, its role declined in moose, and was the smallest in red and roe deer. The opposite

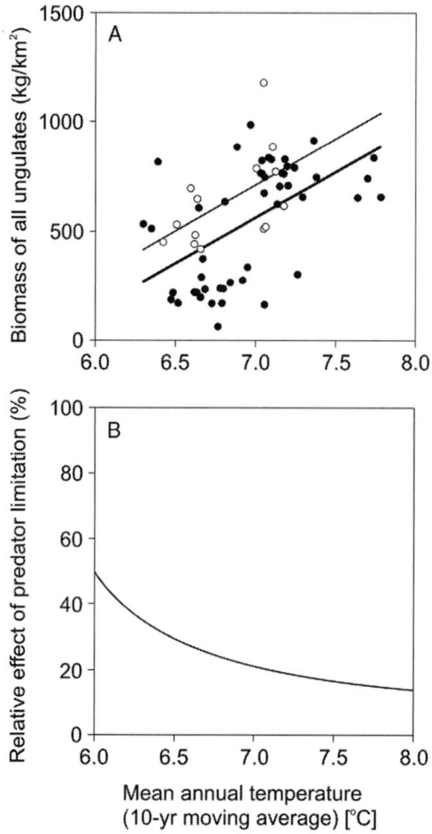

Figure 10.10 The limiting effect of predators on ungulate abundance in relation to habitat productivity (approximated by annual temperature) in Białowieża Primeval Forest (east Poland) in 1890–1993. Graph A: Regression between mean annual temperature and ungulate biomass (European bison, moose, red deer, roe deer, and wild boar combined) for years with predators (wolf and lynx) exterminated or in low densities (open points, upper line), and years with predators present in moderate and high densities (black points, lower line). Graph B: The impact exerted by predators in relation to annual temperature (the difference between the upper and lower line in graph A is presented as percentage of the maximum ungulate biomass, shown by the upper regression line in graph A. Source: Jędrzejewska and Jędrzejewski (2005), modified.

trend was shown in the role of predation, which was most significant in red and roe deer (Jędrzejewska and Jędrzejewski, 2005).

The absolute value of the limitation by predators was relatively stable. Over the whole range of temperatures, the crude biomass of all ungulates was by about 120 kg/km^2 lower in periods with large carnivores present than in years when they had been exterminated. Compared with the maximum attained densities by ungulates (upper line in Figure 10.10), the impact of predation was thus stronger during cooler, less productive periods (prey biomass lowered by 40–50% relative to habitat carrying capacity) than in the warmer, more productive periods (suppression by predators only by 10–20% of ungulate biomass).

The relative roles of habitat productivity, winter severity and predation in limiting the densities of prey were also investigated at the biogeographic scale in European populations of roe deer and wild boar. In the roe deer, productivity of the ecosystem (approximated by the FPAR index, i.e. the fraction of photosynthetically active radiation), presence of predators (the lynx and the wolf), percentage forest cover, and winter harshness were the most important factors shaping the great variation in densities observed in Europe (Melis *et al.*, 2009). The influence of habitat productivity on roe deer density was clearly positive. Increasing forest cover had negative effects on roe deer abundance. Interestingly, the limiting effect of predators varied with habitat productivity. It was strongest in the poorly productive, boreal habitats, and rather small in highly productive, temperate and southern habitats. The same finding applied to the gradient of winter severity: the limiting effect of predation on roe deer density increased in regions with harsh winters (Melis *et al.*, 2009). The biological mechanisms underlying the described phenomena remain to be studied, as they have profound consequences for the management of game, especially the planning of hunting quotas.

A similar study on wild boar showed that the major factors explaining great variation in their densities in Europe and Western Asia were mean temperature of January (acting through elevated mortality in harsh winters) and habitat productivity (operating through food supply). The presence of wolves had only a small, and non-significant, limiting effect on wild boar abundance at the biogeographic scale (Melis *et al.*, 2006). The results of this macroecological study correspond very well to research done at local scales, which usually documented that predation is less important for the wild boar population than abiotic conditions, especially during winter (Okarma *et al.*, 1995, compare also Figure 10.9).

10.8 Is human hunting additive to, or compensatory for, losses to natural predators?

While management culls by human intervention may be expected to reduce and largely replace natural mortality (certainly where levels of natural mortality may be density dependent or related in some way to resource limitation), a number of studies have suggested that, in the specific case of predation, human culls may be directly additive to (rather than compensatory for) mortality due to natural predation – and that thus due allowance may need to be made in setting (human) cull allocations from a given herbivore population where natural predation may play an increasing role.

The study conducted in Białowieża Primeval Forest (eastern Poland) reported on an unintended experiment of greatly elevated hunting harvest in the presence of large predators (Figure 10.5). In the early 1990s, as a result of foresters' complaints about damage caused by overabundant deer, hunting quotas increased two- to threefold compared with earlier years, and intense harvesting continued for 6 years (until 1996). Wolves and lynx have long been present in the forest, but game managers seriously underestimated their impacts on ungulates. Since predation and hunting harvest were additive, and both wolf and lynx continued to kill nearly as many prey as before the heavy hunting by humans, the ungulates declined dramatically within just 5 years (Figure 10.5). Interestingly, despite the fact that heavy hunting ceased after 1996, the populations of roe deer seemed to recover very slowly and red deer continued to decline.

In their studies in multi-predator systems involving foxes, lynx and humans, Melis *et al.* (2009) concluded that lynx killed a higher proportion of roe deer at lower roe deer density; foxes killed a higher proportion of fawns at higher roe deer density. Hunters, however, were shown to take the same proportion of roe deer at any density. Further, while lynx, foxes and hunters did not show biased harvest in relation to sex, lynx showed no specific selection for any age class, foxes preferentially selected young fawns, and hunters took more adults and yearlings than fawns. Melis *et al.* therefore concluded that while the functional response of the different natural predators changed with roe deer density and the impacts of predation by lynx and foxes appeared to be compensatory with respect to each other, mortality through hunting was largely additional to that from natural predation. They also note that since adult survival rates in their study populations were lower by about 35% than those reported in environments without human harvest and with no natural predators, mortality by hunting and predation combined are also likely to be additive to mortality from other causes.

10.9 Ungulate management in areas with large carnivores

The presence of large carnivores in the area therefore does have implications for the harvesting regimes of ungulates. This problem is a particular challenge to game managers now, when wolves are recolonising areas where they have been extinct for decades, and reintroduction programmes of lynx are being performed or planned. Also in regions with long-term presence of these predators, game managers usually lack good estimates of

Table 10.4 *Hunting harvest and predation on ungulates in comparison to summer (seasonally highest) densities of ungulates in four localities in Europe*

Locality/species	Mean density in summer (*N* inds/100 km^2)	Mean annual (*N* inds/100 km^2)		
		Hunting harvest	Wolf predation[a]	Lynx predation
Italy, Western Alps, 2000–2002				
Red deer	302	61	25 (53)	–
Roe deer	512	57	39 (83)	–
Alpine Chamois	495	44	11 (23)	–
Italy, Northern Apennines, 2000–2006				
Wild boar	3100	988	107	–
Roe deer	1800	386	116	–
Switzerland, Jura Mountains, 1988–1997				
Roe deer	?	80	–	50
Chamois	?	8	–	12
Poland, Białowieża Primeval Forest, 1992–1996				
Moose	23	4	1	0
Red deer	653	134	76	61
Roe deer	635	97	22	157
Wild boar	530	81	30	negligible

[a]The lower numbers are estimates based on metabolic rates of wolves, higher numbers (in parentheses) are calculated based on daily food consumption in the wild.
Sources: Italy – Gazzola *et al.* (2007), A. Gazzola *et al.*, unpublished; Poland – Jędrzejewska and Jędrzejewski (1998); Switzerland – Molinari-Jobin *et al.* (2002).

the magnitude of predation. Indeed, the predation impact is most often underestimated, as it is believed to include only kill remains found by humans in the hunting range. In such situations, elevated harvest quotas may lead to declines of game.

Nilsen *et al.* (2005) have treated this problem with modelling moose harvesting strategies in the presence of wolves in Hedmark, south-east Norway. They demonstrated that if human harvests are indeed additive to those of natural predators, continuing to harvest at the rates used prior to wolf recolonisation would soon lead to a decline in the moose population. Thus, managers should reduce the size of harvest quotas to avoid decreases in prey populations.

In Europe, four empirical studies – in Italy, Switzerland and Poland – attempted to compare the magnitude of predation on an ungulate species with the hunting harvest by humans (Table 10.4). The ratios of the hunting harvest to predation varied from 1:0.11 to 1:1.61, mean for all species and localities being 1:0.68 (higher estimates of predation, as more realistic, were taken for the Italian study). Therefore, on average, the magnitude of annual predation was about two thirds of the hunting harvest.

In order accurately to predict predation impact on game species and consider it in harvest plans, it is necessary to have a reliable estimate of productivity of ungulate population and predation rates. Based on the study conducted in Białowieża Primeval Forest (eastern Poland), we propose the following recommendations for planning hunting quotas in Eastern and Central European regions with wolves and lynx. If the aim is to maintain stable populations of ungulates, annual mortality caused by harvest, large carnivores and other factors (e.g. harsh winters, traffic accidents) must not be larger than annual production of young. For red and roe deer that production was assessed based on the assumptions that adult females constituted 46% of population in late winter, and the birth rates were 0.9 calves/adult female in red deer and 1.45 calves/adult female in roe deer. Since the censuses of game are usually done in late winter to early spring, the calculated sustainable harvest was compared with those estimates. To obtain stability in deer populations, we propose that hunting quotas for red and roe deer should be planned as follows: (1) maximally 30% of late winter population numbers in regions with no large carnivores; (2) 20% of red deer and 25% of roe deer in areas with wolves; (3) 15% of late winter numbers of the two deer species in regions where both wolves and lynx occur. It must be noted that those recommendations are not universal. They should be adjusted to local conditions and consider varying productivity of ungulates and numbers of predators, climate harshness, level of poaching, and other factors.

In the past, large carnivores present in the hunting area were typically perceived as unwanted competitors of humans for the same resource. In recent decades, however, the numerous studies conducted in North America and Europe have evidenced the many positive roles and ecological services played by large carnivores. They may include, among others, lowering the density of ungulates below the habitat carrying capacity and thus releasing woody vegetation from heavy browsing pressure (e.g. Ripple and Beschta, 2004), and providing predictable food resources for scavengers, including several rare species (e.g. Wilmers *et al.*, 2003; Selva *et al.*, 2005). Furthermore, in the recent decade a growing concern about biodiversity loss has resulted in placing the top predators in a broader context of conservation of ecosystem biodiversity. There is a growing body of evidence that by limiting the numbers of herbivorous mammals, large carnivores trigger a trophic cascade, which has indirect positive consequences for richness and diversity of plant cover, birds and invertebrates (Ray *et al.*, 2005).

Obviously, in many regions large carnivores do cause problems, mainly by preying on domestic animals in rural areas (e.g. Nowak *et al.*, 2005; Gazzola *et al.*, 2007). Thus, development of good, flexible and rational methods to mitigate human–carnivore conflicts is necessary. In recent decades, the social attitude to wolves and lynxes has changed markedly, and empirical studies have accumulated so much evidence on the positive roles they play in European ecosystems that such methods can no longer be based on predator control alone. If necessary, the control should be focused on problem-causing packs or individuals, and not on the whole population.

There is no doubt that ungulate–large carnivore relationships in Europe need further research. Many questions remain to be answered to provide science-based recommendations to management and those interested in conservation of both ungulates and predators. Kill rates by wolves and lynxes at various densities of prey, and the magnitude of predation in mono- and multi-species communities of ungulates should receive special attention. Understanding the cascading effect of predation (i.e. the role that large carnivores play in the three-level trophic chain – including large herbivores and vegetation cover) is important for game management, forestry practices, and conservation measures of reserves and national parks.

Finally, a fairly dynamic development of large carnivore populations (expansion of wolves and at least regional shrinkage of lynx range) calls for a large-scale recognition of forest corridors that can be used by predators for migration and dispersal. By contrast, the rapidly increasing network of transportation infrastructure (highways and other motorways), especially in the eastern part of Europe where large carnivores abound, makes it necessary to assess the emerging new threats and challenges to their survival.

Acknowledgements

WJ and BJ are grateful for the financial support from Euronatur (Germany), BIOTER Centre of Excellence at MRI PAS (FP5, EVK2-CT-2002–80011) and the budget of the Mammal Research Institute PAS in Białowieża. We thank Tomasz Samojlik and Marzanna Zub for drawing the figures.

Fox box

[*Contributed by Rory Putman; drawing by Catherine Putman*]

The bulk of this chapter deals with the effects on ungulate populations of larger predators such as wolf and lynx – particularly in the wider context of the likely expansion of the distributional range of these large carnivores, both by natural colonisation and through possible reintroduction. Less attention is paid to red fox (*Vulpes vulpes*) since, while this small carnivore is very widespread, it appears more of an opportunist feeder and through much of its range does not appear to have any major impact on ungulate populations.

In Mediterranean environments, for example, published studies would suggest that foxes take a comparatively small number of young ungulates, if any, mainly relying on fruits and rodents (Rosalino and Santos-Reis, 2009), and this is true in a number of other European countries (e.g. Jędrzejewski and Jędrzejewska, 1992; Lanszki *et al.*, 2006; Sidorovich *et al.*, 2006; Webbon *et al.*, 2006; Dell'Arte *et al.*, 2007). In part, however, this low representation of ungulates within the overall annual dietary composition may be due to the fact that even where foxes do take ungulate prey, predation is predominantly targeted on juveniles and thus restricted to a short and markedly seasonal time window. Although red foxes can kill adult roe deer in winter when snow conditions give them an advantage (Borg, 1962; Cederlund and Lindström, 1993) they usually prey primarily on juveniles in their first summer – or indeed within the first 60 days of life (Linnell *et al.*, 1995; Aanes and Andersen, 1996; Kjellander and Nordström, 2003; Jarnemo *et al.*, 2004; Panzacchi, 2007; Melis *et al.* 2009).

In some instances, however, predation by foxes, even during this restricted period, may be highly significant. For example, foxes may take up a significant proportion of sika calves born in UK and the Republic of Ireland (O'Donoghue, 1991) and in a number of countries the impact of fox predation on infant roe deer may also have a very significant impact (see, for example, Kramer, 1990; Liberg *et al.*, 1993; Lindström *et al.*, 1994; Linnell *et al.*, 1995; Aanes and Andersen, 1996; Kjellander and Nordström, 2003; Jarnemo *et al.*, 2004; Panzacchi, 2007; Hewison and Staines, 2008). In Norway, Aanes and Andersen (1996) reported a 48% mortality of radio-collared fawns to fox predation within 60 days of birth, and similar results are reported by subsequent authors with 30% of 107 radio-collared fawns lost in studies of Melis *et al.* (submitted manuscript); an average of 34% of 233 marked neonates were taken by foxes in studies of Jarnemo *et al.* (2004) in Sweden. In 3 out of 14 years in this long-term study mortality rates exceeded 85%.

The highly seasonal and largely opportunistic predation by foxes on neonatal ungulates is thus rather different in nature from predation by wolves or lynx. A number of authors have, however, shown that red fox predation on roe fawns may have a significant effect on population density (Kramer, 1990; Liberg *et al.*, 1993; Lindström *et al.*, 1994; Aanes and Andersen, 1996) and some would argue (e.g. Kjellander and Nordström, 2003; Panzacchi, 2007) that, since the dynamics of large herbivore populations tend in any case to be characterised by relatively constant adult survival, but marked variation in juvenile recruitment, variable rates of predation on neonates may be of major significance in the dynamics of particular populations. Because of the importance of fox predation on the dynamics of some ungulate populations (especially roe deer) in some parts of their range (UK, Scandinavia), some consideration of what we know of these impacts is thus highlighted in this box. This review is taken from Putman (2008); see also Aanes *et al.* (1998).

Throughout, we must emphasise the opportunistic nature of fox predation. It would appear that lynx are to a large degree specialist predators on roe (Breitenmoser and Haller, 1993; Linnell *et al.*, 1995; Okarma *et al.*, 1997; Molinari-Jobin *et al.*, 2002): roe deer constitutes up to 90% of the biomass in lynx diet (Jędrzejewski *et al.*, 1993) and lynx continue to prey predominantly on roe even when roe deer are present only at comparatively low density (see main text).

By contrast, predation by foxes on roe (and on sika calves) is not only strongly seasonal but also varies markedly from year to year, dependent on the density of fox populations, the density of roe or sika populations and the relative availability of alternative prey. It is clear that predation rates vary with landscape type (Aanes and Andersen, 1996; Panzacchi, 2007; Panzacchi *et al.*, 2009: predation is higher in open landscapes) and with the relative availability of both roe deer fawns and alternative prey; high predation rates tend to coincide with reduced availability of alternative prey (such as voles; Kjellander and Nordström, 2003; Panzacchi, 2007; Panzacchi *et al.*, 2008a) and relatively higher proportional availability of roe kids. In consequence, overall mortality, and predation rates amongst roe deer fawns show strong annual variation (Jarnemo *et al.*, 2004; Panzacchi, 2007). In the same way, mortality rates reported for sika calves in the Republic of Ireland (O'Donoghue, 1991; T. Burkitt and N. Raymond, unpublished) show pronounced year to year variation, which has a profound affect on recruitment rates and population dynamics.

As we have already noted, predation of roe kids by foxes seems restricted to a very limited time window, with predation rates lower in the first week of life (while fawns are still 'hiders'; Aanes and Andersen, 1996) and with the bulk of fawns killed before 60 days (Aanes and Andersen, 1996; Jarnemo *et al.*, 2004). Indeed Jarnemo *et al.* report that 85% of fawns are killed before 30 days of age and 98% before 40 days. Panzacchi *et al.* (2008a) note that roe deer remains occurred more frequently in scats found at fox dens than in the scats of adult foxes, indicating that vixens were using fawns primarily to feed their cubs or at least that, during the period they were available, fawns were a profitable and significant food source for vixens raising cubs. This observation that vixens may be taking fawns primarily to feed cubs, and the restricted period of vulnerability of fawns, may help to explain the extremely restricted time window in which fawns are taken.

Timing of predation (and ultimately total amount) may also be affected by relative density of foxes and roe deer. Panzacchi (2007; Panzacchi *et al.*, 2008b) notes that in one study area in Norway where roe deer were at comparatively low abundance, foxes responded to this new source of prey purely opportunistically. Foxes were most successful in hunting roe fawns when surveying open areas, and when conducting less prey specific searches. In consequence roe deer fawns were killed for the most part opportunistically when they were encountered in the peak of their availability (and fawns born at the beginning of the season were comparatively safe).

In a different study site where roe were present at higher density, the higher abundance of fawns within the predator's home range triggered an early prey switch on the part of the fox, promoting a higher degree of specialisation on this prey source. Foxes started to search actively for roe fawns early, from the very beginning of the birth season, and caused a higher predation risk for fawns before or at the beginning of the birth peak. Given the limited time in which fawns are vulnerable to fox predation, the earlier prey switch also resulted in a higher overall impact on neonatal mortality rates in this second area and the proportion of animals killed by foxes was higher at higher density (Melis *et al.*, 2010). While in the latter case, during periods of high abundance of roe deer fawns, foxes become more specialised in actively seeking such prey, this is more of a matter of degree, and in both cases foxes are primarily responding opportunistically to roe as prey. Melis *et al.* conclude that as typical opportunistic feeders, foxes are likely to specialise in predating fawns only when their occurrence in the environment makes it worthwhile to actively spend time in their search.

Since roe deer fawns can only represent a seasonal food for red foxes, it is unlikely that availability or density of fawns will have a significant effect on fox density (since the foxes can readily switch to other prey in periods where few roe fawns are available; see also Aanes *et al.*, 1998; Panzacchi *et al.*, 2008a). Since, in addition, roe fawns (and other juvenile ungulates) are taken only opportunistically, it seems probable that the actual impact on prey population dynamics, as well as the style of fox predation, will be significantly different from that of larger carnivores which are hunting ungulate prey of all ages and throughout the year. Indeed we may expect that while predation by wolves or lynx (as ungulate specialists) is likely to have a limiting, and not a regulating, effect on prey populations (*sensu* Sinclair, 1989), fox predation, where it does impose a significant impact, may genuinely be expected to have a regulating effect on prey populations.

One other feature which may be noted is that the studies of Kramer (1990), Lindström *et al.* (1994), Aanes and Andersen, (1996), Melis *et al.* (2010) all suggest that mortality through human hunting is largely additional to any mortality resulting from natural predation and that thus the varying effects of fox predation on roe deer population dynamics may directly affect availability of roe as a quarry species.

References

Aanes, R. and Andersen, R. (1996) The effects of sex, time of birth, and habitat on the vulnerability of roe deer fawns to red fox predation. *Canadian Journal of Zoology*, **74**, 1857–1865.

Aanes, R., Linnell, J.D.C., Perzanowski, K., Karlsen, J. and Odden, J (1998) Roe deer as prey. In R. Andersen, P. Duncan and J.D.C. Linnell (eds.) *The European Roe Deer: The Biology of Success*. Oslo: Scandinavian University Press, pp. 139–160.

Andersen, R., Duncan, P. and Linnell, J.D.C. (eds.) (1998) *The European Roe Deer: The Biology of Success*. Oslo: Scandinavian University Press, 376 pp.

Andrén, H., Linnell, J.D.C., Liberg, O., *et al.* (2006) Survival rates and causes of mortality in Eurasian lynx (*Lynx lynx*) in multi-use landscapes. *Biological Conservation* **131**, 23–32.

Apollonio, M., Mattioli, L., Scandura, M., *et al.* (2004) Wolves in the Casentinesi Forests: insights for wolf conservation in Italy from a protected area with a rich wild prey community. *Biological Conservation* **120**, 249–260.

Bobek, B., Morow, K., Perzanowski, K. and Kosobucka, M. (1992) *Jeleń.* [Red Deer] (In Polish) Warsaw: Świat, 200 pp.

Boitani, L. (2000) *Action Plan for the Conservation of Wolves in Europe (Canis lupus)*. Nature and Environment 113. Strasbourg, France: Council of Europe Publishing, 86 pp.

Bonenfant, C., Gaillard, J.M., Klein, F., *et al.* (2002) Sex- and age-dependent effects of population density on life history traits of red deer *Cervus elaphus* in a temperate forest. *Ethography*, **25**(4), 446–458.

Borg, K. (1962) Predation on roe deer in Sweden. *Journal of Wildlife Management* **26**, 133–136.

Breitenmoser, U. and Haller, H. (1993) Patterns of predation by reintroduced European lynx in the Swiss Alps. *Journal of Wildlife Management* **57**, 135–144.

Breitenmoser, U., Kaczensky, P., Dotterer, M., *et al.* (1993) Spatial organization and recruitment of lynx (*Lynx lynx*) in a re-introduced population in the Swiss Jura Mountains. *Journal of Zoology, London* **231**, 449–464.

Breitenmoser, U., Breitenmoser-Würsten, C., Okarma, H., *et al.* (2000) *Action Plan for the Conservation of the Eurasian Lynx in Europe (Lynx lynx)*. Nature and Environment 113. Strasbourg, France: Council of Europe Publishing, 69 pp.

Breitenmoser, U., Breirenmoser-Würsten, C., Carbyn, L.N. and Funk, S.M. (2001) Assessment of carnivore reintroductions. In J.L. Gittleman, S.M. Funk, D.W. Macdonald and R.K. Wayne (eds.) *Carnivore Conservation*. Cambridge, UK: Cambridge University Press, pp. 241–281.

Capitani, C., Bertelli, I., Varuzza, P., Scandura, M. and Apollonio, M. (2004) A comparative analysis of wolf (*Canis lupus*) diet in three different Italian ecosystems. *Mammalian Biology* **69**, 1–10.

Capitani, C., Mattioli, L., Apollonio, M., *et al.* (2006) Distribution and population dynamics of a wolf population inhabiting eastern Tuscany, Italy. In C. Capitani, Dynamics of a wolf (*Canis lupus*) population in eastern-central Apennines, PhD thesis in Environmental Biology, University of Sassari, Italy, pp. 8–28.

Cederlund, G. and Lindström, E. (1993) Effects of severe winters and fox predation on roe deer mortality. *Acta Theriologica* **28**, 129–145.

Ciucci, P., Boitani, L., Francisci, F. and Andreoli, G. (1997) Home range, activity and movements of a wolf pack in central Italy. *Journal of Zoology, London* **243**, 803–819.

Danell, A.C., Andrén, H., Segerström, P. and Franzén, R. (2006) Space use by Eurasian lynx in relation to reindeer migration. *Canadian Journal of Zoology* **84**, 546–555.

Del Giudice, G.D., Riggs, M.R., Joly, P. and Pan, W. (2002) Winter severity, survival, and cause-specific mortality of female white-tailed deer in north-central Minnesota. *Journal of Wildlife Management* **66**, 698–717.

Dell'Arte, G.L., Laaksonen, T., Norrdahl, K. and Korpimäki, E. (2007) Variation in the diet composition of a generalist predator, the red fox, in relation to season and density of main prey. *Acta Oecologica* **31**, 276–281.

Dzięciołowski, R. and Pielowski, Z. (1993) *Łoś.* [Moose] (In Polish) Warsaw: Anton-5, 208 pp.

Elgmork, K. and Kaasa, J. (1992) Food habits and foraging of the brown bear *Ursus arctos* in central south Norway. *Ecography* **15**, 101–110.

Filonov, K.P. (1989) *Ungulates and Large Predators in Wildlife Reserves*. (In Russian) Moscow: Nauka, 251 pp.

Findo, S. and Chovancova, B. (2004) Home ranges of two wolf packs in the Slovak Carpathians. *Folia Zoologica* **53**, 17–26.

Frąckowiak, W. and Gula, R. (1992) The autumn and spring diet of brown bear *Ursus arctos* in the Bieszczady Mountains of Poland. *Acta Theriologica* **37**, 339–344.

Fruziński, B. (1992) *Dzik*. [Wild boar] (In Polish) Warsaw: Cedrus, 247 pp.

Gade-Jorgensen, I. and Stagegaard, R. (2000) Diet composition of wolves *Canis lupus* in east-central Finland. *Acta Theriologica* **45**, 537–547.

Gazzola, A., Bertelli, I., Avanzinelli, E., *et al.* (2005) Predation by wolves (*Canis lupus*) on wild and domestic ungulates of the western Alps, Italy. *Journal of Zoology, London* **266**, 205–213.

Gazzola, A., Avanzinelli, E., Bertelli, I., *et al.* (2007) The role of the wolf in shaping a multi-species ungulate community in the Italian western Alps. *Italian Journal of Zoology* **74**, 297–307.

Głowaciński, Z. and Profus, P. (1997) Potential impact of wolves *Canis lupus* on prey populations in eastern Poland. *Biological Conservation* **80**, 99–106.

Grosse, C., Kaczensky, P. and Knauer, F. (2003) Ants: a food source sought by Slovenian brown bears (*Ursus arctos*)? *Canadian Journal of Zoology* **81**, 1996–2005.

Hewison, A.J.M. and Staines, B.W. (2008) Roe deer. In S. Harris and D.W. Yalden (eds.) *Mammals of the British Isles: Handbook*, 4th edn. London: The Mammal Society, pp. 605–617.

Holling, C.S. (1959) Some characteristics of simple types of predation and parasitism. *Canadian Entomologist* **91**, 385–398.

Jarnemo, A., Liberg, O., Lockowandt, S., Olsson, A. and Wahlstrom, K. (2004) Predation by red fox on European red deer fawns in relation to age, sex, and birth date. *Canadian Journal of Zoology* **82**, 416–422.

Jędrzejewska, B. and Jędrzejewski, W. (1998) *Predation in Vertebrate Communities: the Białowieża Primeval Forest as a Case Study*. Berlin: Springer Verlag, 450 pp.

Jędrzejewska, B. and Jędrzejewski, W. (2005) Large carnivores and ungulates in European temperate forest ecosystems: bottom-up and top-down control. In J.C. Ray, K.H. Redford, R.S. Steneck and J. Berger (eds.) *Large Carnivores and the Conservation of Biodiversity*. Washington, DC: Island Press, pp. 230–246.

Jędrzejewska, B., Jędrzejewski, W., Bunevich, A.N., Miłkowski, L. and Okarma, H. (1996) Population dynamics of wolves *Canis lupus* in Białowieża Primeval Forest (Poland and Belarus) in relation to hunting by humans, 1847–1993. *Mammal Review* **26**, 103–126.

Jędrzejewska, B., Jędrzejewski, W., Bunevich, A.N., Miłkowski, L. and Krasiński, Z. A. (1997) Factors shaping population densities and increase rates of ungulates in Białowieża Primeval Forest (Poland and Belarus) in the 19th and 20th centuries. *Acta Theriologica* **42**, 399–451.

Jędrzejewski, W. and Jędrzejewska, B. (1992) Foraging and diet of the red fox *Vulpes vulpes* in relation to variable food resources in Białowieża National Park, Poland. *Ecography* **15**, 212–220.

Jędrzejewski, W., Schmidt, K., Miłkowski, L., Jędrzejewska, B. and Okarma, H. (1993) Foraging by lynx and its role in ungulate mortality: the local (Białowieża Forest) and the Palaearctic viewpoints. *Acta Theriologica* **38**, 385–403.

Jędrzejewski, W., Jędrzejewska, B., Okarma, H., *et al.* (1996) Population dynamics (1869–1994), demography, and home ranges of the lynx in Białowieża Primeval Forest (Poland and Belarus). *Ecography* **19**, 122–138.

Jędrzejewski, W., Jędrzejewska, B., Okarma, H., *et al.* (2000) Prey selection and predation by wolves in Białowieża Primeval Forest, Poland. *Journal of Mammalogy* **81**, 197–212.

Jędrzejewski, W., Schmidt, K., Theuerkauf, J., *et al.* (2002) Kill rates and predation by wolves on ungulate populations in Białowieża Primeval Forest (Poland). *Ecology* **83**, 1341–1356.

Jędrzejewski, W., Niedziałkowska, M., Nowak, S. and Jędrzejewska, B. (2004) Habitat variables associated with wolf (*Canis lupus*) distribution and abundance in northern Poland. *Diversity and Distributions* **10**, 225–233.

Jędrzejewski, W., Niedziałkowska, M., Mysłajek, R.W., Nowak, S. and Jędrzejewska, B. (2005) Habitat selection by wolves *Canis lupus* in the uplands and mountains of southern Poland. *Acta Theriologica* **50**, 417–428.

Jędrzejewski, W., Schmidt, K., Theuerkauf, J., Jędrzejewska, B. and Kowalczyk, R. (2007) Territory size of wolves *Canis lupus*: linking local (Białowieża Primeval Forest, Poland) and Holarctic-scale patterns. *Ecography* **30**, 66–76.

Jonas, T., Geiger, F. and Jenny, H. (2008) Mortality pattern of the Alpine chamois: the influence of snow-meteorological factors. *Annales of Glaciology* **49**, 56–62.

Karlsen, J. (1997) The impact of lynx (*Lynx lynx*) predation and hunting on a roe deer (*Capreolus capreolus*) population in southeastern Norway. Unpublished Candidatus scientiarum thesis, Norges Teknisk-Naturvitenskapelige Universitet, Trondheim, Norway.

Kerechun, S.F. (1979) Vliyanie khishchnikov na pogolove kopytnykh v zone Karpat (Impact of predators on ungulate numbers in the Carpathian zone). In V.E. Sokolov (ed.) *Ekologicheskie osnovy okhrany i ratsionalnogo ispolzovaniya khishchnykh mlekopitayushchikh.* [Ecological fundamentals of protection and rational utilization of predatory mammals] (In Russian) Moscow: Nauka, pp. 43–44.

Kjellander, P. and Nordström, J. (2003) Cyclic voles, prey switching in red fox and roe deer dynamics: a test of the alternative prey hypothesis. *Oikos* **101**, 338–344.

Kojola, I., Huitu, O., Toppinen, K., *et al.* (2004) Predation on European wild forest reindeer (*Rangifer tarandus*) by wolves (*Canis lupus*) in Finland. *Journal of Zoology, London* **263**, 229–235.

Kramer, A. (1990) Uberden einflussder fuchsdichte auf die zuwachsraten des rehwildes in der Schweiz. In S. Myrberget (ed.) *Transactions of the 19th Congress of the International Union of Game Biologists*, held Trondheim 1989. Trondheim, Norway: Norwegian Institute for Nature Research, pp. 53–58.

Kramer-Schadt, S., Revilla, E., Wiegand, T. and Breitenmoser, U. (2004) Fragmented landscapes, road mortality and patch connectivity: modelling influences on the dispersal of Eurasian lynx. *Journal of Applied Ecology* **41**, 711–723.

Kudaktin, A.N. (1978) Selective wolf predation on ungulates in the Caucasian nature reserve. *Byulleten Moskovskogo Obshchestva Ispytatelei Prirody, Otdel Biologii* **83**, 19–28. (In Russian with English summary)

Kusak, J. and Huber, D. (2000) Tracking a wolf pack in livestock-raising areas of Dalmatia, southern Croatia. Paper presented at Beyond 2000: Realities of Global Wolf Restoration symposium, held University of Minnesota, Duluth, USA. Program and Abstracts, p. 32.

Landa, A., Strand, O., Swenson, J.E. and Skogland, T. (1997) Wolverines and their prey in southern Norway. *Canadian Journal of Zoology* **75**, 1292–1299.

Lanszki, J. and Heltai, M. (2002) Feeding habits of golden jackal and red fox in south-western Hungary during winter and spring. *Mammalian Biology* **67**, 129–136.

Lanszki, J., Heltai, M., Szabó, L. (2006) Feeding habits and trophic niche overlap between sympatric golden jackal (*Canis aureus*) and red fox (*Vulpes vulpes*) in the Pannonian ecoregion (Hungary). *Canadian Journal of Zoology* **84**, 1647–1656.

Liberg, O., Johansson, A., Lockowandt, S. and Weahlstrom, K. (1993) Red fox predation as a dominant cause of neonatal mortality in roe deer fawns. Paper presented at Swedish Hunters' Association Conference, Lund. Abstract, p. 12.

Lindström, E.R., Adren, H., Angelstam, P., *et al.* (1994) Disease reveals the predator: sarcoptic mange, red fox predation and prey populations. *Ecology* **75**, 1042–1049.

Linnell, J.D.C., Aanes, R. and Andersen, R. (1995) Who killed Bambi? The role of predation in the neonatal mortality of temperate ungulates. *Wildlife Biology*, **1**, 209–224.

Linnell, J.D.C., Andersen, R., Kvam, T., *et al.* (2001) Home range size and choice of management strategy for lynx in Scandinavia. *Environmental Management* **27**(6), 869–879.

Linnell, J.D.C., Promberger, C., Boitani, L., *et al.* (2005) The linkage between conservation strategies for large carnivores and biodiversity: the view from the 'half-full' forests of Europe. In J.C. Ray, K.H. Redford, R.S. Steneck and J. Berger (eds.) *Large Carnivores and the Conservation of Biodiversity*. Washington, DC: Island Press, pp. 381–399.

Mattioli, L., Capitani, C., Avanzinelli, E., *et al.* (2004) Predation by wolves (*Canis lupus*) on roe deer (*Capreolus caprelous*) in north-eastern Apennine, Italy. *Journal of Zoology, London* **264**, 249–258.

Melis, C., Szafrańska, P., Jędrzejewska, B. and Bartoń, K. (2006) Biogeographic variation in wild boar (*Sus scrofa*) abundance in Europe and western Asia. *Journal of Biogeography* **33**, 803–811.

Melis, C., Jędrzejewska, B., Apollonio, M., *et al.* (2009) Predation has a greater impact in less productive environments: variation in roe deer, *Capreolus capreolus*, population density across Europe. *Global Ecology and Biogeography* **18**, 724–734.

Melis, C., Basille, M., Herfindal, I., *et al.* (2010) Roe deer population growth and lynx predation along a gradient of environmental productivity and climate in Norway. *Ecoscience* **27**, 166–174.

Meriggi, A. and Lovari, S. (1996) A review of wolf predation in southern Europe: does the wolf prefer wild prey to livestock? *Journal of Applied Ecology* **33**, 1561–1571.

Messier, F. (1991) The significance of limiting and regulating factors on the demography of moose and white-tailed deer. *Journal of Animal Ecology* **60**, 377–393.

Messier, F. (1995) On the functional and numerical responses of wolves to changing prey density. In: L.N. Carbyn, S.H. Fritts and D.R. Seip (eds.) *Ecology and Conservation of Wolves in a Changing World*. Canadian Circumpolar Institute, Occasional Publication 35. Edmonton, Alberta, Canada:. Canadian Circumpolar Institute, pp. 187–197.

Mitchell-Jones, A.J., Amori, G., Bogdanowicz, W., *et al.* (1999) *The Atlas of European Mammals*. London: T. and A.D. Poyser, 484 pp.

Molinari-Jobin, A., Molinari, P., Breitenmoser-Wursten, C. and Breitenmoser, U. (2002) Significance of lynx *Lynx lynx* predation for roe deer *Capreolus capreolus* and chamois *Rupicapra rupicapra* mortality in the Swiss Jura Mountains. *Wildlife Biology* **8**, 109–115.

Niedziałkowska M., Jędrzejewski W., Mysłajek R.W., *et al.* (2006) Environmental correlates of Eurasian lynx occurrence in Poland: large scale census and GIS mapping. *Biological Conservation* **133**, 63–69.

Nilsen, E.B., Pettersen, T., Gundersen, H., *et al.* (2005) Moose harvesting strategies in the presence of wolves. *Journal of Applied Ecology* **42**, 389–399.

Nores, C., Llaneza, L. and Álvarez, M.A. (2008) Wild boar *Sus scrofa* mortality by hunting and wolf *Canis lupus* predation: an example in northern Spain. *Wildlife Biology* **14**, 44–51.

Nowak, S., Mysłajek, R.W. and Jędrzejewska, B. (2005) Patterns of wolf *Canis lupus* predation on wild and domestic ungulates in the Western Carpathian Mountains (S Poland). *Acta Theriologica* **50**, 263–276.

Odden, J., Linnell, J.D.C. and Andersen, R. (2006) Diet of Eurasian lynx, *Lynx lynx*, in the boreal forest of south-eastern Norway: the relative importance of livestock and hares at low roe deer density. *European Journal of Wildlife Research* **52**, 237–244.

O'Donoghue, A. (1991) Growth, reproduction and survival in a feral population of Japanese sika deer (*Cervus nippon*). Unpublished PhD thesis, University College, Dublin, Ireland.

Okarma, H. (1995) The trophic ecology of wolves and their predatory role in ungulate communities of forest ecosystems in Europe. *Acta Theriologica* **40**, 335–386.

Okarma, H., Jędrzejewska, B., Jędrzejewski, W., Krasiński, Z.A. and Miłkowski, L. (1995) The roles of predation, snow cover, acorn crop, and man-related factors on ungulate mortality in Białowieża Primeval Forest, Poland. *Acta Theriologica* **40**, 197–217.

Okarma, H., Jedrzejewski, W., Schmidt, K., Kowalczyk, R. and Jedrzejewska, B. (1997) Predation of Eurasian lynx on roe deer *Capreolus capreolus* and red deer *Cervus elaphus* in Białowieża Primeval Forest, Poland. *Acta Theriologica*, **42**, 203–224.

Okarma, H., Jędrzejewski, W., Schmidt, K., *et al.* (1998) Home ranges of wolves in Białowieża Primeval Forest, Poland, compared with other Eurasian populations. *Journal of Mammalogy* **79**, 842–852.

Ornicans, A., Andersone-Lilley, Z., Zunna, A., *et al.* (2005) Radio tracking of Eurasian lynx *Lynx lynx* in Latvia: first results and conclusions. Paper presented at 6th Baltic Theriological Conference, Latvia, 11–15. November 2005. Abstracts, p. 38.

Panzacchi, M. (2007) The ecology of red fox predation on roe deer fawns, with respect to population density, habitat and alternative prey. Unpublished PhD thesis, University of Bologna, Italy.

Panzacchi, M., Linnell, J.D.C., Serrao, G. *et al.* (2008a) Evaluation of the importance of roe deer fawns in the spring-summer diet of red foxes in south-eastern Norway. *Ecological Research* **23**, 889–896.

Panzacchi, M., Linnell, J.D.C., Odden, J., Odden, M. and Andersen, R. (2008b) When a generalist becomes a specialist: patterns of red fox predation on roe deer fawns under contrasting conditions. *Canadian Journal of Zoology* **86**, 116–126.

Panzacchi, M., Linnell, J.D.C., Odden, M., Odden, J. and Andersen, R. (2009) Habitat and roe deer fawn vulnerability to red fox predation. *Journal of Animal Ecology* **78**, 1124–1133.

Pereira, M., Fonseca, F. and Magalhaes, C. (1985) Wolf ecology in Portugal. In *Proceedings of the Simposio de Predadores*. Lisbon: Faculdade de Ciências da Universidade de Lisboa, pp. 122–167.

Promberger, C., Ionescu, O., Mertens, A., *et al.* (1998) *Carpathian Large Carnivore Project: Annual report 1997/1998*. Ettal, Germany: Munich Wildlife Society.

Putman, R.J. (2008) *A Review of Available Data on Natural Mortality of Red and Roe Deer Populations*. Contract report for the Deer Commission for Scotland, Inverness.

Ray, J.C. (2005) Large carnivorous animals as tools for conserving biodiversity: assumptions and uncertainties. In J.C. Ray, K.H. Redford, R.S. Steneck and J. Berger (eds.) *Large Carnivores and the Conservation of Biodiversity*. Washington, DC: Island Press, pp. 34–56.

Ray, J.C., Redford, K.H., Steneck, R.S., Berger, J. (eds.) (2005) *Large Carnivores and the Conservation of Biodiversity*. Washington, DC: Island Press, 526 pp.

Ripple, W.J. and Beschta, R.L. (2004) Wolves, elk, willows, and trophic cascades in the upper Gallatin Range of Southwestern Montana, USA. *Forest Ecology and Management* **200**, 161–181.

Rosalino, L.M. and Santos-Reis, M. (2009) Fruit consumption by carnivores in Mediterranean Europe. *Mammal Review* **39**, 67–78.

Rusakov, O.S. (1979) Khishchniki kak faktor dinamiki chislennosti kopytnykh na Severo-Zapade SSSR. [Predators as factors of the dynamics of ungulate numbers in the North-West USSR] (In Russian) In V.E. Sokolov (ed.) *Ekologicheskie osnovy okhrany i ratsionalnogo ispolzovaniya khishchnykh mlekopitayushchikh* [Ecological fundamentals of protection and rational utilization of predatory mammals]. Moscow: Nauka, pp. 61–62.

Rusakov, O.S. and Timofeeva, E.K. (1984) *Kaban – ekologiya, resursy, khozaistvennoe znachenie na Severo-Zapade SSSR*. [The wild boar: its ecology, resources, and economic importance in the North-West USSR] (In Russian) Leningrad, Russia: Izdatelstvo Leningradskogo Universiteta, 206 pp.

Sand, H., Andren, H., Liberg, O. and Ahlqvist, P. (2000) Telemetry studies of wolves (*Canis lupus*) in Scandinavia: a new research project. Paper presented at Beyond 2000: Realities of Global Wolf Restoration symposium, held University of Minnesota, Duluth, USA. Program and Abstracts, pp. 66–67.

Sand, H., Zimmermann, B., Wabakken, P., Andren, H. and Pedersen, H.C. (2005) Using GPS technology and GIS cluster analyses to estimate kill rates in wolf-ungulate ecosystems. *Wildlife Society Bulletin* **33**, 914–925.

Schmidt, K. (2008) Behavioural and spatial adaptation of the Eurasian lynx to a decline in prey availability. *Acta Theriologica* **53**, 1–16.

Schmidt, K., Jędrzejewski, W. and Okarma, H. (1997) Spatial organization and social relations in the Eurasian lynx population in Białowieża Primeval Forest, Poland. *Acta Theriologica* **42**, 289–312.

Selva, N., Jędrzejewska, B., Jędrzejewski, W. and Wajrak, A. (2005) Factors affecting carcass use by a guild of scavengers in European temperate woodland. *Canadian Journal of Zoology* **83**, 1590–1601.

Sidorovich, V.E. (2006) Relationship between prey availability and population dynamics of the Eurasian lynx and its diet in northern Belarus. *Acta Theriologica* **51**, 265–274.

Sidorovich, V.E., Tikhomirova, L.L. and Jędrzejewska, B. (2003) Wolf *Canis lupus* numbers, diet and damage to livestock in relation to hunting and ungulate abundance in northeastern Belarus during 1990–2000. *Wildlife Biology* **9**, 103–111.

Sidorovich, V.E., Sidorovich, A.A. and Izotova, I.V. (2006) Variations in the diet and population density of the red fox *Vulpes vulpes* in the mixed woodlands of northern Belarus. *Mammalian Biology* **71**, 74–89.

Sinclair, A.R.E. (1989) Population regulation in animals. In J.M. Cherret (ed.) *Ecological Concepts*. Oxford, UK: Blackwell Scientific Publications, pp. 197–241.

Sinclair, A.R.E. (2003) Mammal population regulation, keystone processes and ecosystem dynamics. *Philosophical Transactions of the Royal Society B–Biological Sciences* **358**, 1729–1740.

Śmietana, W. (1998) Drapieżnictwo wilków jako czynnik śmiertelności w zespole kopytnych na terenie Bieszczadzkiego Parku Narodowego i w jego otoczeniu. [Wolf predation as a mortality factor in ungulate community of Bieszczady National Park and its surroundings] (In Polish) Unpublished Doctoral thesis, Instytut Ochrony Przyrody PAN, Kraków, Poland, 74 pp.

Śmietana, W., Okarma, H. and Śnieżko, S. (2000) Bieszczadzka populacja rysia. [Lynx population in the Bieszczady Mountains] (In Polish with English summary) *Monografie Bieszczadzkie* **9**, 147–155.

Stahl, P., Vandel, J.M., Ruette, S., *et al.* (2002) Factors affecting lynx predation on sheep in the French Jura. *Journal of Applied Ecology* **39**, 204–216.

Sunde, P., Kvam, T., Bolstad, J.P. and Bronndal, M. (2000) Foraging of lynxes in a managed boreal-alpine environment. *Ecography* **23**, 291–298.

Timofeeva, E.K. (1974) *Los – ekologia, rasprostranenie, khozaistvennoe znachenie.* [Moose: ecology, distribution and economic importance] (In Russian) Leningrad, Russia: Izdatelstvo Leningradskgo Universiteta, 166 pp.

Vatolin, B.A. (1979) O populatsii volka v Bryanskoi oblasti i regulirovanii ee chislennosti. [On the wolf population and regulation of its numbers in the Bryansk region] (In Russian) In V.E. Sokolov (ed.) *Ekologicheskie osnovy okhrany i ratsionalnogo ispolzovaniya khishchnykh mlekopitayushchikh* [Ecological fundamentals of protection and rational utilization of predatory mammals]. Moscow: Nauka, p. 91.

Vila, C., Urios, V. and Castroviejo, J. (1990) Wolf ecology in Cabrera and in Carballeda (Zamora). In J.C. Blanco, L. Cuesta and S. Reig (eds.) *The Wolf (*Canis lupus*) in Spain: Status, Issues and Notes on his Ecology*. Madrid: Instituto para la Conservacion de la Naturaleza, pp. 95–108. (In Spanish)

Webbon, C.C., Baker, P.J., Cole, N.C. and Harris, S. (2006) Macroscopic prey remains in the winter diet of foxes *Vulpes vulpes* in rural Britain. *Mammal Review* **36**, 85–97.

Wilmers, C.C., Stahler, D.R., Crabtree, R.L., Smith, D.W. and Getz, W.M. (2003) Resource dispersion and consumer dominance: scavenging at wolf- and hunter-killed carcasses in Greater Yellowstone, USA. *Ecology Letters* **6**, 996–1003.

Yevtikov, S.A. (1991) Natural mortality rate in European moose. In B. Bobek, K. Perzanowski and W. Regelin (eds.) *Global Trends in Wildlife Management*. Transactions of the 18th International Union of Game Biologists Congress, Kraków 1987. Krakow, Poland: Swiat Press, pp. 653–654.

11

The role of pathogens in the population dynamics of European ungulates

MARION L. EAST, BRUNO BASSANO AND BJØRNAR YTREHUS

11.1 Introduction

Numerous ecological studies have investigated how key factors such as resource availability and predation may influence the population dynamics of European ungulates (Clutton-Brock *et al.*, 1985; Putman *et al.*, 1996; Sæther, 1997; Forchhammer *et al.*, 1998; Gaillard *et al.*, 1998, 2000; Jędrzejewska and Jędrzejewski, 1998; Coulson *et al.*, 2006). In comparison, research on the consequences of pathogen infection on ungulate population dynamics has been rather limited. In recent years interest in pathogen–host dynamics has increased not only among ecologists, in relation to these dynamics at a population level, but also among scientists in other disciplines such as veterinarians, immunologists and geneticists whose interests focus on the individual, cellular or genetic level.

As we will discuss, pathogen–host dynamics are complex and involve interconnected ecological, behavioural, physiological and genetic pathways. Because of this complexity it is often difficult to unravel these pathways, which is necessary to gain an understanding of the role of pathogens in shaping the population dynamics of ungulates. As the studies discussed in this chapter demonstrate, long-term research is necessary to reveal the role of pathogens on ungulate population dynamics. Unfortunately this requirement does not conform well to the current short-term nature of research funding.

Here we present an overview of pathogen–host dynamics and the long-term consequences of pathogen infection for European ungulates. Although a detailed discussion of the evolutionary role of pathogens is beyond the scope of this chapter, we briefly highlight the importance of pathogens in driving evolutionary processes and physiological trade-offs that modulate immune

Ungulate Management in Europe: Problems and Practices, eds. Rory Putman, Marco Apollonio and Reidar Andersen. Published by Cambridge University Press.

responses. Theoretical models are a useful tool that can provide insights into pathogen transmission and maintenance. We outline the general concepts used to formulate models and discuss empirical studies that illustrate these concepts. We discuss how the apparently simple task of quantifying key parameters for models can pose considerable practical problems, and caution that pathogens can influence population dynamics even when animals show no obvious clinical signs of infection. As the majority of European ungulate populations live in human-altered habitats, we consider how anthropogenic activities may alter pathogen–host dynamics.

We note that there are surprisingly few studies that provide a stringent assessment of pathogen–host dynamics in European ungulate populations, although there are numerous descriptive reports and case studies of pathogen infection and prevalence of exposure to pathogens in European ungulates. Although these studies provide much useful information, they do not generally contain the information necessary to determine the population consequences of pathogen infection and we suggest that there is considerable scope for research in this field.

The current and likely future scale of human-induced habitat and climate changes throughout Europe are cause for concern as these anthropogenic changes are likely to alter pathogen–host dynamics in European ungulate populations; alterations that may be to the benefit of the pathogen or the host. The current shortage of information on pathogen–host dynamics in European ungulate populations may hinder future attempts to rigorously assess the effect of anthropogenic factors on these dynamics.

11.2 Pathogens, evolution and life-history trade-offs

Traditionally, veterinarians and wildlife managers have considered pathogens as unwanted agents that should be controlled or eliminated through the use of medications, mass vaccination or the culling of infected individuals, whereas evolutionary biologists view pathogens as important agents of natural selection (Anderson and May, 1982; Hamilton and Zuk, 1982; Sommer, 2005). As a comprehensive review of the evolutionary importance of pathogens is beyond the scope of this chapter, we restrict ourselves in this section to key factors that modulate immune responses, and discuss how pathogens drive selection for life-history trade-offs and behaviours that may influence population dynamics.

Pathogen infection can influence life-history trade-offs and strategies of hosts (Pontier *et al.*, 1998; Agnew *et al.*, 2000; Lee, 2006; Barrett *et al.*, 2008; Jones *et al.*, 2008) because body resources channelled into fighting infections will not be available for other functions such as body maintenance, growth, costly sexual displays, gestation and lactation (Kirkwood, 1981; Hamilton and Zuk, 1982). For example, because lactation is energetically expensive, higher levels of parasite infection reported in lactating than non-lactating bighorn ewes *Ovis canadensis* are considered by Festa-Bianchet (1989) likely to result from increased investment of body reserves in lactation leading to insufficient investment in immune functions.

Many factors such as age, sex, social status, competition for food resources and seasonal changes in climate may influence an individual's investment in immune function (Lee, 2006; Martin *et al.*, 2008; Martin, 2009). Nutritionally demanding immune responses can deplete body reserves and thus curtail future investment in immune function, making individuals more vulnerable to further infections (Gasbarre, 1997; Beldomenico *et al.*, 2008). Serious infectious and inflammatory diseases can accelerate rate of senescence and curtail lifespan and life-time reproductive success (Aviv, 2008; Ilmonen *et al.*, 2008) and these processes may have population consequences. High pathogen burden can curtail an individual's reproductive success not only because combating infections requires body resources that otherwise could have been devoted to reproduction, but also because of behavioural preferences for mating partners with a low parasite load (Hamilton and Zuk, 1982; Kavaliers *et al.*, 2005).

Pathogens are thought to drive selection in their hosts for polymorphism at loci within both the type I and II regions of the major histocompatibility gene complex (MHC): a gene complex known to play an important role in determining susceptibility to pathogen infection and in individual odour and mate choice (Sommer, 2005). The unmanaged Soay sheep population on the remote island of Hirta in the St Kilda archipelago off the west coast of Scotland has been the focus of studies that consider coevolution between ungulate hosts and pathogens. In this context, Paterson *et al.* (1998) investigated coevolution between MHC in Soay sheep and helminth infection. The abundance of eggs from several helminth species (*Teladorsagia circumcincta*, *Teladorsagia davtiani*, *Ostertagia trifurcate*, *Trichostrongylus axei*, *Trichostrongylus vitrinus*) was assessed in faeces from a large sample of individually known sheep. Five polymorphic microsatellite loci were considered, including loci within both the MHC type II region and the MHC type I region and two flanking markers as controls. Survival of juvenile Soay sheep was associated with all three MHC loci but not with either of the control flanking markers. Allelic variation within the MHC was significantly associated with both survival and resistance to intestinal nematode infection.

In an earlier study on the same population of Soay sheep, Gulland *et al.* (1993) considered the influence of Soay sheep genotype at a specific locus (the diallelic adenosine deaminase locus *Ada*) in relation to nematode *Teladorsagia* sp. infection and survival. Sheep genotypes were either homozygous (FF or SS) or heterozygous (FS) at this locus. Long-term monitoring of the size of the Soay sheep population revealed fluctuations between approximately 600 and 1600 individuals. Population crashes were associated with shortages of winter forage when individuals became nutritionally compromised. Sheep dying in winter population crashes showed abomasal lesions consistent with damage by *Teladorsagia* sp. Using data from 1101 Soay sheep genotyped at the *Ada* locus, no difference was found in genotype frequencies in 10 cohorts before the occurrence of a winter population crash. Prevalence of genotypes after a winter population crash revealed that mortality had been highest in individuals with the homozygous FF genotype, intermediate for individuals with the homozygous SS genotype and lowest for the heterozygous genotypes FS. During five summers of parasite monitoring, prevalence of nematode infection was also influenced by *Ada* genotypes, with significantly lower infections among heterozygous females. Finally, sheep experimentally treated against helminth infection survived winter population crashes better than matched controls, suggesting that infection with nematode parasites contributed to the probability of death during a crash. The results of this study provide strong evidence of the impact of a macroparasite on ungulate host population dynamics as well as parasite-driven selection.

A recent study on the European wild boar *Sus scrofa* also provides evidence that genetic variability is a factor associated with susceptibility to infection with *Mycobacterium bovis*, the causative agent of bovine tuberculosis (bTB). Boar with high heterozygosity were less likely to be infected than individuals with low levels of heterozygosity (Acevedo-Whitehouse *et al.*, 2005).

11.3 Theoretical background

Both predators and pathogens are thought to exert a ‘top-down’ effect on prey/host populations (Tompkins and Begon, 1999; Begon *et al.*, 2006) and predator–prey ecology has laid the foundation for theoretical models of pathogen–host dynamics (Raffel *et al.*, 2008). Mathematical models are useful tools that can help clarify the factors determining the spread and impact of pathogens on host populations (Anderson and May, 1991; Tompkins and Begon, 1999; Diekmann and Heesterbeek, 2000; Ahrens and Pigeot, 2005). Many models are based on the basic SIR concept, in which the host population consists of susceptible (also termed naïve) individuals (S) that can be infected, infected individuals (I) that transmit the pathogen to susceptible animals, and individuals that have recovered from infection (R) which may be considered immune to further infection for a specified period or throughout the remaining life of the recovered individual, depending on the duration of immunity induced by exposure to a specific pathogen. For example, survival of infection by a *Morbillivirus* such as rinderpest (see Section 11.3.2) in cattle or measles in humans is thought to provide life-long immunity to further infection, whereas exposure to other pathogens may only provide transient immunity. Currently the duration of immunity induced by many pathogens is unknown in most wildlife species.

11.3.1 Density-dependent transmission within a single host species

If pathogen transmission increases as host population density increases and declines as host population density declines, then transmission of pathogens is called density dependent. Virulent pathogens cause a proportion of infected individuals to die, thus decreasing host density and reducing pathogen transmission. Density-dependent models that consider pathogen transmission in a single host species predict the local extinction of the pathogen before the pathogen drives the host population to extinction (Anderson and May, 1991; Tompkins and Begon, 1999; Diekmann and Heesterbeek, 2000; McCallum *et al.*, 2001). Pathogen-driven extinction of host populations is more likely when a pathogen infects more than one host species (as described in Section 11.3.5). However, host populations that are dramatically reduced in size by pathogen infection may become more vulnerable to extinction due to other causes.

To examine the impact of the nematode *Ostertagia gruhneri* on a reindeer *Rangifer tarandus plathyrynchus* population in Norway over a period of six years, Albon *et al.* (2002) experimentally manipulated parasite burden in a population with the application of anti-helminth medication. *Ostertagia gruhneri* was one of only two species of parasites from the family Strongylidae known to infect the abomasums of the study population and only *O. gruhneri* was considered to be pathogenic. Reindeer population densities in the study area fluctuated over a twofold range, and birth and death rates in the study population covaried. Low or negative population growth generally followed winters with high precipitation which were associated with poor survival and low birth rate during the subsequent summer. The results of this experimental study revealed that *O. gruhneri* infection decreased fecundity but not survival of infected reindeer and that this parasite-mediated reduction in calf production was density dependent, as fecundity in the population decreased with an increase in the mean estimate of *O. gruhneri* abundance in the host population. The abundance of *O. gruhneri* was also density dependent, with a delayed positive response to increases in host population densities. This nematode appeared to be sufficient to regulate the host population.

11.3.2 Naïve hosts and herd immunity

A high rate of pathogen transmission is expected to lead to an increase in the population density of recovered and immune individuals in a host population (i.e. an increase in herd immunity, and a reduction in the density of susceptible hosts), which in turn will curtail pathogen transmission. When herd immunity is absent, as is the case when a naïve host population is exposed to a pathogen that it has not encountered previously, infection is expected to spread rapidly through the host population. Before the 1880s, cattle and wild ungulates in Africa south of the Sahara had not encountered rinderpest, a *Morbillivirus* from Asia in the family Paramyxoviridae. When rinderpest was introduced (probably by the movement of infected cattle) into naïve ungulate populations without previous exposure to this virus, it swiftly spread throughout sub-Saharan Africa, causing high mortality in cattle and closely related wild herbivores such as the African buffalo *Syncerus caffer*, and famine in many pastoralist communities (Plowright, 1982).

A comparable scenario occurred when a pestivirus spread through chamois *Rupicapra pyrenaica* in the Spanish Pyrenees in 2001 and 2002. Pestiviruses belong to the family Flaviviridae and cause border disease in small ruminants, bovine viral diarrhoea in cattle and classical swine fever in pigs (see Section 11.3.5). Chamois in specific Pyrenean valleys in Spain suffered increased mortality and seroprevalence rapidly increased to 80% in 2002 shortly after the outbreak of the disease, then declined to 56% in 2004 (Marco *et al.*, 2008). Chamois infected with border disease showed various signs including weight loss, alopecia, and a lack of fear of humans (Marco *et al.*, 2007). A detailed study of border disease in chamois in the National Game and Wildlife Reserve of Orlu in the French section of the Pyrenees found serological evidence of exposure to border disease in chamois since 1995 (Pioz *et al.*, 2007). During the period of the study (1995–2004), seroprevalence varied between 43% and 90% (mean 70%). In contrast to other chamois populations infected with border disease, no mass mortality was observed in the Orlu population. This difference was not thought to be due to differences in viral strains, because the Orlu border disease variant was similar to other strains from chamois populations in Spain and Andorra (Pioz *et al.*, 2007).

Infected chamois that manage to clear the virus are thought to gain lifelong immunity. Females infected with the virus during pregnancy play a key role in the epidemiology of the virus because they can produce offspring that are persistently infected and these offspring shed large amounts of virus into the environment. Infected offspring often die prematurely. Pioz *et al.* (2007) constructed a model to investigate factors associated with infection using information on the prevalence of infection (viropositives) and exposure (seropositives) to the virus. The results of the model suggest that herd immunity decreased and virus transmission rates increased during years when the number of young and subadults in the population was high. This was because young animals had a low level of immunity (seroprevalence in young animals was only 18%). As demonstrated for many other diseases, chamois mothers are thought to provide their dependent offspring with protective antibodies via milk, and when this maternally derived protection wanes young become susceptible to infection. In Orlu, protection of offspring is thought to decline in autumn, when virus transmission rates increase, because females and males congregate in large herds to mate. Thus, changes in levels of herd immunity and in spatial behaviour between the mating and non-mating season play an important role in border disease maintenance and transmission. Currently, the majority of adult chamois in the Orlu population are thought to be immune and thus the future effect of the virus on the population is expected to be mediated primarily through reduced birth rates associated with abortions and reduced survival of infected calves. Quantifying the level of abortions and early infant mortality in the population will be problematic, and thus assessing the future impact of border disease on the Orlu populations will be difficult (Section 11.4).

11.3.3 Frequency-dependent transmission

The transmission of vector-borne pathogens and sexually transmitted pathogens is often independent of host population density. Instead, it is determined by the frequency of contacts between individuals (Courchamp *et al.*, 1995; Begon *et al.*, 1999). In many mammal populations, frequency of contact between infectious and susceptible individuals will be strongly influenced by social factors such as social status. For example, transmission of feline immunodeficiency virus (FIV) in the domestic cat *Felis catus* is thought to be through virus-infected saliva in bites received during aggressive contests between males for social status and from males to females when males bite females during coitus. Infection within urban cat colonies ranges between 0% and 30%, with socially dominant, aggressive adult males having the highest probability of infection (Courchamp *et al.*, 1995; Pontier *et al.*, 1998; Natoli *et al.*, 2005). Frequency-dependent transmission has been the focus of several theoretical studies that reveal the capacity for such transmission to drive host populations to extinction or limit host population range (e.g. Thrall *et al.*, 1993; Boots and Sasaki, 2002; Antonovics, 2009). These models have rarely been tested using empirical data from free-ranging mammals (but see Courchamp *et al.*, 1995; Begon *et al.*, 1999).

Recently Smith *et al.* (2009a) used long-term empirical data on the transmission of cowpox (an *Orthopoxvirus*) within a field vole *Microtus agrestis* population to investigate processes of transmission. Their results indicated that both density-dependent and frequency-dependent processes operated in this population and that these processes were influenced by seasonal variation in host density, host susceptibility to infection and contact rates.

11.3.4 Structured host populations

For simplicity, most early models of pathogen transmission assumed that interactions within host populations were homogeneous. These models did not consider structure within a host population, even though this may have a profound impact on pathogen transmission. More recently, mathematical models have been developed that allow for the inclusion of population heterogeneity through the use of sub-communities within populations (e.g. Grenfell and Harwood, 1997; Lindholm and Britton, 2007; Webb *et al.*, 2007). Pathogen transmission is then dependent on the frequency of contact within and between different sub-communities within the population.

11.3.5 Multi-host pathogens

Modelling transmission of pathogens that exploit multiple host species is complex (Anderson and May, 1991; Diekmann and Heesterbeek, 2000; Gandon, 2004) and has received relatively little attention even though multi-host pathogens are numerous, are of considerable economic importance, and are primarily responsible for infectious disease outbreaks in wildlife and emerging diseases in humans, livestock and wildlife (Daszak *et al.*, 2000; Woolhouse, 2002; Gortázar *et al.*, 2007). Pathogens that exploit multi-host species are also of concern to those involved in biological conservation (Smith *et al.*, 2006, 2009b; Gortázar *et al.*, 2008) because multi-host pathogens are more likely to drive local host populations to extinction than pathogens that exploit a single host species (de Castro and Bolker, 2005; Fenton and Pedersen, 2005). This is because theoretically an infection can spread from a host species that occurs at a relatively high density to an endangered species present at a low density.

Many economically important pathogens utilise multiple hosts. Examples include classical swine fever, a pestivirus that infects both domestic pigs and the European wild boar (Kramer-Schadt *et al.*, 2007, 2009), and foot-and-mouth disease, a disease caused by a virus belonging to the family Picornaviridae that can have severe economic implications for intensive livestock producers. As a result, there are strict regulations designed to prevent this disease occurring in European countries. Currently there is no evidence that foot-and-mouth-disease is maintained in any wildlife population in Europe (Thomson *et al.*, 2003).

Bovine tuberculosis (bTB), caused by *Mycobacterium bovis*, is an important multi-host bacterial disease of both livestock and several wildlife species, including European ungulates such as European wild boar, red deer and fallow deer *Dama dama* (Acevedo-Whitehouse *et al.*, 2005; Vincente *et al.*, 2007; Gortázar *et al.*, 2008; McDonald *et al.*, 2008). Brucellosis is another economically important multi-host bacterial disease caused by *Brucella abortus* in cattle and closely related wildlife species, *B. melitensis* in sheep and goats, and *B. suis* in domestic pigs and the European wild boar (Godfroid and Käsbohrer, 2002; Gortázar *et al.*, 2007).

Blood-sucking arthropods such as ticks are very successful agents of pathogen transmission and often feed on several host species. Such arthropod vectors may reduce populations of a rare or vulnerable host species by transmitting pathogens to rare species from a more abundant host (Hudson and Greenman, 1998). Such pathogen-mediated apparent competition may be important in structuring ecological assemblies (Bonsall and Hassell, 1997). When pathogen vectors such as ticks feed on a range of species that vary in their susceptibility to infection by tick-borne pathogens, such as the spirochaete *Borrelia burgdorferi* (Lyme disease), ticks that feed on a rich community of host species are less likely to transmit disease than those that feed on a species-poor community of hosts dominated by the few species that are competent disease reservoirs (LoGiudice *et al.*, 2003). This phenomenon has been termed 'the dilution effect'.

11.3.6 Multiple pathogen infections

In free-ranging ungulate populations, most individuals are likely to be co-infected with more than one pathogen, and co-infections are likely to involve diverse combinations of micropathogens and macropathogens. As the various possible combinations of co-infection are numerous, the development of realistic models is difficult. Lello and Hussell (2008) suggested that one possible solution could be to use the ecological concept of guilds to model the expected outcome of co-infections composed of pathogens from different guilds.

When a host is infected with more than one pathogen, there is the potential for host immune-mediated interactions between co-infecting pathogens. Immune-mediated interactions between parasites may occur through cross-reactivity, in which exposure to one pathogen enhances immunity to a second pathogen. Alternatively, exposure to one pathogen might result in immunosuppression and decreased immunity to another pathogen and such interactions may modulate host susceptibility, influence patterns of co-infection, rates of transmission and pathogen distribution within a host population (Behnke *et al.*, 2001; Graham *et al.*, 2007; Cattadori *et al.*, 2008).

A host's response to infection is determined by the cytokine environment that initiates the development of T cells into one of two types of T helper cells (Th cells). Infection with micropathogens (viruses, intracellular protozoa), induces a pro-inflammatory Th1 response that stimulates the production of cytokines (interferon type 1 and type 2, interleukin (IL)-12) and a tumour necrosis factor. The cytokines induce macrophages to engulf and eliminate micropathogens. If infection cannot be eliminated in this manner, the Th1 response can lead to a dangerous inflammatory response. In contrast, infection by extracellular pathogens such as gastrointestinal helminths leads to a Th2 response, which involves the production of cytokines (IL-4, IL-5, IL-9, IL-13), the production of antibodies, immunoglobulins and class II MHC antigens. IL-4 also suppresses some macrophage functions, thereby preventing potentially damaging inflammatory responses (Graham *et al.*, 2007). Both the Th1 and Th2 response may be induced during infection by macropathogens and micropathogens because many pathogens move through different host tissues during their infectious cycle and immune responses can function at a tissue-specific level.

Cattadori *et al.* (2007) studied co-infection of European rabbits *Oryctolagus cuniculus* with the immunosuppressive myxoma virus that causes myxomatosis and a gastrointestinal helminth *Trychostrongylus retortaeformis*. Myxoma virus infection disrupted the ability of rabbits to clear *T. retortaeformis* and seasonal outbreaks of the virus enhanced the susceptibility of rabbits to infection by the helminth. Myxoma infection might therefore have a significant impact on the transmission and persistence of *T. retortaeformis* in rabbit populations.

11.4 Practical problems

Those aiming to quantify the impact of pathogens on wild ungulate populations face a considerable challenge. The prevalence of infection in the host population ought to be measured during a period of several years. This might require pathogen detection or isolation from specific host tissue and, in the case of many micropathogens, the determination of the phylogenetic relationship of the strain found with other described strains. Phylogenetic analyses use nucleotide sequence data typically from a fragment of the genome and provide a powerful tool for the reconstruction of the evolutionary relationship between different strains of the same micropathogen that spread within populations of the same host species and between host species (Holmes, 1999). A suitable method would have to be identified that could provide accurate estimates of host population size. Finally, adequate samples from different age/sex classes in the host population would need to be obtained so that the prevalence of susceptible, infected and recovered individuals could be quantified during a period of several years. Thus, research to assess the impact of pathogens on host populations requires an interdisciplinary team and is likely to be costly in terms of time and resources.

Although a high prevalence of pathogen infection in apparently healthy adults might suggest that the pathogen has no impact on host population dynamics, this assumption may be false. As mentioned in the discussion of border disease in chamois (Section 11.3.2), pathogens may cause abortion or increase mortality in young animals, and it may be difficult to attribute these losses to pathogen infection (Marco *et al.*, 2007). Similarly, extremely high levels of *Hepatozoon* infection in African large carnivores have resulted in an assumption that this pathogen has no population consequences. However, daily monitoring of young spotted hyenas *Crocuta crocuta* revealed *Hepatozoon*-associated mortality in cubs less than two months of age (East *et al.*, 2008). Another example is the infection of chimpanzees *Pan troglodytes* with simian immunodeficiency virus (SIV), which does not cause acquired immunodeficiency syndrome in this host species. Long-term detailed demographic monitoring of chimpanzees in Gombe National Park, Tanzania, coupled with non-invasive monitoring of SIVcpz infection (the immediate precursor of human immunodeficiency virus), has, however, demonstrated that SIVcpz infection can decrease longevity and fecundity in infected individuals (Keele *et al.*, 2009). This chimpanzee study shows that simply establishing that a pathogen has an effect on infected individuals may not be easy.

Pathogen infection may predispose individuals to predation and may have an indirect effect on herbivore population dynamics through their action on predator population dynamics. For example, sarcoptic mange in red fox *Vulpes vulpes* populations can result in a significant reduction of fox predation of roe deer *Capreolus capreolus* fawns (Kjellander *et al.*, 2004), thereby influencing roe deer population dynamics. Finally, although pathogen-mediated mortality might be expected to affect host population dynamics, this pathogen effect may not necessarily be evident if negative density dependence is sufficiently reduced to allow a substantial increase in fecundity (Forchhammer and Asferg, 2000, Begon *et al.*, 2006). In summary, there is considerable potential for the effects of pathogens on population dynamics to be overlooked or underestimated.

Invasive methods necessary for the collection of blood or other tissue samples to determine prevalence of infection and pathogen exposure require time-consuming methods of trapping and/or anaesthesia which may yield a biased sample of 'trap-happy' or less vigorous individuals. It is also worth bearing in mind that interventions such as trapping and anaesthesia for research purposes may have a variety of effects (see Sapolsky, 1992; Alibhai *et al.*, 2001; Pelletier *et al.*, 2004; Cattet *et al.*, 2008), some of which might be detrimental to the immune status of 'handled' animals (Hofer and East, 1998). Interventions may also increase disease transmission through the use of trapping equipment contaminated with infectious agents and by holding trapped animals together in confined areas before they are blood sampled. Thus researchers should carefully consider how their methods might influence their results and reduce the possibility of such effects by ensuring that interventions have a minimal effect on the health status of animals.

New techniques, such as molecular (RT-PCR) screening methods now permit the detection of pathogen infection using faeces and urine and thus provide a non-invasive method to assess prevalence of infection within host populations (e.g. Keele *et al.*, 2009). Currently, however, molecular analyses are relatively expensive when applied to a large number of samples. It is perhaps not surprising that several studies that have quantified the long-term effect of pathogens on European ungulate population dynamics considered gastrointestinal helminth infections identified through non-invasive, relatively inexpensive methods that quantify helminth eggs in faeces (Gulland *et al.*, 1993; Paterson *et al.*, 1998; Albon *et al.*, 2002). In addition, such macroparasite infections can be experimentally manipulated through the application of anti-helminth medications (Section 11.3.1).

Long-term passive surveillance of mortality in wildlife populations or the use of information obtained from animals shot by hunters may help clarify the impact of a pathogen on host population dynamics (e.g. Forchhammer and Asferg, 2000) but may also suffer from unknown biases. For example the National Health Surveillance Program for Cervids in Norway monitored mortality (excluding hunting and traffic accidents) between 1998 and 2006 (Vikøren *et al.*, 2008). Although only 5% of 2718 cases were categorised to have died from an infectious disease, an unknown number of animals allocated to other mortality categories probably died as a direct or indirect result of pathogen infection (including some of those killed by hunters or in traffic accidents).

Box 11.1
Glossary of terms

There are several terms that are commonly used in studies that discuss pathogen–host dynamics, some of which describe the same phenomenon while others are used by different authors to describe different phenomena. For this reason we define what we mean when we use key terms.

Apparent multiple host pathogens: Pathogens that persist in one host species and are transmitted to another species sufficiently often to cause non-transient infection; however, the pathogen cannot persist in the second species without frequent transmission of infection from the true host.

Emerging pathogens: Novel pathogens able to infect a novel host population and cause an increase in the incidence of infection with time, or those that infect existing pathogen–host associations and cause an increase in the incidence of infection in a host population as a result of long-term changes in epidemiology (Woolhouse, 2002; Anderson *et al.*, 2004).

Host: An individual, population or species infected with a pathogen.

Host-switching: The process where pathogens extend their host range by establishing an association in new host species.

Pathogen: Any organism that lives inside or on a host and causes pathologies. Pathogens may be viruses, bacteria, fungi, protozoa, helminths or arthropods. Although transmissible spongiform encephalopathies (TSEs) do not strictly fit this definition because the transmissible agents that result in the abnormal folding of prion proteins in the brain cells of infected animals are not an organism, the impact of TSEs on infected ungulate populations might be viewed as similar to that caused by pathogens. Since the identification of a form of TSE termed chronic wasting disease (CWD) in ungulate populations in the USA in the 1960s, considerable research has been directed to establishing which species and populations have this neurodegenerative prion disease, how the disease is spread and what management actions are needed to control the disease (e.g. Schauber and Woolf, 2003; Grear *et al.*, 2006; Miller *et al.*, 2006). Large-scale screening of European ungulate species has produced no evidence of TSE infection in free-ranging cervids (Schettler *et al.*, 2006). TSEs include scrapie in sheep, bovine spongiform encephalopathy (BSE), also known as 'mad cow disease', and Creutzfeldt-Jakob disease in humans (Ghani *et al.* 2003).

Reservoir host: A species or population in which a pathogen persists.

Spill-over: Pathogen transmission from a host to another species that does not establish persistent infection in that species.

Box 11.1 (cont.)

True multiple host pathogens: Pathogens that can independently persist in more than one host species.

Virulence: Pathogens differ in their level of virulence, from relatively benign pathogens that have a limited impact on the fitness of their host, to highly virulent pathogens that induce a strong negative effect on host survival. Virulent pathogens typically have a high replication rate coupled to an efficient mechanism to move from the infected host they occupy to a non-infected, naïve host.

11.5 Pathogens with the potential to influence population dynamics

A perusal of veterinary and microbiological journals yields numerous examples of both macropathogens and micropathogens infecting wild European herbivores (Table 11.1). Many of these studies were initiated by an observed increase in disease within a population, or from the need to determine the disease threat posed by wild ungulates to livestock or humans. Certain wild ungulate populations in Europe are of particular cultural or economic importance, and thus likely to be closely monitored for disease. Hunters are thought to be particularly vulnerable to infection by wild ungulate diseases; this has prompted research on specific host species and pathogen types. For all these reasons, our current knowledge of European ungulate diseases is likely to be biased in favour of those pathogens and host populations viewed as important by interest groups and to monitoring periods that may not yield data suitable for the assessment of the possible long-term effects of pathogen infection on ungulate population dynamics.

11.6 Anthropogenic factors

In many European countries, ungulates mostly inhabit landscapes that have been altered by human activities and many ungulate populations are managed to reduce damage to economic crops or for sports hunting. Large carnivores have been removed from most areas of Europe; thus natural predation pressure is low or absent in many areas. Predation is chiefly carried out by hunters that are likely to have a markedly different impact from natural predators on herbivore populations. The availability of food resources for ungulates has increased in many areas in the form of nutritious agricultural crops and artificial forage provided by hunters

Table 11.1 *Examples of pathogens described in European ungulate species and their reported impact on host individuals*

Pathogen type	Host species	Increased adult mortality	Increased juvenile mortality	Decreased fecundity	Reference
Arthropod					
Bot fly *Cephenemyia trompe*	Reindeer				Hagemoen and Reimers, 2002; Weladji *et al.*, 2003
Warble fly *Hypoderma tarandi*	Reindeer				Hagemoen and Reimers, 2002; Weladji *et al.*, 2003
Deer ked *Lipoptena cervi*	Moose; red deer; roe deer; reindeer	✓	✓		Haarløv, 1964; Kadulski, 1996; Kaunisto *et al.*, 2009
Ixodid ticks	Red deer; wild boar; European bison				Ruiz-Fons *et al.*, 2006; Krasińska and Krasiński, 2007
Sarcoptic mange *Sarcoptes scabiei*	Chamois; red deer; Spanish ibex	✓			Rossi *et al.*, 1995; Oleaga *et al.*, 2008 León-Vizcaíno *et al.*, 1999
Chorioptes sp.	Moose; fallow deer				Hestvik *et al.*, 2007; Szczurek and Kadulski, 2004
Helminth					
Teladorsagia sp.; *Ostertagia* sp.; *Trichostrongylus* sp.	Soay sheep	✓	✓		Gulland *et al.*, 1993; Paterson *et al.*, 1998
Ostertagia gruehneri	Reindeer			✓	Albon *et al.*, 2002; Irvine *et al.*, 2000
Viral					
Bluetongue virus	Red deer; roe deer; mouflon; aoudad				Ruiz-Fons *et al.*, 2008a
Aujesky's disease	Wild boar				Ruiz-Fons *et al.*, 2008b
Classical swine fever	Wild boar	✓	✓		Kramer-Schadt *et al.*, 2007; Ruiz-Fons *et al.*, 2008b

Table 11.1 (*cont.*)

Pathogen type	Host species	Increased adult mortality	Increased juvenile mortality	Decreased fecundity	Reference
Bovine viral diarrhoea virus	Alpine chamois; wild boar; red deer; roe deer				Olde Riekerink *et al.*, 2005
Border disease	Chamois	✓	✓	✓	Pioz *et al.*, 2007; Marco *et al.*, 2008
Bacterial					
Mycobacterium bovis	Red deer; fallow deer; wild boar	✓	✓		Gortázar *et al.*, 2008
Mycobacterium avium paratuberculosis	Red deer; alpine ibex	✓			Fraquelli *et al.*, 2005; Ferroglio *et al.*, 2000
Arcanobacterium bialowiezense	European bison			✓	Krasińska and Krasiński, 2007
Mycoplasma conjunctivae	Chamois	✓			Giacometti *et al.*, 2002
Brucella suis	Wild boar				Godfroid and Käsbohrer, 2002
Salmonella enterica serovar *Abortusovis*	Alpine chamois	✓			Pioz *et al.*, 2008
Chlamydophila abortus	Iberian ibex				Ryser-Degiorgis *et al.*, 2009; Salinas *et al.*, 2009
Coxiella burnetii	Red deer; roe deer; wild boar				Ruiz-Fons *et al.*, 2008c; Astobiza *et al.*, 2010

during periods of food shortage. Reduced predation pressure and increased food abundance has resulted in artificially high ungulate densities in many areas. This is a situation likely to impact pathogen–host dynamics by increasing density-dependent transmission of infectious pathogens. Pathogen transmission may also be elevated when animals congregate at artificial feeders and water supplies. For example, artificial feeding of white-tailed deer by hunting clubs aimed at increasing the number and quality of deer in Michigan led to a significant increase in TB in the deer population, most probably through increased direct contact between animals at feeding sites and through contaminated feed sources used by large numbers of animals (Miller *et al.*, 2003). Feeding of European wild boar and the use of artificial water sources by high-density populations of wild boar are also thought to have increased bTB prevalence in some populations in Spain (Gortázar *et al.*, 2008).

Removal of diseased individuals, a traditional method aimed at reducing disease transmission, can change social structure within a population and thereby alter pathogen transmission, leading to some unexpected results. For example, socially mediated changes in the transmission of bTB occurred during the 'randomised culling trials' in badgers *Meles meles* in the UK (McDonald *et al.*, 2008). In Britain, badgers live in stable groups that defend territories. Although dispersal of animals between social groups could increase the incidence of bTB, rates of dispersal between groups is typically low. The randomised culling trials may have radically decreased badger group size but also changed social group composition. Members of culled groups that escaped being culled, and badgers living in areas close to culled areas, responded to the creation of vacant areas and newly created opportunities of reproduction and dispersal by significantly increasing their ranging behaviour. Thus, although culling reduced host density, it also exacerbated the spread of disease to susceptible badgers and cattle (McDonald *et al.*, 2008). The influence of social structure on the transmission of bTB was also demonstrated in free-ranging white-tailed deer in which individuals infected with bTB were significantly closer kin than non-infected deer (Blanchong *et al.*, 2007). The effect of social structure on bTB transmission in Persian fallow deer *Dama mesopotamica* is also considered by Bar-David *et al.* (2006).

Interestingly, habitat structure and quality also influence bTB transmission on farms in the UK, where 'nature friendly' management practices such as the presence of ungrazed wildlife strips and the greater availability, width and continuity of hedgerows were associated with reduced bTB incidence in cattle. This result might be generated by the effect of habitat on cattle contact rates, or by agricultural management practices (Mathews *et al.*, 2006). Various other forms of disturbance caused by humans might also alter pathogen–host dynamics. For example, when disturbance induces sufficient physiological stress to reduce host immunity, susceptibility to infection may be increased (Hofer and East, 1998; Wikelski and Cooke, 2006). Crops and the improvement of pasture may also alter parasite–host interactions. On the one hand, increased food availability may permit wild ungulates to invest more in immune responses (Martin, 2009). On the other hand, herbivores that feed on tannin-rich forage are likely to have fewer helminth parasites because tannin consumption can reduce both the number and fecundity of parasitic helminth worms when compared with the consumption of tannin-poor forage such as ryegrass which is abundant in improved pastures (Hoste *et al.*, 2006).

There is strong evidence that global climate change is already altering host-pathogen dynamics in some ungulate species (see Box 11.2 and Chapter 12).

The exceptionally high densities of livestock that occur in intensive livestock rearing units and the international movement of ungulates (both domestic and wild species) have the potential to spread pathogens to previously uninfected naïve host species or populations, sometimes with devastating effects (Plowright, 1982). All these anthropogenic factors are likely to increase in the coming decades but their effect on pathogen–host dynamics may be difficult to quantify because of the general lack of information on current dynamics.

11.7 Pathogen management

As demonstrated by the case of culling of badgers in the UK, the traditional method used to control pathogens in wildlife populations is culling of host populations, with the assumption that by reducing host density, pathogen transmission will be reduced, leading to local pathogen extinction. As already mentioned, simple density-dependent models do not consider heterogeneity in host populations, or the consequences of disrupting social structure, and for this reason culls may fail to control disease transmission (Gortázar *et al.*, 2007; McDonald *et al.*, 2008). In fact, such practices may theoretically lead to an increase in both disease prevalence and cross-species disease transmission (Choisy and Rohani, 2006). Furthermore, during culling operations that seek to remove diseased animals from a population, distinguishing diseased from non-infected animals may be difficult (Wobeser, 2006). When population size is strongly limited by density-dependent effects, the population effect of culling may also be fully compensated or overcompensated for by increased reproduction and recruitment. For example, the wild boar population in Europe has greatly increased in density since the 1950s and its geographical range has expanded. During this population growth, classical swine fever has become endemic. European wild boar population densities are strongly linked to high-energy food resources (both natural and provisioned), and even when the rate of adult mortality is high because of hunting or classical swine fever infection, increased reproductive success can result in the swift recovery of wild boar populations (Bieber and Ruf, 2005; Kramer-Schadt *et al.*, 2007, 2009).

11.8 Concluding remarks

The studies reviewed in this chapter provide strong evidence that pathogens influence the dynamics of European ungulate populations, and some infectious diseases have the potential to reduce the long-term viability of small, fragmented populations. Clearly there is a need for more research on this topic, particularly for interdisciplinary studies that combine a broad range of expertise and those that apply an experimental approach to help tease apart interconnected ecological, behavioural, genetic and anthropogenic effects. Certainly more information is required on anthropogenic effects on pathogen–host interactions, particularly in relation to major problems such as disturbance, global climate change and the spread of pathogens through international trade. An appreciation of the ecological network in which both pathogens and their hosts exist will improve our knowledge of pathogen–host dynamics, and too narrowly focused pathogen-control measures that ignore these networks and host species behavioural structures may produce unexpected results (Hofer and East, 2010). Scientists need to ensure that funding agencies comprehend the importance of long-term research on this topic and the potential benefits that can be obtained by a broad spectrum of interest groups (hunters, farmers, conservationists, managers of zoological facilities) from a better understanding of the impact of pathogens on the population dynamics of European ungulate populations.

Box 11.2

Global climate change and disease transmission in European ungulates

[*Contributed by Marion L. East*]

There is sufficient evidence to indicate that global climate change is altering ecological processes in habitats worldwide and the increase in average global temperature is already having a marked impact on habitats, particularly those at high latitudes and high elevations (Easterling *et al.*, 2000; Hughes, 2000; Parmesan and Yohe, 2003; Williams *et al.*, 2003; Moritz *et al.*, 2008). As Mysterud and Sæther describe in the next chapter, these, and predicted future, changes in climate are likely to have a significant influence on the behaviour, ecology and population dynamics of European ungulates.

One aspect of ungulate ecology that Mysterud and Sæther mention only briefly in this context is evidence for climate-mediated changes to pathogen–host dynamics. Although such changes may, theoretically, be equally to the benefit of either the host or the pathogen, most research on this topic has focused on changes likely to favour pathogens to the detriment of their hosts. This focus probably arises from concerns that a possible impact of climate change will be decreased agricultural production and biodiversity as a result of an increased prevalence of diseases (Anderson *et al.*, 2004; Kutz *et al.*, 2004; Lips *et al.*, 2008; Harvell *et al.*, 2009). This has led some to suggest that the detrimental effects of pathogens on host populations that are predicted to occur as a result of climate change shifts in pathogen–host dynamics have been exaggerated (Lafferty, 2009a, 2009b).

Box 11.2 (cont.)

As Mysterud and Sæther make clear in the next chapter, not all predicted climate changes in Europe are expected to be detrimental to ungulates or increase the prevalence of or susceptibility to diseases. For example, predicted climatic effects on plant phenology leading to an extension of plant growth in spring and autumn, and thus increased access to higher quality forage, are likely to be beneficial for some European ungulates. However, the predicted increase in temperature extremes and incidences of drought, particularly at lower latitudes, are likely to increase levels of physiological stress and reduce body condition in some ungulate populations. An increase in physiological stress and a decline in body condition would be expected to decrease the immune function of affected animals and increase susceptibility to infection (Patz and Reisen, 2001; Wingfield, 2008). Thus the effects of climate change on pathogen–host dynamics are likely to differ between climatic zones and, where they occur, they will probably be diverse and be mediated by interconnected pathways (Anderson *et al.*, 2004; Kutz *et al.*, 2004; Pedersen and Post, 2008; Wingfield, 2008; Hofer and East, 2010).

The effects of climate on pathogen survival and transmission are perhaps easier to monitor than the more subtle effects that climate change may have on those factors such as nutritional status, physiological stress and immunity which are likely to alter host susceptibility to infection. In this box I outline two basic types of climate-mediated change on pathogen–host dynamics. Firstly, I describe an example that illustrates how increased temperatures, particularly during European winters, has allowed insect vectors of diseases to persist throughout the year and rapidly extend their range (and that of the disease they transmit) in summer. Secondly, I consider how mild, moist weather in far northern habitats can increase parasite populations by decreasing overwinter mortality of aquatic parasite larva and populations of intermediate gastropod hosts.

Bluetongue virus

The bluetongue virus (BTV) is a pathogen that infects both domestic and wild ungulates (Fernández-Pacheco *et al.*, 2008; Mintiens *et al.*, 2008; Ruiz-Fons *et al.*, 2008a) in Europe. BTV is a non-contagious, arthropod-borne virus (arbovirus) belonging to the genus *Orbivirus* (family Reoviridae) that is transmitted by *Culicoides* midges.

There are 24 described serotypes of BTV which are distinguished by their specificity of reactions between proteins on the outer capsid proteins (VP2 and VP5) and neutralising antibodies. Since 1998, six BTV serotypes have been identified in Europe and sequencing of genome segments has helped to identify the geographical origin of different BTV strains (Batten *et al.*, 2008). Historically BTV occurred chiefly in tropical and subtropical regions (Gibbs and Greiner, 1994) but has occasionally been reported from the Iberian Peninsula since 1956 (Gloster *et al.*, 2008).

Box 11.2 (cont.)

More recently, the apparent threat posed by BTV to livestock in Europe has greatly increased and large-scale vaccination programmes have been used in an effort to limit the spread and economic impact of the virus (Batten *et al.*, 2008). In 2000, BTV-2 was recorded in the south-west of mainland Europe, having probably spread from the Mediterranean coast of Africa via islands in the western Mediterranean. In 2003, BTV-4 followed a similar route as did BTV-1 in 2006 (Mintiens *et al.*, 2008). In August 2006, BTV-8 was reported in the Netherlands and then rapidly spread to several northern European countries (France, Luxembourg, Belgium and Germany) by late autumn 2006. In June 2007 BTV-8 was detected again in Germany, suggesting the virus had overwintered in northern Europe, possibly due to an unusually mild winter, and then spread rapidly to all the countries that were infected in 2006 and also to the Czech Republic, Denmark, Switzerland and the UK (Gloster *et al.*, 2008). This rapid spread of BTV-8 was of particular significance since, unlike other BTV strains recorded in Europe, BTV-8 caused overt disease and mortality in domestic cattle (Gould and Higgs, 2009).

It has been proposed that all these recent outbreaks of BTV in Europe were linked to climatic factors associated with climate change (Purse *et al.*, 2005) that are thought to have helped the main midge vector *Culicoides imicola* not only to extend its range in Europe but also the number of months in the year it is active. Furthermore, milder winters appear to allow BTV to persist even in northerly latitudes in Europe (Monteys *et al.*, 2005; Purse *et al.*, 2005).

Outbreaks of BTV in Europe have prompted immunisation programmes using attenuated vaccines (with attenuated vaccine strains of BTV-2, 4, 9,16) which has led to the 'escape' and proliferation of vaccine strains in vectors and hosts, and possible genome reassortment between vaccine strains (Batten *et al.*, 2008). In Italy attenuated vaccine strains BTV-2 and BTV-16 have been recorded in animal populations beyond those actually immunised, possibly because live 'attenuated' strains can be virulent for some European breeds of sheep and thus might cause the onwards transmission of the virus by midge vectors (Monaco *et al.*, 2006; Batten *et al.*, 2008).

Nematode parasites

There is evidence that climate change is affecting the epidemiology of the nematode parasites of arctic ungulates. In the USA, the nematode parasite *Umingmastronylus pallikuukensis* infects musk oxen (Kutz *et al.*, 2004), forming cysts in the lung, and the nematode has an indirect life cycle with an intermediate gastropod host. *Umingmastronylus pallikuukensis* is thought to be an important cause of clinical and subclinical disease and it has been suggested that the survival of heavily infected musk oxen is reduced both directly (by the parasite

Box 11.2 (cont.)

itself) and indirectly through increased predation of individuals with heavy lung infections by grizzly bears *Ursus arctos* (Kutz *et al.*, 2004).

The current high level of infection, with 100% prevalence of infection in some areas, may be due to an accelerated parasite development rate due to increased temperatures. Experiments on the development of the larval forms of the parasite in the gastropod intermediate host suggest that under 'normal' arctic summer temperatures the full development of the larval stages requires two years whereas under the warmer current summer temperatures, larval development can be completed in a year. Larval development may also be favoured by the generally earlier onset of spring and the delayed onset of winter. Overwinter mortality of the gastropod host and larval stages is high and thus the completion of the larval stages within one growth season probably greatly reduces the mortality among early larval stages that previously needed to survive a winter period. There is also evidence from another protostrongylid parasite (*Elaphostrongylus rangiferi*) that warm summers have led to increased infection of both the intermediate gastropod host of this parasite and European reindeer populations (Handeland and Slettbakk, 1994).

Conclusion

The examples described in this box and those outlined by Mysterud and Sæther in Chapter 12 provide sufficient evidence to suggest that global climate change is altering pathogen–host dynamics in ungulate populations and will continue to do so in the coming decades. Long-term monitoring of pathogen–host dynamics is essential if the impact of global climate change on pathogen–host dynamics is to be understood. In particular we need to know more about how climatic change may influence inter-related ecological, behavioural and physiological factors that mediate immune responses, disease transmission and susceptibility to infection. More interdisciplinary research is required if the potential threat of climate change is to be truly assessed and appropriate action taken to diminish any threat climate change may pose to European ungulate populations.

References

Acevedo-Whitehouse, K., Vincente, J., Gortázar, C., *et al.* (2005) Genetic resistance to bovine tuberculosis in the Iberian wild boar. *Molecular Ecology* **14**, 3209–3217.

Agnew, P., Koella, J.C. and Michalakis, Y. (2000) Host life history responses to parasitism. *Microbes and Infection* **2**, 891–896.

Ahrens, W. and Pigeot, I. (2005) *Handbook of Epidemiology*. Berlin: Springer-Verlag.

Albon, S.D., Stien, A., Irvine, R.J., *et al.* (2002) The role of parasites in the dynamics of a reindeer population. *Proceedings of the Royal Society B–Biological Sciences* **269**, 1625–1632.
Alibhai, S.K., Jewell, Z.C. and Towindo, S.S. (2001) Effect of immobilization on fertility in female black rhino *(Diceros bicornis) Journal of Zoology, London* **253**, 333–345.
Anderson, P.K., Cunningham, A.A., Patel, N.G., *et al.* (2004) Emerging infectious diseases of plants: pathogen pollution, climate change and agrotechnology drivers. *Trends in Ecology and Evolution* **19**, 535–545.
Anderson, R.M. and May, R.M. (1982) Coevolution of hosts and parasites. *Parasitology* **85**, 411–426.
Anderson, R.M. and May, R.M. (1991) *Infectious Diseases of Humans: Dynamics and Control*. Oxford, UK: Oxford University Press.
Antonovics, J. (2009) The effect of sterilizing diseases on host abundance and distribution along environmental gradients. *Proceedings of the Royal Society B–Biological Sciences* **276**, 1443–1448.
Astobiza, I., Barral, M., Ruiz-Fons, F., *et al.* (2010) Molecular investigation of the occurrence of *Coxiella burnetii* in wildlife and ticks in an endemic area. *Veterinary Microbiology* [epub ahead of print]
Aviv, A. (2008) The epidemiology of human telomeres: faults and promises. *Journal of Gerontology* **63A**, 979–983.
Bar-David, S., Lloyd-Smith, J.O. and Getz, W.M. (2006) Dynamics and management of infectious diseases in colonizing populations. *Ecology*, **87**, 1215–1224.
Barrett, L.G., Thrall, P.H., Burdon, J.J. and Linde, C.C. (2008) Life history determines genetic structure and evolutionary potential of host-parasite interactions. *Trends in Ecology and Evolution* **23**, 678–685.
Batten, C.A., Maan, S., Shaw, A.E., Maan, N.S. and Mertens, P.P.C. (2008) A European field strain of bluetongue virus derived from two parental vaccine strains by genome segment reassortment. *Virus Research* **137**, 56–63.
Begon, M., Hazel, S.M., Baxby, D., *et al.* (1999) Transmission dynamics of a zootic pathogen within and between wildlife host species. *Proceedings of the Royal Society B–Biological Sciences* **266**, 1939–1945.
Begon, M., Townsend, C.R. and Harper, J.L. (2006) *Ecology: From Individuals to Ecosystems*. Oxford, UK: Blackwell Publishing.
Behnke, J.M., Bajer, A., Sinski, E. and Wakelin, D. (2001) Interactions involving intestinal nematodes of rodents: experimental and field studies. *Parasitology* **122**, S39–S49.
Beldomenico, P.M., Telfer, S., Gebert, S., *et al.* (2008) Poor condition and infection: a vicious cycle in natural populations. *Proceedings of the Royal Society B–Biological Sciences* **275**, 1753–1759.
Bieber, C. and Ruf, T. (2005) Population dynamics in wild boar *Sus scrofa*: ecology, elasticity of growth rate and implications for management of pulsed resource consumers. *Journal of Applied Ecology* **42**, 1203–1213.
Blanchong, J.A., Scribner, K.T., Kravchenko, S.N. and Winterstein, S.R. (2007) TB-infected deer are more closely related than non-infected deer. *Biology Letters* **3**, 103–105.
Bonsall, R.G. and Hassell, M.P. (1997) Apparent competition structures in ecological assemblies. *Nature* **388**, 371–373.
Boots, M. and Sasaki, A. (2002) Parasite driven extinction in spatially explicit host-parasite systems. *American Naturalist* **159**, 706–713.

Cattadori, I.M., Albert, R. and Boag, B. (2007) Variation in host susceptibility and infectiousness generated by co-infection: the myxoma-*Trichostrongylus retortaeformis* case in rabbits. *Journal of the Royal Society Interface* **4**, 831–840.

Cattadori, I.M., Boag, B. and Hudson, P.J. (2008) Parasite co-infection and interaction as drivers of host heterogeneity. *International Journal of Parasitology* **38**, 371–380.

Cattet, M., Boulanger, J., Stenhouse, G., Powell, R.A. and Reynolds-Hogland, M.J. (2008) An evaluation of long-term capture effects in ursids: implications for wildlife welfare and research. *Journal of Mammalogy* **89**, 973–990.

Choisy, M. and Rohani, P. (2006) Harvesting can increase severity of wildlife disease epidemics. *Proceedings of the Royal Society B–Biological Sciences* **273**, 2025–2034.

Clutton-Brock, T.H., Major, M. and Guinness, F.E. (1985) Population regulation in male and female red deer. *Journal of Animal Ecology* **54**, 831–846.

Coulson, T., Benton, T.G., Lundberg, P., *et al.* (2006) Estimating individual contributions to population growth: evolutionary fitness in ecological time. *Proceedings of the Royal Society B–Biological Sciences* **273**, 547–555.

Courchamp, F., Pontier, D., Langlais, M. and Artois, M. (1995) Population dynamics of feline immunodeficiency virus within cat populations. *Journal of Theoretical Biology* **175**, 553–560.

Daszak, P., Cunningham, A.A. and Hyatt, A.D. (2000) Emerging infectious diseases of wildlife: threats to biodiversity and human health. *Science* **287**, 443–449.

de Castro, F. and Bolker, B. (2005) Mechanisms of disease-induced extinction. *Ecology Letters* **8**, 117–126.

Diekmann, O. and Heesterbeek, J.A.P. (2000) *Mathematical Epidemiology of Infectious Diseases: Model Building, Analysis and Interpretation.* Chichester, UK: Wiley.

East, M.L., Wibbelt, G., Lieckfeldt, D., *et al.* (2008) A *Hepatozoon* species genetically distinct from *H. canis* infecting spotted hyenas in the Serengeti ecosystem, Tanzania. *Journal of Wildlife Diseases* **44**, 45–52.

Easterling, D.R., Meehl, G.A., Parmesan, C., *et al.* (2000) Climate extremes: observations, modelling and impacts. *Science* **289**, 2068–2074.

Fenton, A. and Pedersen, A.B. (2005) Community epidemiology framework for classifying disease threats. *Emerging Infectious Diseases* **11**, 1815–1821.

Fernández-Pacheco, P., Fernández-Pinero, J., Agüero, M. and Jiménez-Clavero, M.A. (2008) Bluetongue virus serotype 1 in wild mouflon in Spain. *Veterinary Record* **162**, 659–660.

Ferroglio, E., Nebbia, P., Robino, P., Rossi, L. and Rosati, S. (2000) *Mycobacterium paratuberculosis* infection in two free-ranging Alpine ibex. *Revue Scientifique et Technique* **19**, 859–862.

Festa-Bianchet, M. (1989) Individual differences, parasites and the cost of reproduction for bighorn ewes (*Ovis canadensis*). *Journal of Animal Ecology* **58**, 785–795.

Forchhammer, M.C., and Asferg, T. (2000) Invading parasites cause a structural shift in red fox dynamics. *Proceedings of the Royal Society B–Biological Sciences* **267**, 779–786.

Forchhammer, M.C., Stenseth, N.C., Post, E. and Langvatn, R. (1998) Population dynamics of Norwegian red deer: density-dependence and climate variation. *Proceedings of the Royal Society B–Biological Sciences* **265**, 341–350.

Fraquelli, C., Carpi, G., Bregoli, M., *et al.* (2005) Epidemiology of paratuberculosis in two red deer (*Cervus elaphus*) populations of Trentino (Northern Italy). In E.J.B. Manning and S.S. Nielsen (eds.) *Proceedings of the 8th International*

Colloquium on Paratuberculosis, held Copenhagen, Denmark. Madison, WI: International Association for Paratuberculosis, pp.605–612.

Gaillard, J.-M., Festa-Bianchet, M. and Yoccoz, N.G. (1998) Population dynamics of large herbivores: variable recruitment with constant adult survival. *Trends in Ecology and Evolution* **13**, 58–63.

Gaillard, J.-M., Festa-Bianchet, M., Yoccoz, N.G., Loison, A. and Toïgo, C. (2000) Temporal variation in fitness components and population dynamics of large herbivores. *Annual Review of Ecology and Systematics* **31**, 367–393.

Gandon, S. (2004) Evolution of multihost parasites. *Evolution* **58**, 455–469.

Gasbarre, L.C. (1997) Effects of gastrointestinal nematode infection on the ruminant immune system. *Veterinary Parasitology* **72**, 327–337.

Ghani, A.C., Ferguson, N.M., Donnelly, C.A. and Anderson, R.M. (2003) Factors determining the pattern of the variant Creutzfeldt-Jakob disease (vCJD) epidemic in the UK. *Proceedings of the Royal Society B–Biological Sciences* **270**, 689–698.

Giacometti, M., Janovsky, M., Belloy, L. and Frey, J. (2002) Infectious keratokonjunctivitis of ibex, chamois and other Caprinae. *Review of the Scientific and Technical Office for International Epidemiology* **21**, 335–345.

Gibbs, E.P. and Greiner, E.C. (1994) The epidemiology of bluetongue. *Comparative Immunology, Microbiology and Infectious Diseases* **17**, 207–220.

Gloster, J., Burgin, L., Witham, C., Athanassiadou, M. and Mellor, P.S. (2008) Bluetongue in the United Kingdom and northern Europe in 2007 and key issues for 2008. *Veterinary Record* **162**, 298–302.

Godfroid, J. and Käsbohrer, A. (2002) Brucellosis in the European Union and Norway at the turn of the twenty-first century. *Veterinary Microbiology* **90**, 135–145.

Gortázar, C., Ferroglio, E., Höfle, U., Frölich, K. and Vicente, J. (2007) Diseases shared between wildlife and livestock: a European perspective. *European Journal of Wildlife Research* **53**, 241–256.

Gortázar, C., Torres, M.J., Vincente, J., *et al.* (2008) Bovine tuberculosis in Doñana Biosphere Reserve: the role of wild ungulates as disease reservoirs in the last Iberian lynx strongholds. *PLoS ONE* **3**, e2776.

Gould, E.A. and Higgs, S. (2009) Impact of climate change and other factors on emerging arbovirus diseases. *Transactions Royal Society of Tropical Medicine and Hygiene* **103**, 109–121.

Graham, A., Cattadori, I.M., Lloyd-Smith, J., Ferrari, M. and Bjørnstad, O.N. (2007) Transmission consequences of co-infection: cytokines writ large? *Trends in Parasitology* **23**, 281–291.

Grear, D.A., Samuel, M.D., Langenberg, J.A. and Keane, D. (2006) Demographic patterns and harvest vulnerability of chronic wasting disease infected white-tailed deer in Wisconsin. *Journal of Wildlife Management* **70**, 546–553.

Grenfell, B. and Harwood, J. (1997) (Meta) population dynamics of infectious diseases. *Trends in Ecology and Evolution* **12**, 395–399.

Gulland, F.M.D., Albon, S.D., Pemberton, J.M., Moorcroft, P.R. and Clutton-Brock, T.H. (1993) Parasite-associated polymorphism in a cyclic ungulate population. *Proceedings of the Royal Society B–Biological Sciences* **254**, 7–13.

Haarløv, N. (1964) Life cycle and distribution pattern of *Lipoptena cervi* (L.) (Dipt. Hippobosc.) on Danish deer. *Oikos* **15**, 93–129.

Hagemoen, R.I.M. and Reimers, E. (2002) Reindeer summer activity pattern in relation to weather and insect harassment. *Journal of Animal Ecology* **71**, 883–892.

Hamilton, W.D. and Zuk, M. (1982) Heritable true fitness and bright birds: a role for parasites? *Science* **218**, 384–387.

Handeland, K. and Slettbakk, T. (1994) Outbreaks of clinical cerebrospinal elaphostrongylosis in reindeer (*Rangifer tarandus tarandus*) in Finnmark, Norway, and their relation to climate conditions. *Journal of Veterinary Medicine B* **41**, 407–410.

Harvell, D., Altizer, S., Cattadori, I.M., Harrington, L. and Weil, H. (2009) Climate change and wildlife diseases: when does the host matter the most? *Ecology* **90**, 912–920.

Hestvik, G., Zahler-Rinder, M., Gavier-Widen, D., *et al.* (2007) A previously unidentified *Chorioptes* species infesting outer ear canals of moose (*Alces alces*): characterization of the mite and the pathology of infestation. *Acta Veterinaria Scandinavica* **49**, 21.

Hofer, H. and East, M.L. (1998) Biological conservation and stress. *Advances in the Study of Behavior* **27**, 405–525.

Hofer, H. and East, M.L. (2010). Impact of global climate change on wildlife hosts and their pathogens. *Nova Acta Leopoldina* **NF111** (NR381), 103–110.

Holmes, E.C. (1999) Molecular phylogenies and the genetic structure of viral populations. In S.C. Stearns and J.C. Koell (eds.) *Evolution in Health and Disease*. Oxford, UK: Oxford University Press, pp.173–182.

Hoste, H., Jackson, F., Athanasiadou, S., Thamsborg, S.M. and Hoskin, S.O. (2006) The effect of tannin-rich plants on parasitic nematodes in ruminants. *Trend in Parasitology* **22**, 253–261.

Hudson, P.J. and Greenman, J. (1998) Competition mediated by parasites: biological and theoretical progress. *Trends in Ecology and Evolution* **13**, 387–390.

Hughes, L. (2000) Biological consequences of global warming. *Trends in Ecology and Evolution* **15**, 56–61.

Ilmonen, P., Kotrschal, A. and Penn, D.J. (2008) Telomere attrition due to infection. *Plos One* **3**, 2143.

Irvine, R.J., Stien, A., Halvorsen, O., Langvatn, R. and Albon, S.D. (2000) Life-history strategies and population dynamics of abomasal nematodes in Svalbard reindeer (*Rangifer tarandus platyrrhynchus*). *Parasitology* **120**, 297–311.

Jones, M.E., Cockburn, A., Hamede, R., *et al.* (2008) Life-history change in disease-ravaged Tasmanian devil populations. *Proceedings of the National Academy of Sciences of the USA* **105**, 10 023–10 027.

Jędrzejewska, B. and Jędrzejewski, W. (1998) *Predation in Vertebrate Communities*. Berlin: Springer.

Kadulski, S. (1996) Ectoparasites of Cervidae in north-east Poland. *Acta Parasitologica* **41**, 204–210.

Kaunisto, S., Kortet, R., Härkönen, L., *et al.* (2009) New bedding sites examination-based method to analyse deer ked (*Lipoptena cervi*) infection in cervids. *Parasitological Research* **104**, 919–925.

Kavaliers, M., Choleris, E. and Pfaff, D.W. (2005) Genes, odours and the recognition of parasitized individuals by rodents. *Trends in Parasitology* **21**, 423–429.

Keele, B.F., Jones, J.H., Terio, K.A., *et al.* (2009) Increased mortality and AIDS-like immunopathology in wild chimpanzees infected with SIVcpz. *Nature* **460**, 515–519.

Kirkwood, T.B.L. (1981) Repair and its evolution: survival versus reproduction. In C.R. Townsend and P. Calow (eds.) *Physiological Ecology: An Evolutionary Approach to Resource Use*. Oxford, UK: Blackwell Scientific Press, pp.165–189.

Kjellander, P., Gaillard, J.-M., Hewison, M. and Liberg, O. (2004) Predation risk and longevity influence variation in fitness of female roe deer. *Proceedings of the Royal Society B–Biological Sciences* **271**, S338–S340.

Kramer-Schadt, S., Fernández, N. and Thulke, H.H. (2007) Potential ecological and epidemiological factors affecting the persistence of classical swine fever in wild boar *Sus scrofa* populations. *Mammal Review* **37**, 1–20.

Kramer-Schadt, S., Fernández, N., Eisinger, D., Grimm, V. and Thulke, H.H. (2009) Individual variations in infectiousness explain long-term disease persistence in wildlife populations. *Oikos* **118**, 199–208.

Krasińska, M. and Krasiński, Z.A. (2007) *European Bison: The Nature Monograph.* Białowieża, Poland: Mammal Research Institute, Polish Academy of Sciences.

Kutz, S.J., Hoberg, E.P., Nagy, J., Polley, L. and Elkin, B. (2004) 'Emerging' parasitic infections in arctic ungulates. *Integrative and Comparative Biology* **44**, 109–118.

Lafferty, K. (2009a) The ecology of climate change and infectious diseases. *Ecology* **90**, 888–900.

Lafferty, K. (2009b) Calling for an ecological approach to studying climate change and infectious diseases. *Ecology* **90**, 932–933.

Lee, K. (2006) Linking immune defenses and life history at the levels of the individual and the species. *Integrative and Comparative Biology* **46**, 1000–1015.

Lello, J. and Hussell, T. (2008) Functional group/guild modelling of inter-specific pathogen interactions: a potential tool for predicting the consequences of co-infections. *Parasitology* **135**, 825–839.

León-Vizcaíno, L., Ruiz de Ybáñez, M.R., Cubero, M.J., *et al.* (1999) Sarcoptic mange in Spanish ibex from Spain. *Journal of Wildlife Diseases* **35**, 647–659.

Lindholm, M. and Britton, T. (2007) Endemic persistence or disease extinction: the effect of separation into sub-communities. *Theoretical Population Biology* **72**, 253–263.

Lips, K.R., Diffendorfer, J., Mendelson III, J.R. and Sears, M.W. (2008) Riding the wave: reconciling the role of disease and climate change in amphibian declines. *PLoS Biology* **6**, 72.

LoGiudice, K., Ostfeld, R.S., Schmidt, K.A. and Keesing, F. (2003) The ecology of infectious disease: Effects of host diversity and community composition on Lyme disease risk. *Proceedings of the National Academy of Sciences USA* **100**, 567–571.

Marco, I., López-Olvera, J.R., Rosell, R., *et al.* (2007) Severe outbreak of disease in the southern chamois (*Rupicapra pyrenaica*) associated with border disease virus infection. *Veterinary Microbiology* **120**, 33–41.

Marco, I., Rosell, R., Cabezón, O., *et al.* (2008) Epidemiological study of border disease virus infection in southern chamois (*Rupicapra pyrenaica*) after an outbreak of disease in the Pyrenees (NE Spain). *Veterinary Microbiology* **127**, 29–38.

Martin, L.B. (2009) Stress and immunity in wild vertebrates: timing is everything. *General and Comparative Endocrinology* **163**, 70–76.

Martin, L.B., Weil, Z.M. and Nelson, R.J. (2008) Seasonal changes in vertebrate immune activity: mediation by physiological trade-offs. *Philosophical Transactions of the Royal Society B–Biological Sciences* **363**, 321–339.

Mathews, F., Lovett, L., Rushton, S. and Macdonald, D.W. (2006) Bovine tuberculosis in cattle: reduced risk on wildlife-friendly farms. *Biology Letters* **2**, 271–274.

McCallum, H., Barlow, N.D. and Hone, J. (2001) How should pathogen transmission be modelled? *Trends in Ecology and Evolution* **16**, 295–300.

McDonald, R.A., Delahay, R.J., Carter, S.P., Smith, G.C. and Cheeseman, C.L. (2008) Perturbing implications of wildlife ecology for disease control. *Trends in Ecology and Evolution* **23**, 53–56.

Miller, M.W., Hobbs, N.T. and Tavener, S.J. (2006) Dynamics of prion disease transmission in mule deer. *Ecological Applications* **16**, 2208–2214.

Miller, R., Kaneene, J.B., Fitzgerald, S.D. and Schmitt, S.M. (2003) Evaluation of the influence of supplemental feeding of white-tailed deer (*Odocoileus virginianus*) on the prevalence of bovine tuberculosis in the Michigan wild deer population. *Journal of Wildlife Diseases* **39**, 84–95.

Mintiens, K., Méroc, E., Mellor, P.S., *et al.* (2008) Possible routes of introduction of bluetongue virus serotype 8 into the epicentre of 2006 epidemic in north-western Europe. *Preventative Veterinary Medicine* **87**, 131–144.

Monaco, F., Camma, C., Serini, S. and Savini, G. (2006) Differentiation between field and vaccine strain of bluetongue virus serotype 16. *Veterinary Microbiology* **116**, 45–52.

Monteys, V.S.I., Ventura, D., Pages, N., Aranda, C. and Escosa, R. (2005) Expansion of *Culicoides imicola*, the main bluetongue virus vector in Europe, into Catalonia, Spain. *Veterinary Record* **156**, 415–417.

Moritz, C., Patton, J.L., Conroy, C.J., *et al.* (2008) Impact of a century of climate change on small-mammal communities in Yosemite National Park, USA. *Science* **10**, 261–264.

Natoli, E., Say, L., Cafazzo, S., *et al.* (2005) Bold attitude males make urban feral domestic cats more vulnerable to feline immunodeficiency virus. *Neuroscience and Biobehavioral Reviews* **29**, 151–157.

Olde Riekerink, R.G.M., Dominici, A., Barkema, H.W. and de Smit, A.J. (2005) Seroprevalence of pestivirus in four species of alpine wild ungulates in the High Valley of Susa, Italy. *Veterinary Microbiology* **108**, 297–303.

Oleaga, A., Casais, R., González-Quirós, P., Prieto, M. and Gortázar, C. (2008) Sarcoptic mange in red deer from Spain: improved surveillance or disease emergence? *Veterinary Parasitology* **154**, 103–113.

Parmesan, C. and Yohe, G. (2003) A globally coherent fingerprint of climate change impacts across natural systems. *Nature* **421**, 37–42.

Paterson, S., Wilson, K. and Pemberton, J.M. (1998) Major histocompatibility complex variation associated with juvenile survival and parasite resistance in a large unmanaged ungulate population (*Ovis aries* L.). *Proceedings of the National Academy of Sciences of the USA* **95**, 3714–3719.

Patz, J.A. and Reisen, W.K. (2001) Immunology, climate change and vector-borne diseases. *Trends in Immunology* **22**, 171–172.

Pedersen, C. and Post, E. (2008) Interactions between herbivory and warming in aboveground biomass production of arctic vegetation. *BMC Ecology* **8**, 17.

Pelletier, F., Hogg, J.T. and Festa-Bianchet, M. (2004) Effect of chemical immobilization on the social status of bighorn rams. *Animal Behaviour* **67**, 1163–1165.

Pioz, M., Loison, A., Gibert, P., *et al.* (2007) Transmission of a pestivirus infection in a population of Pyrenean chamois. *Veterinary Microbiology* **119**, 19–30.

Pioz, M., Loison, A., Gauthier, D., *et al.* (2008) Diseases and reproductive success in a wild mammal: example in the alpine chamois. *Oecologia* **155**, 691–704.

Plowright, W. (1982) The effect of rinderpest and rinderpest control on wildlife in Africa. *Symposia of the Zoological Society, London* **50**, 1–28.

Pontier, D., Fromont, E., Courchamp, F., Artois, M. and Yoccoz, N.G. (1998) Retroviruses and sexual size dimorphism in domestic cats (*Felis catus* L.). *Proceedings of the Royal Society B–Biological Sciences* **265**, 167–173.

Purse, B.V., Mellor, P.S., Rogers, D.J., *et al.* (2005) Climate change and the recent emergence of bluetongue in Europe. *Nature Reviews Microbiology* **3**, 171–181.

Putman, R.J., Langbein, J., Hewison, A.J.M. and Sharma, S.K. (1996) Relative roles of density-dependent and density-independent factors in the population dynamics of British deer. *Mammal Review* **26**, 81–101.

Raffel, T.R., Martin, L.B. and Rohr, J.R. (2008) Parasites as predators: unifying natural enemy ecology. *Trends in Ecology and Evolution* **23**, 610–618.

Rossi, L., Meneguz, P.G., DeMartin, P. and Rodolfi, M. (1995) The epizootiology of sarcoptic mange in chamois, *Rupicapra rupicapra*, from the Italian eastern Alps. *Parasitologia* **37**, 233–240.

Ruiz-Fons, F., Fernández-de-Mera, I.G., Acevedo, P., *et al.* (2006) Ixodid ticks parasitizing Iberian red deer (*Cervus elaphus hispanicus*) and European wild boar (*Sus scrofa*) from Spain: geographical and temporal distribution. *Veterinary Parasitology* **140**, 133–142.

Ruiz-Fons, S., Reyes-Garcia, A.R., Alcaide, V. and Gortázar, C. (2008a) Spatial and temporal evolution of bluetongue virus in wild ruminants, Spain. *Emerging Infectious Diseases* **14**, 951–953.

Ruiz-Fons, S., Segalés, J. and Gortázar, C. (2008b) A review of the viral diseases of the European wild boar: effect on population dynamics and reservoir role. *Veterinary Journal* **176**, 158–169.

Ruiz-Fons, F., Rodríguez, O., Torina, A., Gortázar, C. and de la Fuente, J. (2008c) Prevalence of *Coxiella burnetii* infection in wild and farmed ungulates. *Veterinary Microbiology* **126**, 282–286.

Ryser-Degiorgis, M.P., Bischof, D.F., Marreros, N., *et al.* (2009) Detection of *Mycoplasma conjunctivae* in the eyes of healthy, free-ranging Alpine ibex: possible involvement of Alpine ibex as carriers for the main causing agent of infectious keratoconjunctivitis in wild Caprinae. *Veterinary Microbiology* **2**, 368–374.

Salinas, J., Caro, M.R., Vincent, J., *et al.* (2009) High prevalence of antibodies against Chlamydiaceae and *Chlamydophila abortus* in wild ungulates using two 'in house' blocking-ELISA tests. *Veterinary Microbiology* **16**, 46–53.

Sapolsky, R.M. (1992) Neuroendocrinology of the stress response. In J.B. Becker S.M. Breedlove and D. Crews (eds.) *Behavioral Endocrinology*. Cambridge, MA: MIT Press, pp.287–324.

Schauber, E.M. and Woolf, A. (2003) Chronic wasting disease in deer and elk: a critique of current models and their applications. *Wildlife Society Bulletin* **31**, 610–616.

Schettler, E., Steinbach, F., Eschenbacher-Kaps, I., *et al.* (2006) Surveillance for prion disease in cervids, Germany. *Emerging Infectious Diseases* **12**, 319–322.

Smith, K.F., Sax, D.F. and Lafferty, K.D. (2006) Evidence for the role of infectious disease in species extinction and endangerment. *Conservation Biology* **20**, 1349–1357.

Smith, K.F., Acevedo-Whitehouse, K. and Pedersen, A.B. (2009b) The role of infectious diseases in biological conservation. *Animal Conservation* **12**, 1–12.

Smith, M., Telfer, S., Kallio, E.R., *et al.* (2009a). Host-pathogen time series data in wildlife support a transmission function between density and frequency dependence. *Proceedings of the National Academy of Sciences of the USA* **106**, 7905–7909.

Sommer, S. (2005) The importance of immune gene variability (MHC) in evolutionary ecology and conservation. *Frontiers in Zoology* **2**, 16.

Szczurek, B. and Kadulski, S. (2004) Ectoparasites on fallow deer, *Dama dama* (L.) in Pomerania, Poland. *Acta Parasitologica* **49**, 80–86.

Sæther, B.-E. (1997) Environmental stochasticity and population dynamics of large herbivores: a search for mechanisms. *Trends in Ecology and Evolution* **12**, 143–149.

Thomson, G.R., Vosloo, W. and Bastos, A.D.S. (2003) Foot and mouth disease in wildlife. *Virus Research* **91**, 145–161.

Thrall, P.H., Antonovics, J. and Hall, D.W. (1993) Host and pathogen coexistence in sexually transmitted and vector-borne diseases characterized by frequency-dependent disease transmission. *American Naturalist* **142**, 543–552.

Tompkins, D.M. and Begon, M. (1999) Parasites can regulate wildlife populations. *Parasitology Today* **15**, 311–313.

Vikøren, T., Ytrehus, B. and Handeland, K. (2008) *Årsrapport for 2006 og 2007. Helseovervakingsprogrammet for hjortevilt (HOP)*. Report Series 19. Oslo: National Veterinary Institute.

Vincente, J. Höfle, U., Garrido, J.M., *et al.* (2007) Risk factors associated with the prevalence of tuberculosis-like lesions in fenced wild boar and red deer in south central Spain. *Veterinary Research* **38**, 451–464.

Webb, S.D., Keeling, M.J. and Boot, M. (2007) Spatially extended host-parasite interactions: the role of recovery and immunity. *Theoretical Population Biology* **71**, 251–266.

Weladji, R.B., Holand, Ø. and Almøy, T. (2003) Use of climatic data to assess the effect of insect harassment on the autumn weight of reindeer (*Rangifer tarandus*) calves. *Rangifer* **22**, 33–50.

Wikelski, M. and Cooke, S.J. (2006) Conservation physiology. *Trends in Ecology and Evolution* **21**, 38–46.

Williams, S.E., Bolitho, E.E. and Fox, S. (2003) Climate change in Australian tropical forests: an impending environmental catastrophe. *Proceedings of the Royal Society B–Biological Sciences* **270**, 1887–1892.

Wingfield, J.C. (2008) Comparative endocrinology, environment and global change. *General and Comparative Endocrinology* **157**, 207–216.

Wobeser, G.A. (2006) *Essentials of Disease in Wild Animals*. Oxford, UK: Wiley-Blackwell Publishing.

Woolhouse, M.E. (2002) Population biology of emerging and re-emerging pathogens. *Trends in Microbiology* **10**, 3–7.

12

Climate change and implications for the future distribution and management of ungulates in Europe

ATLE MYSTERUD AND BERNT-ERIK SÆTHER

12.1 Introduction

There is little doubt that predicted changes in climate (IPCC, 2007) are of an order of magnitude that they are likely to affect large herbivores in Europe in a number of ways. Whichever predictive models are used, the general expectation for future climate is for greater levels of precipitation, warmer temperatures, and perhaps most significantly an overall increase in variability (IPCC, 2007). However, at the regional level, both variation and uncertainty is expected to be much higher.

It is important to realise that climate in itself is not necessarily the only limitation for the current distribution ranges of all ungulate species, whose range may also be affected by both natural and artificial barriers, by patterns of land use (and urbanisation), by direct management, or by the fact that they are still colonising (Groot Bruinderink *et al.*, 2003). This, combined with the absence of published assessments of how the distribution of different large herbivore species may be affected by climate, make our attempt here a risky business in terms of accuracy of predictions. However, with that said, what might we nevertheless expect? We focus, in the following, mainly on global distribution patterns.

In general, large herbivores can be both directly and indirectly affected by climate. Direct effects of climate are mainly related to thermoregulation either due to extreme heat or cold (Parker and Robbins, 1985), water limitation (Wallach *et al.*, 2007), and costs of moving in snow (Parker *et al.*, 1984). Indirect effects are largely mediated through access to forage (snow or ice) or actual vegetational productivity and quality (e.g. Hebblewhite *et al.*, 2008). Climate does not in general affect distribution ranges directly. Global climatic conditions in any given regional area interact with factors such as

Ungulate Management in Europe: Problems and Practices, eds. Rory Putman, Marco Apollonio and Reidar Andersen. Published by Cambridge University Press.

topography to determine the local weather, which is what may influence the individual. Individual responses of behaviour or physiology in turn affect the life history and thereby the demography and population dynamics of a given species.

Changes in distribution of any animal species are influenced by the dynamics of local populations as well as the pattern of movement of individuals (Maurer and Taper, 2002). Variation in population growth rates of ungulates will mainly be influenced by how summer foraging conditions affect productivity and recruitment, how many animals survive the winter, and, in populations so large that demographic stochasticity can be ignored, how environmental variability may affect temporal fluctuations in population growth rates. Such changes determine whether or not any individual local population is growing or declining. No population can persist if the expected population growth rate is less than zero. In the face of climate change, environmental conditions necessary to allow positive long-term population growth rates of most species (i.e. the niche) will be moved in space (Holt *et al.*, 2005).

Due to local declines or extinctions, together with increases in productivity of yet other populations in different areas, overall distributional range may change through time. More immediately, distributional ranges may be altered if animals shift their home ranges through time in response to altered environmental conditions, gradually abandoning areas which are becoming less favourable and directly colonising areas whose suitability is increasing. The rate of this response in distributional range will, however, also be dependent on the magnitude of the demographic stochasticity (Holt and Keitt, 2005).

Future distribution of large herbivores, within individual countries and within their European range, will result from whether they are able to persist in their historical distribution ranges or move to new ranges. They may persist in the historical ranges if the new conditions still allow positive expected population growth rates. This may be facilitated by plastic responses in demography or by developing evolutionary adaptations to the new environmental conditions. Most studies of climate effects derive from single populations, and we have a fair understanding of how climate variability affects large herbivores in a given area (reviews in Sæther, 1997; Gaillard *et al.*, 2000; Weladji *et al.*, 2002; Mysterud *et al.*, 2003). This background enables us to assess the local dynamics and thus likelihood of continued persistence within historical ranges. The behavioural ecology of the species (mainly dispersal) interacting with geographic barriers will determine if the species is capable of moving to previously hostile environments that become suitable habitat during climate change.

Within this general framework, one of the particular factors which puts a special constraint on large herbivore populations in temperate latitudes is seasonality (mainly by affecting the timing and duration of plant growth). It is fairly clear that large herbivores at the northern distribution limits face different challenges from those at the southern distribution limits, which can be related to differing seasonal limitations. We will therefore review climate factors known to affect the population dynamics during winter and summer separately.

In this chapter, we aim to highlight what we regard as the most likely climate components to affect distribution ranges of ungulates. At northern latitudes, these are mainly changes in snow depth (a lowering will increase range size) and more frequent icing events (an increase may reduce range size) during winter. At the southern limits, increased frequencies of droughts may similarly restrict range sizes.

We are less certain about how the overall seasonal structure might affect ranges. We do know that the growing season will be extended in many areas, but this positive effect may be countered by increased summer heat. The strict division of northern and southern distributions can be questioned, as the relative role of different seasonal limitations for a given area may change in the future. For one of the best studied populations of large herbivores, the Soay sheep, winter conditions have been regarded as the critical period with a limited effect of summer conditions (Clutton-Brock *et al.*, 1997). However, the first ever summer drought appeared in summer 2008, becoming a new, potentially important factor for future dynamics (T. Coulson, pers. comm.). The appearance of new population-limiting factors make any population prediction uncertain (Coulson *et al.*, 2001b; Festa-Bianchet *et al.*, 2006).

There are a number of limiting factors in addition to climate that will determine the future distribution of large herbivore populations. These include large carnivores, diseases, land use, habitat loss and, not the least, human management. These factors may increase or decrease the effect of future climate changes, and they may frequently interact. Climate may increase likelihood of diseases and parasites, which are increasingly becoming apparent as a very important field of which we know little (Murray *et al.*, 2006; Brooks and Hoberg, 2008; Chapter 11 of this volume). Large carnivores are also expanding in much of Europe (Enserink and Vogel, 2006), and the effects of predation often depend on climate interactions (Cederlund and Lindström, 1983; Hebblewhite, 2005; see also Chapter 10, this volume). These interactions will undoubtedly be important for prediction of how the future distribution ranges of large herbivores will be affected by the expected changes in climate.

Let us then review what we do know, and try to identify the possible climate winners and losers – which ones are likely to be correlated with the possibility of expanding or restricting their distributional ranges.

12.2 Global climate, local weather and regional variation

In Europe, much of the climate variation, especially during winter, can be linked to the state of an atmospheric phenomenon across the North Atlantic Ocean called the North Atlantic Oscillation (NAO; Hurrell, 1995; Hurrell *et al.*, 2003). During high phases of the NAO, there is lowered surface pressure over the Arctic and increased surface pressure over the subtropical North Atlantic. This pressure configuration causes a northward shift in the Atlantic storm activity, with stronger westerly winds from southern Greenland across Iceland into northern Europe (especially Scandinavia) and a modest decrease in activity towards the south. This leads in turn to increased precipitation and temperatures occurring over Iceland and Scandinavia, and drier conditions over much of central and southern Europe, the Mediterranean and parts of the Middle East (Stenseth *et al.*, 2003). During low phases of the NAO, fewer winter storms hit northern Europe, resulting in drier and colder conditions in the north, but wetter conditions in the Mediterranean region. To enable more accurate predictions of large herbivore distribution with climate change in Europe, we need to know the relationship between the NAO and climate change and between the NAO and regional variation in local weather in Europe. The link between the NAO and climate change is complicated (review in Gillett *et al.*, 2003). However, most climate models suggest some increase in the winter NAO index in response to increasing concentrations of greenhouse gases (Gillett *et al.*, 2003).

The NAO as a pressure system can be indexed in several different ways and for different periods. The simplest NAO index (the so-called 'station-based index') quantifies the average deviation from the long-term mean sea-level surface pressure difference between Lisbon in Portugal and Stykkisholmur in Iceland. The reason for using a simple index is to get a long time series (back to 1864), but the problem is that small variations in the exact situation of the Arctic or subtropical pressure centres can have very large effects on the index if based on only two locations. More recent indices (which can be calculated from raw data available from 1899) use the first axis of a principal component analysis (PCA-based, equivalent to EOF-based; Gillett *et al.*, 2003) of sea-level pressure anomalies over the Atlantic sector (20°–80° N, 90° W–40° E). The latter indices are more optimal representations of the full NAO spatial pattern and better capture the NAO dynamics on an annual scale (Stenseth *et al.*, 2003).

The NAO index most commonly used is for the months of December through March, but the index for spring has also found to be linked to local temperature and precipitation and to have an effect on large herbivores (Pettorelli *et al.*, 2005a; Mysterud *et al.*, 2008). The NAO is interesting due to its simplicity, and that a single index of the NAO can say something about the state of climate in much of Europe. It happens to capture variation of climate in time and space at a scale relevant in many ecological studies (Hallett *et al.*, 2004, Stenseth and Mysterud, 2005). However, there is large regional variation in the impact of the NAO in Europe – related to altitude, latitude, topography and distance to coast (continentality), to name the most important axes of variation (Beniston, 2006).

On the west coast of Norway, the NAO is positively correlated with temperature and precipitation during winter. However, since temperatures in this region often are around 0 °C during winter, the relationship between the NAO and snow depth depends on altitude (Mysterud *et al.*, 2000). During high phases of the NAO, there is more snow at high altitude and less at low altitude, due to both increased precipitation and warmer temperature, and since it is colder at higher altitude. In the Alps, a higher NAO in contrast tends to give below-average levels of precipitation, but above-average temperatures (Beniston and Jungo, 2002; Beniston, 2006).

There is a much lower impact of the NAO for inland areas in Europe. In inland locations in Poland (Mysterud *et al.*, 2007), PCA-based indices of local weather outperformed the NAO in predicting population dynamics of European bison. Population (harvest) growth rates of roe deer in Norway, most of them in more inland locations, were positively related to the NAO in 94.7% of the populations, but local indices of snow depth explained more of the variation (Grøtan *et al.*, 2005). No effect of the NAO has been reported for body mass variation of moose in Estonia (Veeroja *et al.*, 2008) or roe deer in Sweden and France (Kjellander *et al.*, 2006). The NAO therefore does not capture the most relevant climate variability for all of Europe. There has also been suggested a so-called phase shift in the link between a given value of the NAO and local weather; that is, that the correlation between the NAO and local weather might change over decadal scales (Mysterud *et al.*, 2003; Durant *et al.*, 2004), which will limit the predictive value. This issue is controversial and not all climatologists are convinced this is the case (Gillett *et al.*, 2003).

It is therefore likely that the NAO will be a useful starting point to give some expectation towards what regional patterns to expect in large herbivore dynamics as well as in scale of regional synchrony. So far, there have been few studies looking at spatial synchrony in the effects of climate on large herbivore populations. According to the Moran effect (Moran, 1953) we would expect this correlation to be equal to the spatial autocorrelation in the environmental noise, provided that the density regulation is loglinear and of similar strength in all populations (Engen and Sæther, 2005).

The population sizes of Soay sheep on two islands 3.5 km apart in the St Kilda archipelago was remarkably synchronised over a 40-year period (Grenfell *et al.*, 1998; Blasius and Stone, 2000). For roe deer in Norway, the density dynamic was synchronous up to a scale of about 200 km (Grøtan *et al.*, 2005). For Svalbard reindeer, only two neighbouring populations were synchronised, but not a population situated about 140 km north-west of the other two areas (Aanes *et al.*, 2003). The dynamics of caribou and musk oxen on each side of Greenland were reported to be highly synchronised from a minimum of 1000 km up to about 1700 km (Post and Forchhammer, 2002), but this result is controversial (Vik *et al.*, 2004). This indicates that climate covariation is able to induce spatial synchrony in the population fluctuations of ungulate populations over quite large distances (but see Grøtan *et al.*, 2008).

There is a considerable effort among climatologists in downscaling the global climate predictions to scales relevant for large herbivores. This has so far been done for the Alps (N.G. Yoccoz, pers. comm.), and will likely be available for most of Europe in the near future, allowing for much more accurate and relevant predictions of the effects of climate change on large herbivore populations.

12.3 Winter limitation at northern distribution ranges: snow depth and icing

There is a large literature from northern areas of what climatic conditions might limit large herbivore populations (reviews Sæther, 1997; Weladji *et al.*, 2002; Mysterud *et al.*, 2003). In a classical study, Formozow (1946) stated that northern distribution ranges of large herbivores are mainly limited by snow depth and duration of snow cover. Though cold temperature and strong winds add a negative factor to the energy budget, later quantitative research also shows that snow is the main problem (Sæther *et al.*, 1996; Grøtan *et al.*, 2005).

Severe winters increase loss of body mass (Sæther and Gravem, 1988; Cederlund *et al.*, 1991; Herfindal *et al.*, 2006a), which results mainly in reduced survival of calves (Cederlund and Lindström, 1983; Loison and Langvatn, 1998). Climatic conditions during gestation (i.e. late winter) was the most important period to red deer (Sims *et al.*, 2007). The future snow depth maps will therefore be decisive for large herbivore distribution at northern latitudes and at high altitude. As stated above, since both temperature and precipitation will increase at northern latitudes, temperature at time of precipitation will be critical and determine whether or not snow depths are likely to increase or decrease. At present, maximum snow depths in the mountains of Norway are increasing, though the duration of snow cover has decreased (Kausrud *et al.*, 2008). One can therefore expect that with future stronger warming, it may also begin to rain more in the mountains.

A higher frequency of alternating warm and cold spells (Shabbar and Bonsal, 2003) can cause problems for grazers, especially, by 'locking pastures'. The importance of icing events can most clearly be seen in the ungulate species inhabiting the most extreme environment in Europe, the Svalbard reindeer living at 80 °N. Occasional winters induce massive die-offs due to winter rains and subsequent freezing (Aanes *et al.*, 2002; Putkonen and Roe, 2003; Kohler and Aanes, 2004; Grenfell and Putkonen, 2008). The mortality may reach levels as high as 60% of the total population.

Snow depth, quality and duration of snow cover are therefore the most likely drivers of northern distribution ranges. Decreased snow depth and cover are likely to increase northern distribution ranges of roe deer, red deer, fallow deer, wild boar, and possibly also the mouflon (depending on management action) in Sweden. Changes in the spatial distribution of snow are also likely to affect the distribution patterns of alpine ungulate species in Central Europe such as the ibex and chamois (Sæther *et al.*, 2002; Jacobson *et al.*, 2004; Grøtan *et al.*, 2008). These processes are already ongoing, though there are no case studies documenting a direct link of distribution range increase to climate.

12.4 Changes in growing season length and indirect effects of snow depth

The productivity of large herbivores will also be affected by summer range conditions, but it is less clear whether it will yield changed distribution ranges. The length of growing season is probably important, and together with timing of the onset of spring defines the critical period for productivity. Early spring conditions indexed by the Normalised Difference Vegetation Index (NDVI), a measure of photosynthetic activity (often based on sensing from satellites; Pettorelli *et al.*, 2005b), was the most important predictor for individual performance (body mass) of roe deer in France (Pettorelli *et al.*, 2006).

Early plant growth is determined mainly by spring temperature, but for alpine and northern populations also by snow accumulation patterns during winter (Pettorelli *et al.*, 2005a). Duration of snow can impact plant phenology and quality for the rest of the grazing season (Borner *et al.*, 2008; Cebrian *et al.*, 2008). Effects of the winter NAO operating through snow depths

(Mysterud *et al.*, 2000) are shown to affect soil moisture (Kettlewell *et al.*, 2006) and in turn plant phenology, as shown directly on plant first-flowering dates (Post and Stenseth, 1999; Aanes *et al.*, 2002) or indirectly with NDVI (Pettorelli *et al.*, 2005a, 2005c, 2006; Herfindal *et al.*, 2006b). A variable plant phenology is predicted to be favourable to most large herbivores, at least grazing species, due to an extended period with access to young, high-quality forage (Albon and Langvatn, 1992; Mysterud *et al.*, 2001a; Hebblewhite *et al.*, 2008).

In heterogeneous, mountainous landscapes, phenology has so far become more variable with recent warming favouring red deer in Norway (Pettorelli *et al.*, 2005a). In contrast, less variable plant phenology was reported in the more flat landscape on Greenland, likely detrimental to caribou (Post and Forchhammer, 2008; Post *et al.*, 2008). With increased warming, less snow will result also at higher elevation.

These same factors seem important also for continental Europe. For chamois in the Alps and the closely related isard in the eastern Pyrenees, survival decreased following high phases of the NAO and low spring degree-days (Loison *et al.*, 1999b). There was no direct impact of winter weather, so global changes in the NAO seemed also here to operate indirectly through plants. Lower snow cover as predicted, especially below 1700–2400 m a.s.l., would therefore affect populations negatively (Loison *et al.*, 1999b). The study suggested opposite effects of increased spring temperature in the two populations. As adult survival was affected, having a large impact on dynamics (Gaillard *et al.*, 1998), this may have severe consequences. For alpine and northern populations, decreased amount of snow fields during summer may speed up plant phenology, reducing production markedly. Whether the beneficial effect of lower snow depth during winter may be countered or enhanced by the indirect effect of snow on plant growth during summer will likely depend on topography. Accordingly, although large snow depth during late winter generally affects the population growth of the Swiss ibex negatively (Sæther *et al.*, 2002; Jacobson *et al.*, 2004), large differences were found over short geographical distances in the relative impact on the population dynamics of the same climate variable (Grøtan *et al.*, 2008). This is also likely to affect grazers more than browsers.

12.5 Summer/spring limitations: drought and southern distribution ranges

When moving from north to south, the factors limiting plant growth are changing (Loe *et al.*, 2005). In the north, the main limitation is low temperature and a very short growing season, while precipitation takes over as the most important factor further south. For example, while red deer in Norway and Scotland decrease in mass during winter and grow from May–June until plant material senesces in late autumn (around October), Spanish red deer can grow during autumn and winter, but often do not grow during summer due to drought (Martínez-Jauregui *et al.*, 2009). Similarly, water becomes a clear limitation for roe deer habitat selection in the southern distribution ranges of Europe (Telleria and Virgos, 1997; Virgos and Telleria, 1998; Wallach *et al.*, 2007). At lower latitudes, summer drought may become more important, and may be a key factor to future distribution of large herbivores in southern Europe even in the mountains. A severe drought hit southern Europe in 2003 (Ciais *et al.*, 2005). The mortality of lambs in the mouflon population of Caroux-Espinouse in southern France increased markedly, from $<10\%$ to 25.7% (Garel *et al.*, 2004). Survival depended on the amount of rainfall recorded at a given 14 day period and in the previous 14–21 day period, but was not influenced by the exceptionally high mean daily temperature recorded.

12.6 Buffering of (winter) climate effects: morphology and behaviour

Ungulate species differ greatly in their adaptations. Some of them are much more prone to severe conditions, and it may affect which season is the more limiting. An old paradigm for large herbivores is that the winter season determines number of animals, while summer foraging conditions determine body condition and thus quality (Klein, 1965; Sæther and Heim, 1993; Pettorelli *et al.*, 2005c). A few studies report interspecific synchrony in body mass for a given location. In Norway, annual variation in body mass of domestic sheep (grazing during summer only) were synchronised with body mass of moose (Sæther, 1985), red deer (Mysterud *et al.*, 2001b) and reindeer (Weladji *et al.*, 2003b). These studies at least indicate consistent interspecific responses to summer foraging conditions.

A comparative framework for species' ability to cope with snowy conditions was derived for North American species (Telfer and Kelsall, 1984). The ability to cope with snow was suggested to be related to both morphological factors and behavioural strategies. The chest height and foot loading were used to derive a morphological index, while a behavioural index included scoring use of trails, specific habitat selection, feeding below or above snow, cratering, migration and technique of locomotion (Telfer and Kelsall, 1984). No formal comparative work has been done at a population level to see whether such a scheme works well to predict stability of large herbivore populations facing severe climate conditions. The comparative work of Gaillard *et al.* (2000) found no link between annual pattern of age-specific survival and body size, but was not season specific.

12.6.1 Morphology

The single most important factor is body size; the larger, the better the chances of winter survival. Due to surface:volume ratios, larger animals need less energy per unit of body mass. Since rumen size scales isometrically with body size, larger ruminants can survive on a lower quality diet, which is termed the Jarman–Bell principle (Demment and Van Soest, 1985). Since plant quality is overall lower during winter, and since the fat reserves are too small to last the entire winter, this is an extremely critical component for large herbivores. The fur and fat reserves are also regarded as important. Larger deer have fat reserves to survive longer periods than smaller deer, since fat reserves scale approximately isometrically with size (Calder, 1984), while energy use per kilogram body mass is lower for larger deer (Demment and Van Soest, 1985). The higher and more variable mortality of male compared with female ungulates likely results from exhaustion of males' fat reserves due to rutting just prior to winter (Toïgo and Gaillard, 2003). Direct empirical data on the importance of the fat reserves are limited and body mass is most often the index used. For roe deer, dynamics was best explained by variation in November snow fall, indicating that duration of winter (and fat reserves) might be decisive (Grøtan *et al.*, 2005). Climate change is expected to shorten duration of snow cover, which will benefit roe deer and other small ungulates with limited fat reserves.

12.6.2 Behaviour

Behaviour is a key factor in buffering climate impacts (review in Moen, 1973; Telfer and Kelsall, 1984). The single most important behavioural buffer for climate effects is migration to lower elevation by northern large herbivores (references in Fryxell and Sinclair, 1988; Mysterud, 1999). This is a widespread phenomenon, as lower elevation areas have lower snow depths, though snow quality may be similar (Lundmark and Ball, 2008). Any factor reducing the possibility for migration may thus reduce the animals' ability to buffer such effects of climate change. Building of infrastructures at low elevation can have a negative effect (Andersen, 1991); also, building of railways and major roads can either cut off migration routes completely (fences) or lead to increased number of traffic accidents if they are built in the animals' main wintering areas. Indeed, increased numbers of moose–vehicle collisions are found in harsh winters (Gundersen and Andreassen, 1998).

There is also a challenge related to how the behavioural strategies to buffer climate effects may be influenced by predation (Hebblewhite *et al.*, 2008). Predators may frequently lead to suboptimal use of plants by large herbivores (e.g. Fortin *et al.*, 2005), and moose and reindeer, having calves that are more prone to predation, may be deterred from migrating to areas with the best foraging conditions (Edwards, 1983; Bergerud *et al.*, 1984). Therefore, both building of human infrastructure and recolonisation of large carnivores can reduce the behavioural buffering of climate effects (see also below regarding dispersal).

12.7 Likely demographic consequences of increased frequency of extreme climatic events

Due to large body size and limited time for offspring to grow sufficiently large to meet the winter conditions, annual reproduction by large herbivores is the rule with few exceptions (such as wild boar in some cases). All large herbivore species are iteroparous, and the key to female fitness is thought to be mainly related to number of reproductive attempts (Weladji *et al.*, 2006). A general feature of the life history is thus that adult females are conservative in their resource use (Gaillard and Yoccoz, 2003). Females will decrease reproductive effort rather than risking their own survival if there is an unusually poor year (fewer resources available for reproduction). Female reindeer forced to breed late raised offspring that were smaller in autumn, but maintained the same body condition as females breeding at normal times (Holand *et al.*, 2006). Similarly, many adult female red deer in Scotland fail to breed at all in years of low resource availability (Mitchell, 1973; Ratcliffe, 1984).

A general pattern resulting from such a conservative life history strategy of adult females, is that juvenile survival is generally lower and much more variable than adult survival (Putman *et al.*, 1996; Gaillard *et al.*, 1998). Adult survival is therefore usually above 90% per year in areas with no large predators or hunting, and with very low variation between years (e.g. Toïgo *et al.*, 1997). At very high density and in poor habitat, particularly harsh years may have a slight penalty even on adult survival (Albon *et al.*, 2000). The adult stage is typically between 2 and 7 years of age; older individuals again have higher mortality (Loison *et al.*, 1999a). There is thus a fairly predictable sequence of traits affected as climate gets more severe (Eberhardt, 2002). This conservative life history strategy has important implications for predicting future plastic responses to climate changes (Morris *et al.*, 2008).

The resulting demographic pattern of mortality is typically that the young and very old die under extreme climatic conditions (Gaillard *et al.*, 1998), most clearly shown for the Soay sheep in the St Kilda archipelago off Scotland (Coulson *et al.*, 2001a). Massive die-offs of young result in cohort effects that are indeed widespread in ungulates (red deer: Albon *et al.*, 1992;

Post *et al.*, 1997; Mysterud *et al.*, 2002; moose: Solberg *et al.*, 2004, 2007; roe deer: Kjellander *et al.*, 2006), affecting both the quantity and quality of offspring produced (review in Gaillard *et al.*, 2003). Animals from poor cohorts typically delay onset of maturation by a year, causing lagged effects (Solberg *et al.*, 1999; Langvatn *et al.*, 2004).

Clearly, therefore, the frequency of extreme events will be important for population persistence; a single extreme year is not enough to drive most populations to extinction (Reed *et al.*, 2003). High values of the NAO will lead to a higher frequency of cold and warm spells in the Arctic (Shabbar and Bonsal, 2003). Exactly how frequent such years must be to be a threat to persistence has not been modelled: it will also depend on the level of mortality rates and also on the reproductive potential of the species. Several of the alpine bovids have rather low population growth rates, and may therefore be particularly vulnerable.

12.8 Interactions with other limiting factors may buffer or increase climate effects

It is now well documented that the effects of severe winter climate are more pronounced at high population density (Sauer and Boyce, 1983; Coulson *et al.*, 2001a), as shown for Svalbard reindeer (Aanes *et al.*, 2000; Solberg *et al.*, 2001), red deer (Albon *et al.*, 2000) and ibex (Sæther *et al.*, 2002; Jacobson *et al.*, 2004). In comparison, there is only one study (on moose) reporting an interaction between summer climate and density (Herfindal *et al.*, 2006a), suggesting climate and density interaction to be somewhat less important during summer (Mobæk *et al.*, 2009). Lower effects of climate at low population density are likely to buffer effects of climate change on large herbivore distributions to some extent (see, for example, Hallett *et al.*, 2004).

Apart from the effects of local population density, however, there are a number of other factors that will strongly affect the outcome of climate change on large herbivore numbers and distribution patterns. These factors – disease, predators and competitors – may buffer or increase the effects of climate change. In addition, land use changes may have a marked impact on some species, for example if new crops are grown. We have no overview of such changes at the present time. One feature that we regard as particularly compelling is the possible link between climate and disease (see Chapter 11 and Box 11.2). In Minnesota in the USA, the lowered food quality available after increased warming resulted in increased infestation of parasites, and this caused a decline in the moose population (Murray *et al.*, 2006). This potential influence of changes in disease risk may thus mean that even slight changes in temperature can affect large herbivore populations to a much larger degree than anticipated if, for example, effects were operating through plant quality alone.

In Norway there has been growing concern regarding an increased frequency of tick-borne diseases, and the recent spread of deer ked to moose. In reindeer, the largest effects of summer climate are expected to be linked to activity of both flies and warble flies (Hagemoen and Reimers, 2002; Weladji *et al.*, 2003a). The infestations of warble flies are negatively correlated with density (Fauchald *et al.*, 2007), so that if populations decline, increased effects can be expected. Also, reindeer typically use snow fields during insect harassment, which may melt away giving no refuge. In the Alps, chamois fecundity was affected by bacterial infections, which in turn was related to climate (negatively to the NAO, more after cold and snowy winters: Pioz *et al.*, 2008).

More severe winters can also lead to higher rates of predation, as shown for red fox predation on roe deer (Cederlund and Lindström, 1983) and wolves for moose (Vucetich and Peterson, 2004). Predators can therefore, in combination with winter, set the northern distribution limits at lower latitudes than if climate alone was operating. Though not well documented, roe deer seem to have reduced the extent of their distribution in the north of Norway after recolonisation of lynx in recent decades. It has also been argued that wolf predation in Russia causes roe deer to have a lower northern distribution range than expected based on snow depth alone, but here human hunting also played a role (Danilkin, 1996).

Another indirect way in which climate change may influence ungulate populations is through competition (Case *et al.*, 2005). For birds, it has been shown that climate can affect the balance of interspecific competition (Sætre *et al.*, 1999). More generally it is well established that in any competitive interaction, competitive advantage is greatly influenced by environmental conditions. If climatic change is likely to affect the balance of competition amongst ungulate species, competitive relationships may be altered even among species which already have overlapping distributions. In addition, if some species increase their distributional range, they may become sympatric with new species or species combinations with which they have to establish new competitive relationships, with unpredictable outcome. Animals forced into suboptimal habitat through competition may subsequently be more vulnerable to other effects of climate change. In Spain, the Iberian ibex is forced into suboptimal habitat by goats (Acevedo *et al.*, 2008a) and possibly the invasive exotic aoudad (Cassinello *et al.*, 2004; Acevedo *et al.*, 2008b). In the UK, the red deer seem to be competitively superior to the roe deer (Latham *et al.*, 1996, 1997; reviews in Putman, 1996; Latham, 1999).

At northern latitudes, there is also competition between rodents and larger herbivores in peak rodent years (Steen *et al.*, 2005). The fading of the cyclic dynamics of voles and lemmings due to climate change (review in Ims *et al.*, 2008) may relieve competition with alpine reindeer. However, in forested areas in Sweden, the generalist red fox predates more on roe deer fawns during years with a low abundance of rodents (Kjellander and Nordström, 2003). Consistent low populations of rodents might therefore markedly elevate predation rates on roe and thus lower population productivity in roe deer. Indeed, such ecosystem effects might be strong, but are often difficult to predict.

12.9 What role will evolution play?

A central focus in the literature on climate effects on animals in general is whether animals are sufficiently phenotypically plastic to respond or whether evolution will be rapid enough to track changes in the environment (e.g. Przybylo *et al.*, 2000; Nussey *et al.*, 2005c; Skelly *et al.*, 2007; Charmantier *et al.*, 2008). Theories for evolution of quantitative characters (Lande 1982) have shown that the response to selection in the mean of the life history trait z_i is

$$d\bar{z}_i/dt = \sum_{j=1}^{n} G_{ij}\partial r/\partial \bar{z}_j,$$

where G_{ij} is the genetic variance–covariance between characters i and j and $\partial r/\partial \bar{z}_j$ is the selection gradient of the j'th character. This means that the rate of evolution of the mean life history equals approximately the additive variance–covariance matrix of the characters, times the selection gradient, which is a gradient vector of the intrinsic rate of increase of the population with respect to changes in the mean history. Thus, rapid evolutionary responses can only occur in characters under strong selection gradient with a substantial additive genetic variance or covariance component. Several studies have in fact indicated significant heritability components on life history traits of large herbivores (e.g. Réale *et al.*, 1999; Réale and Festa-Bianchet, 2000; Pelletier *et al.*, 2007a, 2007b) and relative strong selection (Coulson *et al.*, 2003; Pelletier *et al.*, 2007a, 2007b, but see Wilson *et al.*, 2006).

These evolutionary responses can be especially important to consider when understanding how the demography and thus changes in distributions will be affected by changes in climate. The life history of large herbivores is adapted to the timing of plant growth and aims to target the most energy-demanding period of lactation to periods with rapid plant growth to ensure high-quality food for the neonates (Klein, 1965; Sæther, 1997). For example, red deer in France rut in September so that calves are born in April, while red deer at northern latitudes in Norway, with a later spring, are rutting in October and giving birth in May (Loe *et al.*, 2005, see also Fletcher, 1974). Similar geographical gradients in the timing of ovulation are also found in the moose (Garel *et al.*, 2009).

We also know that the phenological development of plants is likely to change with climate. Thus, since large herbivores appear to use cues to initiate reproduction well in advance of the actual time of birth, a change in the relationship between the cue and the plant can severely affect the populations and induce strong selection on the timing of birth. According to this match–mismatch hypothesis (Cushing, 1990; Visser *et al.*, 1998; Stenseth and Mysterud, 2002), the key point is that there are adaptations to the seasonal cycle of plant growth that can be broken down with the expected changes in climate. With advancing spring emergence of plants occurring with climate change (Post and Stenseth, 1999), the question arises of whether large herbivores are able to track these changes with appropriate changes in mating dates to advance birth dates.

Will ungulates adapt to these changes in spring flowering? Data are limited to studies of red deer on the Isle of Rum, Scotland (Coulson *et al.*, 2003; Nussey *et al.*, 2005a, 2005b). Here, the effect of spring temperature on birth mass of offspring was for individual females dependent on whether or not they had experienced high density early in life. Only females born at low population density showed any plasticity for these traits (Nussey *et al.*, 2005a, 2005b). Birth date had become 10 days earlier (0.37 days/year), which was suggested to be due to selection (Coulson *et al.*, 2003). Based on these analyses it would seem likely that there will be selection with future climate change. These are probably not the only aspects in which evolutionary change might happen. Examples are, however, few. Different genotypes of red deer on Rum, Scotland, survived better under different climatic conditions (Coulson *et al.*, 1998). In locations with more severe snow conditions, moose had significantly larger hooves and longer legs than would be expected from their size (Lundmark, 2008). We suspect the limited focus on these issues to date is likely to alter in the near future, and in particular there may be more research on how evolutionary change might interact with the way we manage these populations.

Many species of large herbivores are subject to intense harvesting over larger parts of their distributional ranges. The rate of evolution of adaptive life history variation to climate changes may in those species be closely related to choice of harvest strategies. Theoretical studies have shown that rate of change in the effective population size, which determines the rate of genetic drift (Wright, 1931), in harvested populations increases rapidly with a sex-biased harvest and small harvest rates of calves (Sæther *et al.*, 2009). Thus, the probability of fixation of a slightly advantageous mutation, which can contribute to adaptations to a changing climate, may be affected by the choice of harvest strategy.

12.10 Conclusion

The impending climate change will undoubtedly bring about changes in the potential distribution ranges of large ungulates, which is the main focus of our review above. Large herbivores are a rather wide taxonomic group and differ greatly in their adaptations. This makes some of them much more susceptible to severe climatic conditions, and it may influence which climatic season is the more limiting. A few studies report interspecific synchrony in body mass for a given location. In Norway, annual variations in body mass of domestic sheep (grazing during summer only) were synchronised with body mass of moose (Sæther, 1985), red deer (Mysterud *et al.*, 2001b) and reindeer (Weladji *et al.*, 2003b). These studies at least indicate consistent interspecific responses to summer foraging conditions, and that good and poor years might be so both for browsers and mixed-feeders/grazers. Highly specialised species are likely more affected by severe climate than generalists, but empirical evidence is absent. Species with a high reproductive output have a more variable annual survival (Gaillard *et al.*, 2000), suggesting they react more to severe conditions, but by contrast they are likely to recover more quickly as well. Formal comparative analysis to understand climate susceptibility in more detail is clearly needed.

For some large herbivore species in Europe to persist under the coming climate change, some of them might have to colonise new areas. Also, climate change opens possibilities for range expansion for species limited in particular by snow depth, quality and duration. Large herbivores are mobile species compared with many other taxonomic groups, and should therefore be less susceptible to climate change compared with plants and insects, although even some birds have tracked recent climate change in Europe (albeit with a time lag) (Devictor *et al.*, 2008). How soon will changes in large herbivore distribution happen? Habitat persistence is a major determinant for the evolution of dispersal (Travis and Dytham, 1999). In cervids, a comparative analysis suggested that those species evolved for utilising less stable habitats (typically forest opened by fire or storm) have larger litter sizes and greater inherent propensity to disperse (Liberg and Wahlström, 1995). For Europe, these are roe deer and moose, and the best documented case of colonisation speed in large herbivores comes from the recolonisation of Scandinavia by roe deer from 1850 onwards. The speed was up to 20–30 km a year in low quality habitat and 2–3 km a year in good habitat (Andersen *et al.*, 2004). Dispersal in roe deer at the yearling stage is similar among the sexes, density independent and often long distance (Wahlström and Liberg, 1995; Gaillard *et al.*, 2008). Colonisation speed of sika deer in Argyll, Scotland, was 3–5 km a year (Ratcliffe, 1987).

The behavioural ecology of the species plays a key role in determining colonisation speed. Male-biased dispersal is the rule for most large, polygynous herbivores, making colonisation a slower process since a dispersing male is less likely to encounter a partner, which will increase the demographic stochasticity and reduce the growth rate of the recolonising populations (Engen *et al.*, 2003). Typical examples of this are red deer (Clutton-Brock *et al.*, 2002) and chamois (Loison *et al.*, 2008). Climate variation can also affect dispersal rates, but not to a large degree (Clutton-Brock *et al.*, 2002).

For large herbivores encountering more or less continuous ranges, moving their range is not likely to be problematic, in the sense that climate change is not so rapid that they will be extinct before moving. Also, many species already have a wide distribution, and it is unlikely that climate change will affect conditions other than at the margin of current ranges. For these species, it is fairly easy to predict range expansion further north (and up in elevation), due to reduced snow depths and duration of snow cover, and contraction in the south due to summer droughts. These processes are already ongoing.

But while roe deer, red deer, wild boar and fallow deer show more or less continuous distributions, most species of ungulates do not (Groot Bruinderink *et al.*, 2003). Alpine populations such as those of reindeer, ibex, chamois and mouflon face some of the same limiting factors as northern large herbivore populations in general. However, many alpine ranges are fairly isolated, making dispersal difficult. Reindeer in Norway are already fragmented in 23 populations, partly due to natural mountain borders, partly due to infrastructure. Southern populations of alpine species may therefore be particularly vulnerable.

Habitat destruction together with climate changes may further limit possibilities for colonisation (Travis, 2003). Deer around the world face challenges related to habitat loss through forest cutting or agricultural expansion (Klein, 1992). On a global scale, there are reported problems for migratory species of ungulates (Berger, 2004; Bolger *et al.*, 2008), and protecting migration corridors is considered important (Berger *et al.*, 2008). The extension of cities, farmland and infrastructure has resulted in increasingly fragmented landscapes in Europe. Small populations in isolated habitats are at a higher risk of extinction due to stochastic demographic and environmental variation. Habitat networks will be important for future distribution of large herbivores in Europe (Groot Bruinderink *et al.*, 2003). Or do we need to adopt a framework for assisted colonisation for large herbivores in Europe to mitigate effects of future climate change (Hoegh-Guldeberg *et al.*, 2008)?

The history of large herbivores in Europe shows many examples of successful reintroductions following extinction due to overhunting etc. (Chapter 2). It is also clear that future, reduced or increased, productivity can perhaps, at least to some extent, be buffered with changes in human harvesting off-take.

References

Aanes, R., Sæther, B.-E. and Øritsland, N.A. (2000) Fluctuations of an introduced population of Svalbard reindeer: the effects of density dependence and climatic variation. *Ecography* **23**, 437–443.

Aanes, R., Sæther, B.-E., Smith, F.M., *et al.* (2002) The Arctic Oscillation predicts effect of climate change in two trophic levels in a high-arctic ecosystem. *Ecology Letters* **5**, 445–453.

Aanes, R., Sæther, B.-E., Solberg, E.J., *et al.* (2003) Synchrony in Svalbard reindeer population dynamics. *Canadian Journal of Zoology* **81**, 103–110.

Acevedo, P., Cassinello, J. and Gortázar, C. (2008a) The Iberian ibex is under an expansion trend but displaced to suboptimal habitats by the presence of extensive goat livestock in central Spain. *Biodiversity and Conservation* **16**, 3361–3376.

Acevedo, P., Cassinello, J., Hortal, J. and Gortázar, C. (2008b) Invasive exotic aoudad (*Ammotragus lervia*) as a major threat to native Iberian ibex (*Capra pyrenaica*): a habitat suitability model approach. *Diversity and Distributions* **13**, 587–597.

Albon, S.D. and Langvatn, R. (1992) Plant phenology and the benefits of migration in a temperate ungulate. *Oikos* **65**, 502–513.

Albon, S.D., Clutton-Brock, T.H. and Langvatn, R. (1992) Cohort variation in reproduction and survival: implications for population demography. In R.D. Brown (ed.) *The Biology of Deer*. New York: Springer Verlag, pp.15–21.

Albon, S.D., Coulson, T.N., Brown, D., *et al.* (2000) Temporal changes in key factors and key age groups influencing the population dynamics of female red deer. *Journal of Animal Ecology* **69**, 1099–1110.

Andersen, R. (1991) Habitat deterioration and the migratory behaviour of moose (*Alces alces* L.) in Norway. *Journal of Applied Ecology* **28**, 102–108.

Andersen, R., Herfindal, I., Sæther, B.-E., *et al.* (2004) When range expansion rate is faster in marginal habitats. *Oikos* **107**, 210–214.

Beniston, M. (2006) Mountain weather and climate: a general overview and a focus on climatic change in the Alps. *Hydrobiologia* **562**, 3–16.

Beniston, M. and Jungo, P. (2002) Shifts in the distributions of pressure, temperature and moisture and changes in the typical weather patterns in the Alpine region in

response to the behavior of the North Atlantic Oscillation. *Theoretical and Applied Climatology* **71**, 29–42.
Berger, J. (2004) The last mile: how to sustain long-distance migration in mammals. *Conservation Biology* **18**, 320–331.
Berger, J., Young, J.K. and Berger, K.M. (2008) Protecting migration corridors: challenges and optimism for Mongolian saiga. *PLoS Biology* **6**, 1365–1367.
Bergerud, A.T., Butler, H.E. and Miller, D.R. (1984) Antipredator tactics of calving caribou: dispersion in mountains. *Canadian Journal of Zoology* **62**, 1566–1575.
Blasius, B. and Stone, L. (2000) Nonlinearity and the Moran effect. *Nature* **406**, 846–847.
Bolger, D.T., Newmark, W.D., Morrison, T.A. and Doak, D.F. (2008) The need for integrative approaches to understand and conserve migratory ungulates. *Ecology Letters* **11**, 63–77.
Borner, A.P., Kielland, K. and Walker, M.D. (2008) Effects of simulated climate change on plant phenology and nitrogen mineralization in Alaskan arctic tundra. *Arctic, Antarctic, and Alpine Research* **40**, 27–38.
Brooks, D.R. and Hoberg, E.P. (2008) How will global climate change affect parasite–host assemblages? *Trends in Parasitology* **23**, 571–574.
Calder, W.A. (1984) *Size, Function, and Life History*. Cambridge, MA: Harvard University Press.
Case, T.J., Holt, R.D., McPeek, M.A. and Keitt, T.H. (2005) The community context of species' borders: ecological and evolutionary perspectives. *Oikos* **108**, 28–46.
Cassinello, J., Serrano, E., Calabuig, G. and Pérez, J.M. (2004) Range expansion of an exotic ungulate (*Ammotragus lervia*) in southern Spain: ecological and conservation concerns. *Biodiversity and Conservation* **13**, 851–866.
Cebrian, M.R., Kielland, K. and Finstad, G. (2008) Forage quality and reindeer productivity: multiplier effects amplified by climate change. *Arctic, Antarctic, and Alpine Research* **40**, 48–54.
Cederlund, G. and Lindström, E. (1983) Effects of severe winters and fox predation on roe deer mortality. *Acta Theriologica* **28**, 129–145.
Cederlund, G., Sand, H. and Pehrson, Å. (1991) Body mass dynamics of moose calves in relation to winter severity. *Journal of Wildlife Management* **55**, 675–681.
Charmantier, A., McCleery, R.H., Cole, L.R., *et al.* (2008) Adaptive phenotypic plasticity in response to climate change in a wild bird population. *Science* **230**, 800–803.
Ciais, P., Reichstein, M., Viovy, N., *et al.* (2005) Europe-wide reduction in primary productivity caused by the heat and drought in 2003. *Nature* **437**, 529–533.
Clutton-Brock, T.H., Illius, A.W., Wilson, K., *et al.* (1997) Stability and instability in ungulate populations: an empirical analysis. *American Naturalist* **149**, 195–219.
Clutton-Brock, T.H., Coulson, T.N., Milner-Gulland, E.J., Thomson, D. and Armstrong, H.M. (2002) Sex differences in emigration and mortality affects optimal management of deer populations. *Nature* **415**, 633–637.
Coulson, T.N., Albon, S.D., Pemberton, J.M., *et al.* (1998) Genotype by environment interactions in winter survival in red deer. *Journal of Animal Ecology* **67**, 434–445.
Coulson, T., Catchpole, E.A., Albon, S.D., *et al.* (2001a) Age, sex, density, winter weather, and population crashes in Soay sheep. *Science* **292**, 1528–1531.

Coulson, T., Mace, G.M., Hudson, E. and Possingham, H. (2001b) The use and abuse of population viability analysis. *Trends in Ecology and Evolution* **16**, 219–221.

Coulson, T., Kruuk, L.E.B., Tavecchia, G., Pemberton, J.M. and Clutton-Brock, T. H. (2003) Estimating selection on neonatal traits in red deer using elasticity path analysis. *Evolution* **57**, 2879–2892.

Cushing, D.H. (1990) Plankton production and year-class strength in fish populations: an update of the match/mismatch hypothesis. *Advances in Marine Biology* **26**, 249–293.

Danilkin, A.A. (1996) *Behavioural Ecology of Siberian and European Roe Deer*. London: Chapman and Hall.

Demment, M.W. and Van Soest, P.J. (1985) A nutritional explanation for body-size patterns of ruminant and nonruminant herbivores. *American Naturalist* **125**, 641–672.

Devictor, V., Julliard, R., Couvet, D. and Jiguet, F. (2008) Birds are tracking climate warming, but not fast enough. *Proceedings of the Royal Society B–Biological Sciences* **275**, 2743–2748.

Durant, J.M., Anker-Nilssen, T., Hjermann, D.Ø. and Stenseth, N.C. (2004) Regime shifts in the breeding of an Atlantic puffin population. *Ecology Letters* **7**, 388–394.

Eberhardt, L.L. (2002) A paradigm for population analysis of long-lived vertebrates. *Ecology* **83**, 2841–2854.

Edwards, J. (1983) Diet shifts in moose due to predator avoidance. *Oecologia* **60**, 185–189.

Engen, S. and Sæther, B.-E. (2005) Generalizations of the Moran effect explaining spatial synchrony in population fluctuations. *American Naturalist* **166**, 603–612.

Engen, S., Lande, R. and Sæther, B.-E. (2003) Demographic stochasticity and Allee effects in populations with two sexes. *Ecology* **84**, 2378–2386.

Enserink, M. and Vogel, G. (2006) The carnivore comeback. *Science* **314**, 746–749.

Fauchald, P., Rodven, R., Bardsen, B.J., *et al.* (2007) Escaping parasitism in the selfish herd: age, size and density-dependent warble fly infestation in reindeer. *Oikos* **116**, 491–499.

Festa-Bianchet, M., Coulson, T., Gaillard, J.-M., Hogg, J.T. and Pelletier, F. (2006) Stochastic predation events and population persistence in bighorn sheep. *Proceedings of the Royal Society B–Biological Sciences* **273**, 1537–1543.

Fletcher, T.J. (1974) The timing of reproduction in red deer (*Cervus elaphus*) in relation to latitude. *Journal of Zoology, London* **172**, 363–367.

Formozow, A.N. (1946) *Snow Cover as an Integral Factor of the Environment and its Importance in the Ecology of Mammals and Birds* (In Russian, translated to English). Edmonton, Canada: Boreal Institute, University of Alberta.

Fortin, D., Beyer, H.L., Boyce, M.S., *et al.* (2005) Wolves influence elk movements: behavior shapes a trophic cascade in Yellowstone National Park. *Ecology* **86**, 1320–1330.

Fryxell, J.M. and Sinclair, A.R.E. (1988) Causes and consequences of migration by large herbivores. *Trends in Ecology and Evolution* **3**, 237–241.

Gaillard, J.-M. and Yoccoz, N.G. (2003) Temporal variation in survival of mammals: a case of environmental canalization? *Ecology* **84**, 3294–3306.

Gaillard, J.-M., Festa-Bianchet, M. and Yoccoz, N.G. (1998) Population dynamics of large herbivores: variable recruitment with constant adult survival. *Trends in Ecology and Evolution* **13**, 58–63.

Gaillard, J.-M., Festa-Bianchet, M., Yoccoz, N.G., Loison, A. and Toigo, C. (2000) Temporal variation in fitness components and population dynamics of large herbivores. *Annual Review of Ecology and Systematics* **31**, 367–393.

Gaillard, J.-M., Loison, A., Toïgo, C., Delorme, D. and Van Laere, G. (2003) Cohort effects and deer population dynamics. *Ecoscience* **10**, 412–420.

Gaillard, J.-M., Hewison, A.J.M., Kjellander, P., *et al.* (2008) Population density and sex do not influence fine-scale natal dispersal in roe deer. *Proceedings of the Royal Society of London B–Biological Sciences* **275**, 2025–2030.

Garel, M., Loison, A., Gaillard, J.-M., Cugnasse, J.-M. and Maillard, D. (2004) The effects of a severe drought on mouflon lamb survival. *Proceedings of the Royal Society B–Biological Sciences, Suppl.* **271**, S471–S473.

Garel, M., Solberg, E.J., Sæther, B.-E. *et al.* (2009) Age, size and spatio-temporal variation in ovulation patterns of a seasonal breeder, the Norwegian moose (*Alces alces*). *American Naturalist* **173**, 89–104.

Gillett, N.P., Graf, H.F. and Osborn, T.J. (2003) Climate change and the North Atlantic Oscillation. In J.W. Hurrell, Y. Kushnir, G. Ottersen and M. Visbeck (eds.) *The North Atlantic Oscillation*. Washington, DC: American Geophysical Union, pp.193–210.

Grenfell, B.T., Wilson, K., Finkenstädt, B.F., *et al.* (1998) Noise and determinism in synchronized sheep dynamics. *Nature* **394**, 674–677.

Grenfell, T.C. and Putkonen, J. (2008) A method for the detection of the severe rain-on-snow events on Banks island, October 2003, using passive microwave remote sensing. *Water Resources Research* **44**, (article number W03425).

Groot Bruinderink, G., van der Sluis, T., Lammertsma, D., Opdam, P. and Pouwels, R. (2003) Designing a coherent ecological network for large mammals in northwestern Europe. *Conservation Biology* **17**, 549–557.

Grøtan, V., Sæther, B.-E., Engen, S., *et al.* (2005) Climate causes large-scale spatial synchrony in population fluctuations of a temperate herbivore. *Ecology* **86**, 1472–1482.

Grøtan, V., Sæther, B.-E., Filli, F. and Engen, S. (2008) Effects of climate on population fluctuations of ibex. *Global Change Biology* **14**, 218–228.

Gundersen, H. and Andreassen, H.P. (1998) The risk of moose *Alces alces* collision: a predictive logistic model for moose-train accidents. *Wildlife Biology* **4**, 103–110.

Hagemoen, R.I.M. and Reimers, E. (2002) Reindeer summer activity pattern in relation to weather and insect harassment. *Journal of Animal Ecology* **71**, 883–892.

Hallett, T.B., Coulson, T., Pilkington, J.G., *et al.* (2004) Why large-scale climate indices seem to predict ecological processes better than local weather. *Nature* **430**, 71–75.

Hebblewhite, M. (2005) Predation by wolves interacts with the North Pacific Oscillation (NPO) on a western North American elk population. *Journal of Animal Ecology* **74**, 226–233.

Hebblewhite, M., Merrill, E. and McDermid, G. (2008) A multi-scale test of the forage maturation hypothesis in a partially migratory ungulate population. *Ecological Monographs* **78**, 141–166.

Herfindal, I., Sæther, B-E., Solberg, E.J., Andersen, R. and Høgda, K.-A. (2006a) Population characteristics predict responses in moose body mass to temporal variation in the environment. *Journal of Animal Ecology* **75**, 1110–1118.

Herfindal, I., Solberg, E.J., Sæther, B.-E., Høgda, K.-A. and Andersen, R. (2006b) Environmental phenology and geographical gradients in moose body mass. *Oecologia* **150**, 213–224.

Hoegh-Guldeberg, O., Hughes, L., McIntyre, S.M., *et al.* (2008) Assisted colonization and rapid climate change. *Science* **321**, 345–346.

Holand, Ø., Mysterud, A., Røed, K.H., *et al.* (2006) Adaptive adjustment of offspring sex ratio and maternal reproductive effort in an iteroparous mammal. *Proceedings of the Royal Society B–Biological Sciences* **273**, 293–299.

Holt, R.D. and Keitt, T.H. (2005) Species' borders: a unifying theme in ecology. *Oikos* **108**, 3–6.

Holt, R.D., Keitt, T.H., Lewis, M.A., Maurer, B.A. and Taper, M.L. (2005) Theoretical models of species' borders: single species approaches. *Oikos* **108**, 18–27.

Hurrell, J.W. (1995) Decadal trends in the North Atlantic Oscillation: regional temperatures and precipitation. *Science* **269**, 676–679.

Hurrell, J.W., Kushnir, Y., Ottersen, G. and Visbeck, M. (2003) *The North Atlantic Oscillation: Climatic Significance and Environmental Impact*. Washington, DC: American Geophysical Union.

Ims, R.A., Henden, J.-A. and Killengreen, S.T. (2008) Collapsing population cycles. *Trends in Ecology and Evolution* **23**, 79–86.

IPCC (2007) *Climate Change 2007: The Physical Science Basis*. Contribution of Working Group I to the Fourth Assessment Report of the Intergovernmental Panel on Climate Change. Cambridge, UK: Cambridge University Press.

Jacobson, A.R., Provenzale, A., von Hardenberg, A., Bassano, B. and Festa-Bianchet, M. (2004) Climate forcing and density dependence in a mountain ungulate population. *Ecology* **85**, 1598–1610.

Kausrud, K.L., Mysterud, A., Steen, H., *et al.* (2008) Linking climate change to lemming cycles. *Nature* **456**, 93–98.

Kettlewell, P.S., Easey, J., Stephenson, D.B. and Poulton, P.R. (2006) Soil moisture mediates association between the winter North Atlantic Oscillation and summer growth in the Park Grass Experiment. *Proceedings of the Royal Society B–Biological Sciences* **273**, 1149–1154.

Kjellander, P. and Nordström, J. (2003) Cyclic voles, prey switching in red fox, and roe deer dynamics: a test of the alternative prey hypothesis. *Oikos* **101**, 338–344.

Kjellander, P., Gaillard, J.-M. and Hewison, A.J.M. (2006) Density-dependent responses of fawn cohort body mass in two contrasting roe deer populations. *Oecologia* **146**, 521–530.

Klein, D.R. (1965) Ecology of deer range in Alaska. *Ecological Monographs* **35**, 259–284.

Klein, D.R. (1992) The status of deer in a changing world environment. In R.D. Brown (ed.) *Proceedings of the International Symposium on Biology of Deer*. New York: Springer, pp.3–12.

Kohler, J. and Aanes, R. (2004) Effect of winter snow and ground-icing on a Svalbard reindeer population: results of a simple snowpack model. *Arctic, Antarctic, and Alpine Research* **36**, 333–341.

Lande, R. (1982) A quantitative genetic theory of life-history evolution. *Ecology* **63**, 607–615.

Langvatn, R., Mysterud, A., Stenseth, N.C. and Yoccoz, N.G. (2004) Timing and synchrony of ovulation in red deer constrained by short northern summers. *American Naturalist* **163**, 763–772.

Latham, J. (1999) Interspecific interactions of ungulates in European forests: an overview. *Forest Ecology and Management* **120**, 13–21.

Latham, J., Staines, B.W. and Gorman, M.L. (1996) The relative densities of red (*Cervus elaphus*) and roe (*Capreolus capreolus*) deer and their relationship in Scottish plantation forests. *Journal of Zoology, London* **240**, 285–299.

Latham, J., Staines, B.W. and Gorman, M.L. (1997) Correlations of red (*Cervus elaphus*) and roe (*Capreolus capreolus*) deer densities in Scottish forests with environmental variables. *Journal of Zoology,London* **242**, 681–704.

Liberg, O. and Wahlström, L.K. (1995) Habitat stability and litter size in the Cervidae; a comparative analysis. In L.K. Wahlström, Natal dispersal in roe deer: an evolutionary perspective. PhD thesis, University of Stockholm.

Loe, L.E., Bonenfant, C., Mysterud, A., *et al.* (2005) Climate predictability and breeding phenology in red deer: timing and synchrony of rutting and calving in Norway and France. *Journal of Animal Ecology* **74**, 579–588.

Loison, A. and Langvatn, R. (1998) Short- and long-term effects of winter and spring weather on growth and survival of red deer in Norway. *Oecologia* **116**, 489–500.

Loison, A., Festa-Bianchet, M., Gaillard, J.-M., Jorgenson, J.T. and Jullien, J.-M. (1999a) Age-specific survival in five populations of ungulates: evidence of senescence. *Ecology* **80**, 2539–2554.

Loison, A., Jullien, J.-M. and Menaut, P. (1999b) Relationship between chamois and isard survival and variation in global and local climate regimes: contrasting examples from the Alps and Pyrenees. *Ecology Bulletin* **47**, 126–136.

Loison, A., Darmon, G., Cassar, S., Jullien, J.-M. and Maillard, D. (2008) Age- and sex-specific settlement patterns in chamois (*Rupicapra rupicapra*) offspring. *Canadian Journal of Zoology* **86**, 588–593.

Lundmark, C. (2008) Morphological and behavioural adaptations of moose to climate, snow, and forage. PhD thesis, Swedish University of Agricultural Sciences, Umeå, Sweden.

Lundmark, C. and Ball, J.P. (2008) Living in snowy environments: quantifying the influence of snow on moose behavior. *Arctic, Antarctic, and Alpine Research* **40**, 111–118.

Martínez-Jauregui, M., Miguel-Ayanz, A.S., Mysterud, A., *et al.* (2009) Are local weather, NDVI or NAO consistent determinants of female red deer weight across Europe? *Global Change Biology* **15**, 1727–1738.

Maurer, B.A. and Taper, M.L. (2002) Connecting geographical distributions with population processes. *Ecology Letters* **5**, 223–231.

Mitchell, B. (1973) The reproductive performance of wild Scottish red deer, *Cervus elaphus*. *Journal of Reproduction and Fertility, Supplement* **19**, 271–285.

Mobæk, R., Mysterud, A., Loe, L.E., Holand, Ø. and Austrheim, G. (2009) Density dependent and temporal variability in habitat selection by a large herbivore: an experimental approach. *Oikos* **118**, 209–218.

Moen, A.N. (1973) *Wildlife Ecology*. San Fransciso, CA: W.H. Freeman and Company.

Moran, P.A.P. (1953) The statistical analysis of the Canadian lynx cycle. II. Synchronisation and meteorology. *Australian Journal of Zoology* **1**, 291–298.

Morris, W.F., Pfister, C.A., Tuljapurkar, S., *et al.* (2008) Longevity can buffer plant and animal populations against changing climatic variability. *Ecology* **89**, 19–25.

Murray, D.L., Cox, E.W., Ballard, W.B., *et al.* (2006) Pathogens, nutritional deficiency, and climate influences on a declining moose population. *Wildlife Monographs* **166**, 1–30.

Mysterud, A. (1999) Seasonal migration pattern and home range of roe deer (*Capreolus capreolus*) in an altitudinal gradient in southern Norway. *Journal of Zoology, London* **247**, 479–486.

Mysterud, A., Yoccoz, N.G., Stenseth, N.C. and Langvatn, R. (2000) Relationships between sex ratio, climate and density in red deer: the importance of spatial scale. *Journal of Animal Ecology* **69**, 959–974.

Mysterud, A., Langvatn, R., Yoccoz, N.G. and Stenseth, N.C. (2001a) Plant phenology, migration and geographic variation in body weight of a large herbivore: the effect of a variable topography. *Journal of Animal Ecology* **70**, 915–923.

Mysterud, A., Stenseth, N.C., Yoccoz, N.G., Langvatn, R. and Steinheim, G. (2001b) Nonlinear effects of large-scale climatic variability on wild and domestic herbivores. *Nature* **410**, 1096–1099.

Mysterud, A., Langvatn, R., Yoccoz, N.G. and Stenseth, N.C. (2002) Large-scale habitat variability, delayed density effects and red deer populations in Norway. *Journal of Animal Ecology* **71**, 569–580.

Mysterud, A., Stenseth, N.C., Yoccoz, N.G., Ottersen, G. and Langvatn, R. (2003) The response of the terrestrial ecosystems to climate variability associated with the North Atlantic Oscillation. In J.W. Hurrell, Y. Kushnir, G. Ottersen and M. Visbeck (eds.) *The North Atlantic Oscillation*. Washington, DC: American Geophysical Union, pp.235–262.

Mysterud, A., Baron, K., Jedrzejewska, B., *et al.* (2007) Population ecology and conservation of endangered megafauna: the case of European bison (*Bison bonasus*) in Bialowieza Primeval Forest, Poland. *Animal Conservation* **10**, 77–87.

Mysterud, A., Yoccoz, N.G., Langvatn, R., Pettorelli, N. and Stenseth, N.C. (2008) Hierarchical path analysis of deer responses to direct and indirect effects of climate in northern forest. *Philosophical Transactions of the Royal Society* **363**, 2359–2368.

Nussey, D.H., Clutton-Brock, T.H., Albon, S.D., Pemberton, J. and Kruuk, L.E.B. (2005a) Constraints on plastic responses to climate variation in red deer. *Biology Letters* **1**, 457–460.

Nussey, D.H., Clutton-Brock, T.H., Elston, D.A., Albon, S.D. and Kruuk, L.E.B. (2005b) Phenotypic plasticity in a maternal trait in red deer. *Journal of Animal Ecology* **74**, 387–396.

Nussey, D.H., Postma, E., Gienapp, P. and Visser, M.E. (2005c) Selection on heritable phenotypic plasticity in a wild bird population. *Science* **310**, 304–306.

Parker, K.L. and Robbins, C.T. (1985) Thermoregulation in ungulates. In R.J. Hudson and R.G. White (eds.) *Bioenergetics of Wild Herbivores*. Boca Raton, FL: CRC Press, pp.161–182.

Parker, K.L., Robbins, C.T. and Hanley, T.A. (1984) Energy expenditure for locomotion by mule deer and elk. *Journal of Wildlife Management* **48**, 474–488.

Pelletier, F., Clutton-Brock, T., Pemberton, J., Tuljapurkar, S. and Coulson, T. (2007a) The evolutionary demography of ecological change: linking trait variation and population growth. *Science* **315**, 1571–1574.

Pelletier, F., Réale, D., Garant, D., Coltman, D.W. and Festa-Bianchet, M. (2007b) Selection on heritable seasonal phenotypic plasticity of body mass. *Evolution* **61**, 1969–1979.

Pettorelli, N., Mysterud, A., Yoccoz, N.G., Langvatn, R. and Stenseth, N.C. (2005a) Importance of climatological downscaling and plant phenology for red deer in heterogeneous landscapes. *Proceedings of the Royal Society of London B–Biological Sciences* **272**, 2357–2364.

Pettorelli, N., Vik, J.O., Mysterud, A., *et al.* (2005b) Using the satellite-derived Normalized Difference Vegetation Index (NDVI) to assess ecological responses to environmental change. *Trends in Ecology and Evolution* **20**, 503–510.

Pettorelli, N., Weladji, R.B., Holand, Ø., *et al.* (2005c) The relative role of winter and spring conditions: linking climate and landscape-scale plant phenology to alpine reindeer performance. *Biology Letters* **1**, 24–26.

Pettorelli, N., Gaillard, J.-M., Mysterud, A., *et al.* (2006) Using a proxy of plant productivity (NDVI) to find key periods for animal performance: the case of roe deer. *Oikos* **112**, 565–572.

Pioz, M., Loison, A., Gauthier, D., *et al.* (2008) Diseases and reproductive success in a wild mammal: example in the alpine chamois. *Oecologia* **155**, 691–704.

Post, E. and Forchhammer, M.C. (2002) Synchronization of animal population dynamics by large-scale climate. *Nature* **420**, 168–171.

Post, E. and Forchhammer, M.C. (2008) Climate change reduces reproductive success of an arctic herbivore through trophic mismatch. *Philosophical Transactions of the Royal Society* **363**, 2369–2375.

Post, E. and Stenseth, N.C. (1999) Climatic variability, plant phenology, and northern ungulates. *Ecology* **80**, 1322–1339.

Post, E., Stenseth, N.C., Langvatn, R. and Fromentin, J.-M. (1997) Global climate change and phenotypic variation among red deer cohorts. *Proceedings of the Royal Society B–Biological Sciences* **264**, 1317–1324.

Post, E., Pedersen, C., Wilmers, C.C. and Forchhammer, M.C. (2008) Warming, plant phenology and the spatial dimension of trophic mismatch for large herbivores. *Proceedings of the Royal Society B–Biological Sciences* **275**, 2005–2013.

Przybylo, R., Sheldon, B.C. and Merilä, J. (2000) Climatic effects on breeding and morphology: evidence for phenotypic plasticity. *Journal of Animal Ecology* **69**, 395–403.

Putkonen, J. and Roe, G. (2003) Rain-on-snow events impact soil temperatures and affect ungulate survival. *Geophysical Research Letters* **30**, 1118–1122.

Putman, R.J. (1996) *Competition and Resource Partitioning in Temperate Ungulate Assemblies*. London: Chapman and Hall.

Putman, R.J., Langbein, J., Hewison, A.J.M. and Sharma, S.K. (1996) Relative roles of density-dependent and density-independent factors in population dynamics of British deer. *Mammal Review* **26**, 81–101.

Ratcliffe, P.R. (1984) Population dynamics of red deer (*Cervus elaphus* L.) in Scottish commercial forests. *Proceedings of the Royal Society of Edinburgh B* **82B**, 291–302.

Ratcliffe, P.R. (1987) Distribution and current status of sika deer, *Cervus nippon*, in Great Britain. *Mammal Review* **17**, 39–58.

Réale, D. and Festa-Bianchet, M. (2000) Mass-dependent reproductive strategies in wild bighorn ewes: a quantitative genetic approach. *Journal of Evolutionary Biology* **13**, 679–688.

Réale, D., Festa-Bianchet, M. and Jorgenson, J.T. (1999) Heritability of body mass varies with age and season in wild bighorn sheep. *Heredity* **83**, 526–532.

Reed, D.H., O'Grady, J.J., Ballou, J.D. and Frankham, R. (2003) The frequency and severity of catastrophic die-offs in vertebrates. *Animal Conservation* **6**, 109–114.

Sæther, B.-E. (1985) Annual variation in carcass weight of Norwegian moose in relation to climate along a latitudinal gradient. *Journal of Wildlife Management* **49**, 977–983.

Sæther, B.-E. (1997) Environmental stochasticity and population dynamics of large herbivores: a search for mechanisms. *Trends in Ecology and Evolution* **12**, 143–149.

Sæther, B.-E. and Gravem, A.J. (1988) Annual variation in winter body condition of Norwegian moose calves. *Journal of Wildlife Management* **52**, 333–336.

Sæther, B.-E. and Heim, M. (1993) Ecological correlates of individual variation in age at maturity in female moose (*Alces alces*): the effects of environmental variability. *Journal of Animal Ecology* **62**, 482–489.

Sæther, B.-E., Andersen, R., Hjeljord, O. and Heim, M. (1996) Ecological correlates of regional variation in life history of the moose *Alces alces*. *Ecology* **77**, 1493–1500.

Sæther, B.-E., Engen, S., Filli, F., *et al.* (2002) Stochastic population dynamics of an introduced Swiss population of the ibex. *Ecology* **83**, 3457–3465.

Sætre, B.-E., Engen, S. and Solberg, E.J. (2009) Effective size of harvested ungulate populations. *Animal conservation* **12**, 488–495.

Sætre, G.-P., Post, E. and Král, M. (1999) Can environmental fluctuation prevent competitive exclusion in sympatric flycatchers? *Proceedings of the Royal Society B–Biological Sciences* **266**, 1247–1251.

Sauer, J.R. and Boyce, M.S. (1983) Density dependence and survival of elk in northwestern Wyoming. *Journal of Wildlife Management* **47**, 31–37.

Shabbar, A. and Bonsal, B. (2003) An assessment of changes in winter cold and warm spells over Canada. *Natural Hazards* **29**, 173–188.

Sims, M., Elston, D.A., Larkham, A., Nussey, D.H. and Albon, S.D. (2007) Identifying when weather influences life-history traits of grazing herbivores. *Journal of Animal Ecology* **76**, 761–770.

Skelly, D.K., Joseph, L.N., Possingham, H.P., *et al.* (2007) Evolutionary responses to climate change. *Conservation Biology* **21**, 1353–1355.

Solberg, E.J., Sæther, B.-E., Strand, O. and Loison, A. (1999) Dynamics of a harvested moose population in a variable environment. *Journal of Animal Ecology* **68**, 186–204.

Solberg, E.J., Jordhøy, P., Strand, O., *et al.* (2001) Effects of density-dependence and climate on the dynamics of a Svalbard reindeer population. *Ecography* **24**, 441–451.

Solberg, E.J., Loison, A., Gaillard, J.-M. and Heim, M. (2004) Lasting effects of conditions at birth on moose body mass. *Ecography* **27**, 677–687.

Solberg, E.J., Heim, M., Grøtan, V., Sæther, B.-E. and Garel, M. (2007) Annual variation in maternal age and calving date generate cohort effects in moose (*Alces alces*) body mass. *Oecologia* **154**, 259–271.

Steen, H., Mysterud, A. and Austrheim, G. (2005) Sheep grazing and rodent populations: evidence of negative interactions from a landscape scale experiment. *Oecologia* **143**, 357–364.

Stenseth, N.C. and Mysterud, A. (2002) Climate, changing phenology, and other life history traits: non-linearity and match-mismatch to the environment. *Proceedings of the National Academy of Sciences of the USA* **99**, 13379–13381.

Stenseth, N.C. and Mysterud, A. (2005) Weather packages: finding the right scale and composition for climate in ecology. *Journal of Animal Ecology* **74**, 1195–1198.

Stenseth, N.C., Ottersen, G., Hurrell, J.W., *et al.* (2003) Studying climate effects on ecology through the use of climate indices: the North Atlantic Oscillation, El Niño Southern Oscillation and beyond. *Proceedings of the Royal Society B–Biological Sciences* **270**, 2087–2096.

Telfer, E.S. and Kelsall, J.P. (1984) Adaptation of some large North American mammals for survival in snow. *Ecology* **65**, 1828–1834.

Telleria, J.L. and Virgos, E. (1997) Distribution of an increasing roe deer population in a fragmented Mediterranean landscape. *Ecography* **20**, 247–252.

Toïgo, C. and Gaillard, J.-M. (2003) Causes of sex-biased adult survival in ungulates: sexual size dimorphism, mating tactic or environment harshness? *Oikos* **101**, 376–684.

Toïgo, C., Gaillard, J.-M. and Michallet, J. (1997) Adult survival pattern of the sexually dimorphic Alpine ibex (*Capra ibex ibex*). *Canadian Journal of Zoology* **75**, 75–79.

Travis, J.M.J. (2003) Climate change and habitat destruction: a deadly anthropogenic cocktail. *Proceedings of the Royal Society B–Biological Sciences* **270**, 467–473.

Travis, J.M.J. and Dytham, C. (1999) Habitat persistence, habitat availability and the evolution of dispersal. *Proceedings of the Royal Society B–Biological Sciences* **266**, 723–728.

Veeroja, R., Tilgar, V., Kirk, A. and Tõnisson, J. (2008) Climatic effects on life-history traits of moose in Estonia. *Oecologia* **154**, 703–713.

Vik, J.O., Stenseth, N.C., Tavecchia, G., Mysterud, A. and Lingjerde, O.C. (2004) Living in synchrony on Greenland coasts? *Nature* **427**, 697–698.

Virgos, E. and Telleria, J.L. (1998) Roe deer habitat selection in Spain: constraints on the distribution of a species. *Canadian Journal of Zoology* **76**, 1294–1299.

Visser, M.E., Van Noordwijk, A.J., Tinbergen, J.M. and Lessells, C.M. (1998) Warmer springs lead to mistimed reproduction in great tits (*Parus major*). *Proceedings of the Royal Society B–Biological Sciences* **265**, 1867–1870.

Vucetich, J.A. and Peterson, R.O. (2004) The influence of top-down, bottom-up and abiotic factors on the moose (*Alces alces*) population of Isle Royale. *Proceedings of the Royal Society B–Biological Sciences* **271**, 183–189.

Wahlström, L.K. and Liberg, O. (1995) Patterns of dispersal and seasonal migration in roe deer (*Capreolus capreolus*). *Journal of Zoology, London* **235**, 455–467.

Wallach, A.D., Inbar, M., Scantlebury, M., Speakman, J.R. and Shanas, U. (2007) Water requirements as a bottleneck in the reintroduction of European roe deer to the southern edge of its range. *Canadian Journal of Zoology* **85**, 1182–1192.

Weladji, R.B., Klein, D.R., Holand, Ø. and Mysterud, A. (2002) Comparative response of *Rangifer tarandus* and other northern ungulates to climatic variability. *Rangifer* **22**, 33–50.

Weladji, R.B., Holand, Ø. and Almøy, T. (2003a) Use of climatic data to assess the effect of insect harassment on the autumn weight of reindeer (*Rangifer tarandus*) calves. *Journal of Zoology, London* **260**, 79–85.

Weladji, R.B., Steinheim, G., Holand, Ø., *et al.* (2003b) Temporal patterns of juvenile body weight variability in sympatric reindeer and sheep. *Annales Zoologica Fennica* **40**, 17–26.

Weladji, R.B., Gaillard, J.-M., Yoccoz, N.G., *et al.* (2006) Good reindeer mothers live longer and become better in raising offspring. *Proceedings of the Royal Society of London B–Biological Sciences* **273**, 1239–1244.

Wilson, A.J., Pemberton, J.M., Pilkington, J.G., *et al.* (2006) Environmental coupling of selection and heritability limits evolution. *PLoS Biology* **4**, 1270–1275.

Wright, S. (1931) Evolution in Mendelian populations. *Genetics* **16**, 97–159.

13

Ungulate management in Europe: towards a sustainable future

ROBERT KENWARD AND RORY PUTMAN

13.1 Management of ungulate populations: resource and conflict

As noted by Apollonio *et al.* (2010), the management of large ungulates in Europe is no easy task. There are some 20 different species, each living in a great variety of environmental conditions across their full distributional range; populations are increasing in nearly all European countries and, in consequence, they are having a profound effect on the ecological dynamics of both natural and human-created ecosystems of which they are a part.

These ungulates represent in themselves an immense potential resource – in terms of biodiversity and also in economic terms. More than 5.2 million animals harvested each year represents more than 120 000 tonnes of meat, and a potential hunting revenue of several hundred million euros (Apollonio *et al.*, 2010); in addition these animals have inestimable aesthetic and cultural value as country-specific carriers of a whole range of cultural and hunting traditions. At the same time, while they may be exploited in this way as sources of food and recreation, they may also have many negative impacts through damage to forests or agricultural crops, damage through heavy impacts on natural habitats (Chapter 6), as vectors of disease (Chapter 7), or through implication in collisions with vehicles (Chapter 8).

They have significant impact on human interests (whether positive or negative), but are also themselves strongly influenced by human actions. By introduction and reintroduction, humans have established a number of exotic species, have markedly altered the distribution and abundance of many native European species, and through translocations have had an impact on the population genetics of many local populations (Chapter 2). Direct anthropogenic changes to the landscape also affect the availability,

Ungulate Management in Europe: Problems and Practices, eds. Rory Putman, Marco Apollonio and Reidar Andersen. Published by Cambridge University Press.

quality and distribution of suitable habitats for shelter and foraging (affecting ungulate distributions and demographics – and in turn potentially increasing the risk of conflict through impacts on silvicultural or agricultural crops) and may influence population dynamics further through introduction of competitors and, more recently, reintroduction of their predators (Chapter 10).

In effect their distribution, their status and their impacts are, one way or another, inextricably intertwined with human activities. They affect us in a variety of ways, both positive and negative; by the same token, we affect them. **And ultimately humans must assume overall responsibility for the future management of those ungulate populations and their impacts**. Which equally means that we must plan and undertake that management with due consideration to ensure that, as far as possible, we deliver effective and appropriate management.

In the past, perhaps, management systems for ungulates were rather inflexible (often rather unimaginative), and often directed towards achieving a clear, single objective. Some of the successes and failures of past management were rehearsed by Apollonio *et al.* (2010) who also consider some of the factors associated with several failures in the past. Many of the issues which must be addressed by managers are explored in more detail in the various chapters of this volume.

13.1.1 The need for holistic, multi-objective and adaptive management

Overall it is clear that effective management can no longer focus on delivery of some single aim in isolation (Putman and Watson, 2009; Apollonio *et al.*, 2010). Any management decision taken in any given context (to increase or reduce numbers of a specific ungulate population, to influence their distribution, or simply to alter relative utilisation of different parts of their range) will by definition affect densities, affect range use and thus of necessity have implications on other ecosystem processes of which ungulate populations are an integral part.

Of necessity, therefore, effective management must consider all relevant ecosystem processes in which these ungulate populations may play a role, and must seek to develop integrated management systems which take account of *all* these effects and take due account of other, potentially competing land-use interests which are affected by the same populations of ungulates. Rather than adopting management approaches which seek to solve a single problem, or deliver a single objective, management must pay due heed to all aspects of the interactions between ungulate populations and their wider environment, to develop management strategies which take due account of all impacts and all needs, balancing conflicting or complementary interests in a much more *holistic* way. This emphasises that managing ungulates isn't just about managing the animals themselves but has a number of 'knock-on' effects ... so that management decisions must be made within a more inclusive understanding of land-use interests overall and how management decisions may affect each part of that integrated whole. Management of deer and other ungulates shouldn't just be about managing *them* in isolation, but needs to be considered (and integrated) within a wider framework of how they themselves, and their management, relate to other land-management aims and objectives in more general terms.

Putman and Watson (2009), for example, present a decision-making framework to help managers integrate information about a wide range of possible impacts when making decisions about what actions should be taken in management (emphasising a need to integrate information about damage to agriculture, forestry and vulnerable natural habitats, scale of involvement of ungulates in vehicle collisions and disease status, *inter alia*). Recognising that the objective is to alert managers to a potential need for management intervention, or alteration to existing management policy, this particular matrix focuses inevitably on negative impacts associated with ungulates. Any decision to intervene, or alter existing management practices to address any negative impacts should, however, be taken in consideration of a parallel evaluation of the *positive* impacts which may be derived from ungulate populations in a given area: sporting and recreation, as above, beneficial impacts on certain open habitat types by suppression of scrub, etc. (for further details, see Putman and Watson, 2009; Putman *et al.*, in press).

In much the same way, it is clear that, to be effective, management effort needs to be coordinated over a wide area. Management at the level of an individual site, or individual landholding, will not be effective, if that landholding is by definition smaller than the effective home-range area of the biological population to be managed (again see Apollonio *et al.*, 2010). Management, to be effective, must be carried out at the landscape scale, or at the very least at a geographical scale equivalent to that of the range of the population to be managed. We suggest that the minimum required area for management constitutes the known or estimated population home range for the species of ungulates present in an area, while some impacts (such as deer–vehicle collisions, potential risk of deer as vectors for diseases) may need to be managed at regional or ecosystem level. That would be consistent with the Ecosystem Approach of the Convention on Biological Diversity, recognising the need to take supporting and regulating ecosystem services into account, as well as the more obvious provisioning and cultural services (Millennium Ecosystem Assessment, 2005).

Finally, it is clear that management systems need to incorporate some flexibility, some ability to show responsiveness. The development of any management plan is inevitably an attempt to hit a moving target and none of its prescriptions should be considered rigid or inflexible. Above all, management must be responsive and capable of being adjusted along the way. There should be careful monitoring both of abundance and distribution of ungulate populations themselves, as well as those factors which are the main targets of management (impacts on agriculture, forestry, harvest rates, etc.), to ensure that management policies adopted are actually working to deliver objectives, or to inform appropriate shifts in management to make that delivery more effective.

As noted by Morellet *et al.* (2007) (and see Chapter 5; Section 5.7):

> Managers need to set out some expectations or goals to monitor and manage ungulate populations. Whatever monitoring is carried out after that point must be assessed against those initial aims and objectives. Then, the approach consists of monitoring change over years in both individual performance, population productivity, and habitat quality and/or herbivore impact on the habitat. The temporal variation of this set of ecological indicators can be quantified and compared to predefined goals to assess if a change in management is required or not. This approach is more and less equivalent to a trial and error process during the first years of monitoring, but the understanding of the population-environment system increases with the accumulation of information over the years.
>
> This process bears some resemblance to **adaptive management**. Indeed, in adaptive management, the information on the system response to management is gathered continuously so that this information is used to improve biological understanding and to inform future decision-making (Nichols *et al.*, 1995; Shea *et al.*, 1998; Williams *et al.*, 2002). We believe that the management of ungulates should take advantage of this sort of approach by improving the monitoring of the population-environment system.

13.2 The effect of governance and administrative structures on sustainable management for ungulates

One of the other main factors highlighted by Apollonio *et al.* (2010) as contributing to difficulties in delivering effective management of ungulate populations or their impacts was the particular constraints on management which might be set by national or local legislative controls, or by the actual administrative system under which hunting and management are regulated (see Chapter 3 in this volume). Bureaucratic structures within which hunting is regulated, the degree to which ungulate management is considered a state responsibility or is devolved to individual level, and finally particular legislative controls may equally constrain, or empower, effective management. Different cultural attitudes (Chapter 1) may also influence effectiveness of management both directly (affecting the number of active hunters involved in management activity, affecting the acceptability of hunting to the general public) and indirectly (through cultural differences in willingness to accept different levels of state intervention).

The extent to which these factors might influence management and its effectiveness in delivering

- the ability to control ungulate populations
- the ability to control damaging impacts
- the ability to generate sustainability of harvest from sustainable ungulate populations
- the ability to manage ungulate populations in balance with their habitats

can be explored by combining data collected in this volume with much additional material from a recent EU study (Kenward and Sharp, 2008). The EU project on the Use Nationally of Wild Resources across Europe (UNWIRE) was part of a larger project on Governance and Ecosystem Management for the Conservation of Biodiversity (GEMCONBIO; Manos and Papathanasiou, 2008), co-funded (as #028827) by the European Commission in its 6th Research Framework. The project explored how conservation of resources (using the term in the sense that includes the sustainable use of resources) may be affected not just by local management, but also governance, which is the way societies manage their affairs (through government and other institutions such as regulations, markets and social constructs).[1] Because of its relevance here in helping to explore how governance and administrative systems might constrain or empower management, results from this study are summarised briefly below.

GEMCONBIO developed an analytic framework (Figure 13.1) to look at how environmental and institutional *Capacities* combined with particular *Objectives*, through a variety of governance *Processes*, to create *Impacts*, on the status of different animal and plant species, as well as considering the impacts of management systems on wider biodiversity, on the quality of ecosystem services and on their sustainability. Extensive questionnaires to

[1] Throughout this chapter we use the term 'conservation' in its wider sense, as embraced by the International Union for Conservation of Nature, of maintaining wild resources and their ecosystems for the future, rather than in terms of a more restricted focus on the simple protection or preservation of rare species.

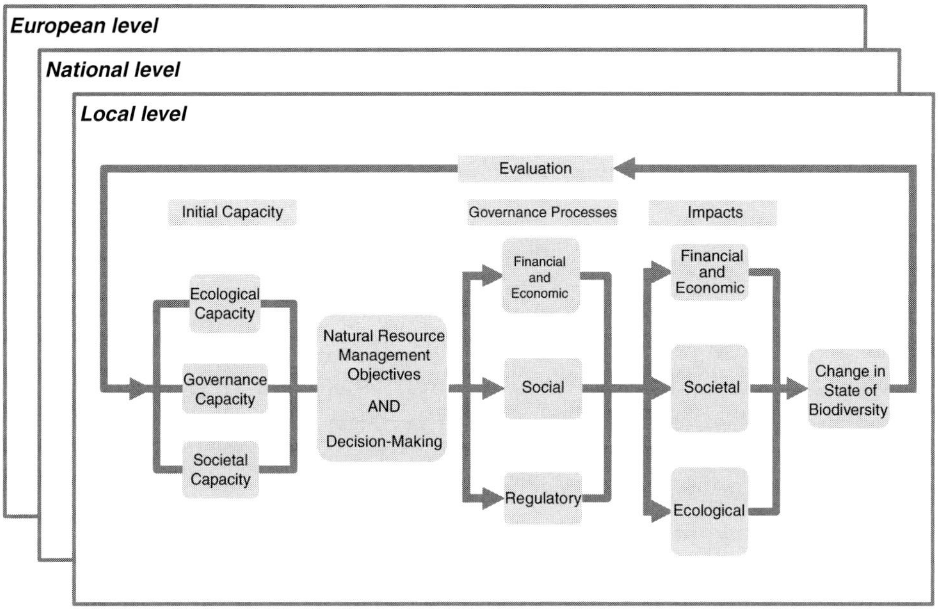

Figure 13.1 The COPIE framework for impacts of governance on natural resources.

national organisations and government administrators gathered data on *Capacities*, *Objectives* and *Processes*, each considered in terms of ecological, economic and social components. Questions included a final *Evaluation* of whether respondents considered economic and social processes beneficial for conserving the resource, giving a COPIE analysis framework (Figure 13.1).

GEMCONBIO as a whole analysed governance and management across more than 30 local case studies and international topics. Within this umbrella, UNWIRE specifically explored six key topics across all 27 EU states. These topics were all concerned with cultural ecosystem services provided from wild resources. The important aspect for this book is that they included hunting of ungulates, as well as hunting birds, angling, gathering fungi, collecting wild plant products, and bird watching.

Conveniently, the best survey response rate was from national organisations for hunting, from 25 of 27 states (including two without ungulate populations suitable for hunting). The data from GEMCONBIO–UNWIRE offer an elegant complement to the material in Chapter 3 on different administrative systems for controlling hunting and use of natural resources, as well as on differences between countries in legislative control of hunting. These are governance factors which may affect the sustainability of hunting, and hence may affect the ability of hunting to control or regulate ungulate populations, and ultimately may affect objectives and outcomes in relation to maintaining wider biodiversity (in terms of habitat quality). The resulting analysis and synthesis, in this concluding chapter, can perhaps go some way towards helping us understand how governments can take positive action to help us manage this valuable resource.

13.2.1 UNWIRE for ungulates

The data collected from across Europe in UNWIRE provide an important resource for exploring relationships between ungulate populations, their biotopes, hunters, management and governance. Variables reported and recorded in the UNWIRE project included several on governance and management *capacity* of each country. These included the proportions of state, private and community-owned land represented within areas over which ungulate hunting is carried out; how much of the management of ungulates is organised or controlled by state, private sector or community; what was the human population density in each country at the time of survey; number of hunters (and proportion of the population engaged in hunting activity).

Aspects of social capacity also included an assessment, by government officials and national organisations, of how much trust occurred between organisations at national level, and between groups at local level, and between government and local level. Other 'administrative factors' included in the analysis were factors such as the degree to which artificial (supplementary) feeding of ungulates was encouraged – or discouraged – by government, whether or not hunters were obliged to pay compensation to landowners for damage caused by ungulates, etc.

As a whole, the data provide an opportunity, for example, to see whether the *Process* of increased devolvement of management to local level, shown by the five administrative models described in Table 3.5 (Chapter 3), is linked to aspects of sustainability of hunting, of ungulate populations or of their habitats. In the same way, the effect of factors such as artificial feeding practices and responsibility for compensation, in helping or hindering effective management, can be explored. Finally, the UNWIRE analysis included an *Evaluation* by respondents of whether licences, access restrictions, other regulations, markets, state payments and other economic incentives benefited conservation of the resource.

In effect we may divide the different factors for which data were collected into a series of potential 'driving variables' and a separate series of 'response variables'. The candidate variables are shown below, coded C for the so-styled 'capacity variables', O for objectives, P for process, E for evaluation variables. Some of these variables constitute a judgement assessed subjectively on a five-point scale; these are noted in the list below with a (5).

C1 Human population density in each country
C2 Urbanisation: proportion of human population in large towns
C3 Number of different habitats used by ungulates (farmland, forest, montane, etc.)
C4 Proportions of those habitats under (i) public, (ii) private or (iii) community ownership
C5 Trust between stakeholders (government, other organisations) at national level (5)
C6 Trust between stakeholders at local level (5)
C7 The awareness among hunters of economic benefits from hunting (5)
C8 Whether or not hunters paid compensation for damage caused by ungulates

O1 Proportion of government management effort for ecological objectives
O2 Proportion of government management effort for economic objectives
O3 Proportion of government management effort for social objectives

P1 Extent of devolvement of management to local level
P2 Adaptive management: whether management changes by monitoring species and habitats
P3 Local knowledge: whether or not hunting organisations used this in management
P4 Leadership: whether management involves regular consultation of a single authority
P5 The number of types of licences and access restrictions and other regulations

E1 Whether hunting organisations considered regulations to benefit resource conservation (5)
E2 Whether hunting organisations deemed economic factors to benefit resource conservation (5)
E3 How highly government rated ungulates as a cost to the public (5)
E4 How highly government rated ungulates as benefit to the public (5).

The response variables considered in relation to these different factors are:

1. Percentage of ungulate hunters in population of each country
2. Change in numbers of ungulate hunters in each country during 1996–2006
3. Change in numbers of ungulates (average with equal weighting for each hunted species)
4. Assessment by hunting organisations of condition of habitats used by ungulates.

Two of these response variables are fairly 'hard' and readily quantified: percentage of ungulate hunters in the human population and the change in their numbers between 1996–2006. One, the change in ungulate populations, is less robust, partly because in some states this was based on estimations for at least some species and partly as the overall estimate was based on summing overall the change (weighted equally) for each species. Finally, the change in quality of habitats used by ungulates was a purely subjective assessment and scored by hunting organisation representatives from −2 for 'much worse' to +2 for 'much better'.

When assessing whether relationships between variables are meaningful, it is usual to estimate how likely they are to have occurred by chance. If something is only 5% likely to have happened as a random occurrence, it is considered statistically significant at the 5% level, but it must be remembered that if 20 relationships are examined, one will be estimated as significant by chance alone. For a more credible analysis process, all variables that did not show significance at the 1% level in one of the UNWIRE topics or across all cases in GEMCONBIO were discarded. This left about 20, several of which showed consistent and significant effects.

Table 13.1 shows the governance and management variables, in capacity, process and evaluation categories, listed as C1–E3 above, that had univariate relationships to the four response variables with less than 1 in 100 likelihood that the effect was due to chance ($P < 0.01$). The direction of the relationship is indicated by a + or − sign.

Relationships summarised in the table suggest that the percentage of the human population actively engaged in hunting was greatest when population density was low, there was least urbanisation and there was good trust between hunters and others at local level. Hunting activity was also more common when management was least devolved by the state and where administrative organisations considered that income made from hunting was good for conservation. Number of hunters was seen to be increasing where regulations limiting hunting access in time (hunting seasons) and place (hunting rights and areas) were generally not considered too restrictive or actively damaging from the point of view of sustainable

Table 13.1 *The main univariate relationships between governance and management variables (in capacity, process and evaluation) and the four response variables of: hunter numbers as % of population, change in hunter numbers over time, trends in ungulate stocks and habitat improvement*

	Capacity	Process	Evaluation
Hunters as % of population	− **C1**. Human population density + **C6**. Trust at local level − **C2**. Urbanisation	− **P1**. Devolved management	+ **E2**. Benefit of taxes for wildlife
Increase in number of hunters			+ **E1**. Benefit of limited-access − **E3**. Perceived cost of biodiversity
Increase in ungulate stocks	+ **C7**. Awareness of state payments		+ **E2**. Benefit of economic factors + **E2**. Benefit of state payments + **E2**. Benefit of markets
Improvement of habitat quality for ungulates	− **C3**. Number of ungulate habitats − **C1**. Human population density	+ **P3**. Use of local knowledge − **P5**. Requirement for licences	− **E1**. Benefit of licensing − **E2**. Benefit of markets

management, and where government officers perceived least negative impact or cost to the general public posed by increasing numbers of ungulates.

Changes in ungulate numbers overall were strongly related to awareness and availability of financial incentives available from the state, with trends showing *increasing* ungulate populations where hunters were most aware of financial incentives available from the state, and where these payments, markets and economic factors were considered beneficial for sustainable harvest/management. Finally, habitat quality for ungulates was perceived to be deteriorating most obviously where human population density was high, where ungulates were using many different habitats and where administrative burden was such that hunters required many different licences in order to hunt. Improvements in habitat quality were noted to be associated specifically with situations where ungulate management systems made significant use of local knowledge.

Table 13.2 *Multivariate relationships between governance and management variables (in capacity, process and evaluation) and the response variables. In this case we consider also effects of artificial feeding*

	Capacity	Process	Evaluation	*P*
Hunters as % of population	+ **C6**. Trust at local level	– **P1**. Devolved management		< 0.001
Increase in number of hunters	+ **C5**. Trust at national level		+ **E1**. Benefit of limited-access	0.001
Increase in ungulate stocks	+ **C7**. Awareness of state payments		+ **E2**. Benefit of economic factors	< 0.001
Improvement of habitat quality	– **C1**. Human population density	+ **P3**. Use of local knowledge	– **E1**. Benefit of licensing	< 0.001
Policy on artificial feeding	– **C4**. Private land ownership			0.001
	– **C2**. Urbanisation			< 0.001

Variables from Table 13.1 which showed significant univariate relationship were carried forward into multivariate analyses to explore further the importance and interaction of the different driving variables; such multivariate analyses are designed to focus on the major variables and tend to exclude the variables with least effect. Table 13.2 shows the governance and management variables, in capacity, process and evaluation categories, listed as C1–E4 above, that in combination related most strongly to the response variables. In this analysis we also consider the role and effect of attitudes to supplementary feeding.

Factors affecting the number and prevalence of ungulate hunters

The proportion of hunters was up to 3% of the human population of each country, though in only five cases was much more than about 1% (Figure 13.2). The proportion declined strongly with increase in population density (Figure 13.2) and degree of urbanisation (Table 13.1).

However, the strongest relationship among two variables was for a low prevalence of ungulate hunters in states where the management model for deer was highly devolved away from state level (Figure 13.3), and where trust between groups at local level was also poor (as shown by small bubbles in the figure). Thus, although prevalence of hunters in the population declines with population density and urbanisation, the strongest relationship is for numbers of hunters to be high where there is most centralised state

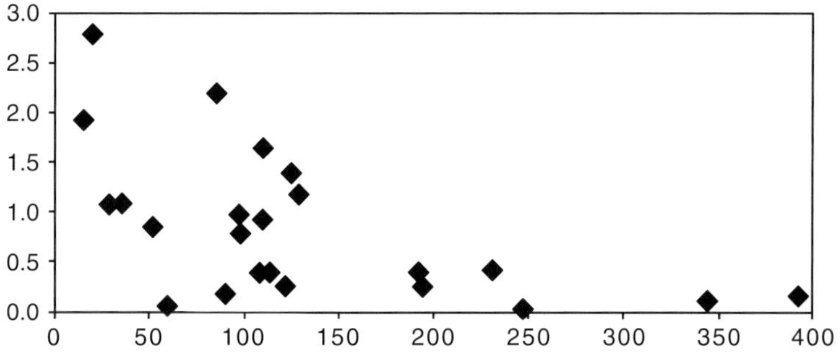

Figure 13.2 Prevalence of ungulate hunters (vertical axis: as a percentage of the total population in each country) declines with population density across EU states (horizontal axis: population number per km^2).

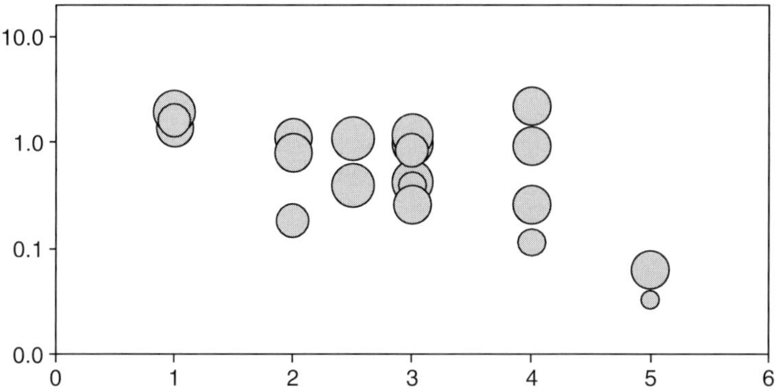

Figure 13.3 The prevalence of ungulate hunters as a percentage of total population (vertical axis: log scale) was least in those EU states where models for deer management were most devolved (horizontal axis, left to right) and national representatives for hunting registered worst relations between ungulate hunters and other groups at local level (bubble size).

management of ungulate stocks and a perception by hunting organisations of best relationships between hunters and other groups at local level.

Factors affecting the trend in number of ungulate hunters

Numbers of ungulate hunters were reported to be increasing or stable in most European states during 1996–2006, with an average change of +9% (Figure 13.4). The numbers showed most tendency to increase where their representatives estimated that there was most trust between institutions at national level. The tendency for hunter numbers to increase was strengthened somewhat where hunting organisations were most inclined to favour restrictions on land access for conserving ungulate stocks (as opposed to licences, quotas or other regulations), as shown by large bubbles in Figure 13.4). Hunter numbers increased most where there was most trust between

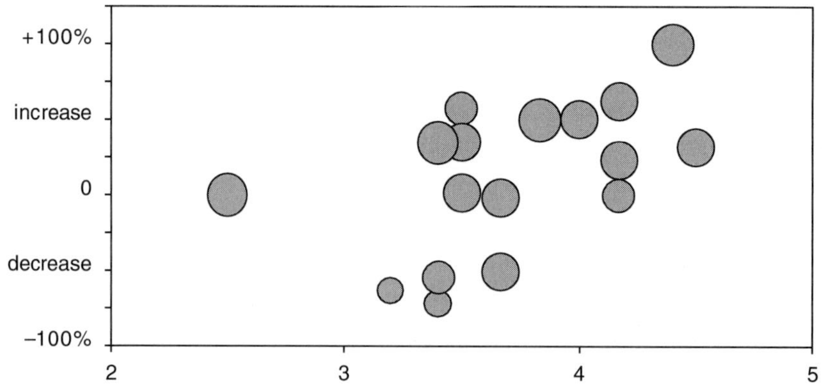

Figure 13.4 There was most increase in ungulate hunters (vertical axis, as the change recorded in the number of ungulate hunters from 1996 to 2006) where hunting representatives perceived most trust across institutions at national level (horizontal axis) and most conservation benefit from access regulations (bubble size).

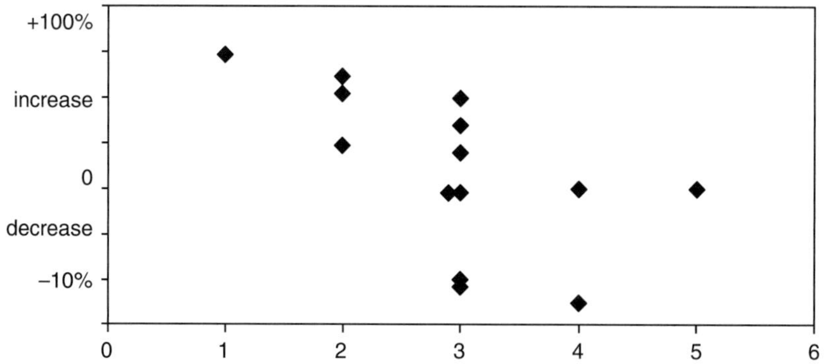

Figure 13.5 There was least increase in numbers of ungulate hunters between 1996 and 2006 (vertical axis) where government officers gave highest scores for the cost to society of biodiversity (horizontal axis).

institutions and the most favourable attitudes to managing hunting by limiting access to land.

In 15 states where government officials responsible for hunting answered a questionnaire, they gave a score for how much cost the public had to bear from biodiversity. The lowest assessments of costs from biodiversity were where numbers of ungulate hunters were increasing most strongly ($F = 8.35$, $n = 15$, $P = 0.01$, Figure 13.5).

Factors related to ungulate numbers

Ungulate numbers were recorded as stable in one EU state, increasing slowly in 3 and increasing strongly in 18 of 22 others (Figure 13.6). Increases

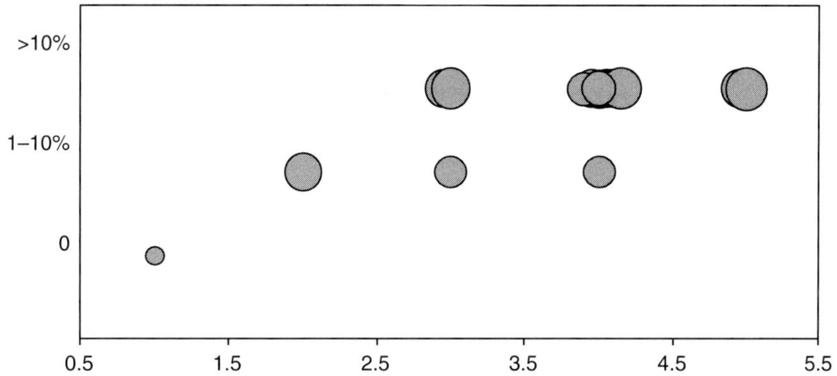

Figure 13.6 Greatest increases in ungulate stocks (1996–2006: vertical axis) were recorded in countries where national representatives for hunting assessed that hunters had greatest awareness of economic benefits (horizontal axis) and perceived most benefit from state payments for conservation (bubble size).

in ungulate populations were consistently linked to perception by hunting representatives at national level that economic measures, including state payments and market opportunities, were beneficial for conserving stocks (Table 13.1). The strongest relationship (Table 13.2) was for increase in ungulate stocks where hunters were considered to be most aware of economic opportunities from hunting (the trend in Figure 13.6), especially where state payments decreed at national level (probably as compensation for damage) were also considered most beneficial (large bubbles). In all states where ungulate populations increased on average by more than 10% during 1996–2006, payments decreed by the state were considered moderately to highly beneficial and hunters were considered moderately to strongly aware of economic benefits from hunting, although only 14 states recorded all three variables.

Factors related to quality of habitat for ungulates

In contrast to the change in numbers of ungulates, the perceived change in quality of preferred habitats for ungulates was strongly negative where market opportunities were deemed most beneficial (Table 13.1). Thus, where markets were considered most beneficial, number of ungulates increased and quality of the habitats declined, perhaps due to the impact of large stocks. Habitats were also perceived to change positively where there was much use of local knowledge for management, and negatively with human population density (Table 13.1). The strongest combination of two variables was an association of habitat decline with high human density and failure to use local knowledge in management (Figure 13.7),

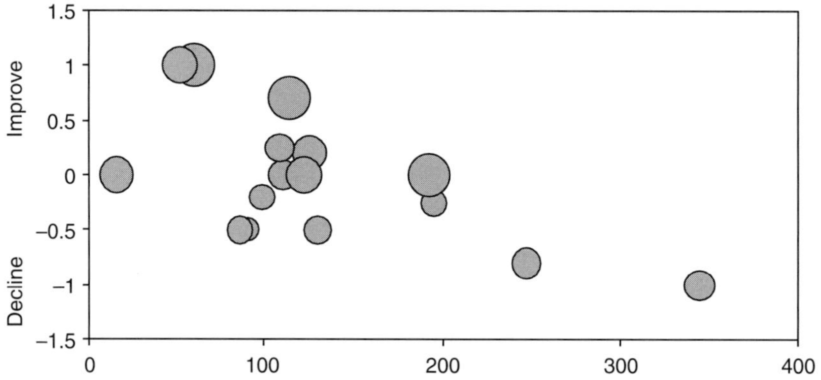

Figure 13.7 Assessments of the change in quality of habitats used by ungulates (vertical axis) indicated decline in habitat quality where the density of humans in each country was high (horizontal axis: population number per km^2) and local knowledge was little used in management of ungulates (bubble size).

with no further effect of markets once these factors were taken into account.

However, another strong combination was of local knowledge with attitudes to licensing requirements (Table 13.2), to which relationship the human population density would have contributed significantly if there had been enough data to justify including a third variable. Remarkably, both the strength of licensing requirements and the approval of licences as a conservation tool were associated with reports of declining habitat quality (Table 13.1).

Artificial or supplementary feeding

The propensity to feed wild deer, as reported in Chapter 3 was quantified using a scale from 0, for where feeding is forbidden, through 1 for sporadic feeding, 2 for common feeding and 3 for obligatory feeding in winter. Among the suite of variables tested, by far the strongest was a negative relationship between feeding and private ownership of hunting areas in each country, which relationship was further improved by inclusion of the urbanisation in the states concerned (Table 13.2). Feeding was prevalent in countries where there was least private ownership of land used to hunt ungulates and least urbanisation (Figure 13.8).

The tendency to feed deer was also linked to the requirement for hunters to compensate landowners for damage from deer. Deer were fed most in countries where compensation was paid by hunters (logistic regression, $n = 20$, $Z = 2.32$, $P = 0.02$), perhaps because feeding game was seen as a way to try and reduce damage that might otherwise require compensation.

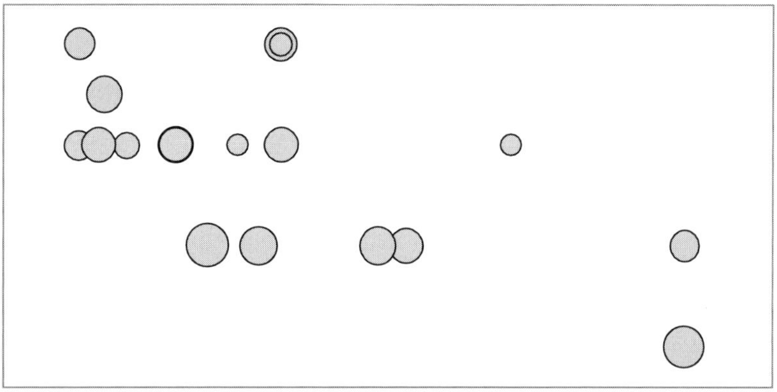

Figure 13.8 Wild deer were offered supplementary food over winter most consistently (vertical axis) where there was little private ownership of land (horizontal axis) and least urbanisation (bubble size).

13.2.2 Some conclusions

Despite the strong significance in many of these relationships, it is important to be cautious about them, not least because multivariate analyses in GEM-CONBIO were based on regression analyses that are being surpassed by information-theory-based randomisation tests (Burnham and Anderson, 2002; Whittingham *et al.*, 2005). Caution is also needed due to possibilities of bias. Thus, the tendency for greatest appreciation of access regulations and trust between institutions when hunter numbers were increasing (Figure 13.4) could have been biased by optimism of officials in hunting organisations. Nevertheless, if good conditions made them optimistic, it was clearly good institutional relations and access regulation that they considered most important among many other factors, such as licensing regulations or economic factors. Likewise, if economic benefits and awareness of them were linked to improvement in ungulate stocks (Figure 13.6) by optimistic opinions of some officials, they were singling out those aspects rather than others for their optimism.

Finally, caution is also needed when considering cause and effect. For any association between variables, the arrow of causation can go from A to B, or from B to A. It could also go strongly from a third factor to both A and B without A being related to B at all; this relationship would not be detected if the third factor was not measured. Thus one may wonder why ungulate hunters are least prevalent in countries where human density is high (Figure 13.2). Is it because in countries with high human density there is scope for relatively few to hunt, or because high human populations render areas unfit for ungulates (as suggested by Figure 13.7) or because hunting becomes unpopular in such countries because of increased opposition from largely urbanised populations, or perhaps due to a combination of all these factors?

A similar consideration applies to the tendency for hunters to be most prevalent where the state was most involved in management of ungulates and there was also good trust between hunters and other groups at local level (Figure 13.3). Trust and state control could tend to support hunting, or alternatively be a response of society when there are many hunters. The fact that numbers of hunters also show most growth where there is most trust (Figure 13.4) supports a view that social trust benefits hunting. Moreover, it is clear that hunting ungulates is not harmed by management models that emphasise state responsibility (Figure 13.3).

Questions of causality are especially pertinent for relationships with evaluation variables, such as the tendency for hunter numbers to grow when their representatives favour regulations that limit land access for hunting (Figure 13.4). For example, are access regulations favoured because of pressure from growing numbers of hunters, or has a system based on access rather than other regulations enabled increasing numbers to hunt? It is also interesting that the disapproval of licence-based regulation is strongest where there is most damage to ungulate habitats (Table 13.2). At the least, it can reasonably be concluded that investigating the relative benefits of land access and licensing should be a priority for regulating the hunting of ungulates.

It is interesting that trends in ungulate numbers are stable or increasing throughout Europe (Figure 13.6). This provides a good example of hunting being sustainable over wide areas, but is it also an example of good management? That question is a hard one to answer without having recorded whether management objectives in each country were to maintain stable ungulate populations or engineer increases or decreases. However, whatever the objectives, it does seem that ungulate populations increased most where economic incentives were viewed favourably and where there was most awareness of them (Tables 13.1 and 13.2).

Questions about management objectives also apply to habitats used by ungulates. Habitats were deemed to have deteriorated where human population density was highest and to be worst where there was least use of local knowledge for managing ungulates (Figure 13.7). However, their quality was also linked negatively to use of licences for regulating hunting, and to approval of licences and market tools for conservation (Table 13.1). As these are the same factors that associate with growth in ungulate populations, their relationships with habitat quality are consistent with habitats being damaged by increasing ungulate populations. Further support for this view is a perception by hunting officials of decline in habitat quality where ungulate populations had increased ($r = -0.56$, $n = 16$, $P = 0.035$). And with hunting as the means of managing ungulate populations, an impact of ungulates on their habitats would explain why government officials saw the highest cost from wildlife where the number of hunters was declining (Figure 13.5).

13.3 Some thoughts for the future

If ungulate populations in Europe are to be a maximally valuable resource in terms of natural food and recreation, while at the same time not a nuisance for crops, transport and disease transmission, they need to be managed very carefully. In a Europe with little non-human predation on ungulates, that involves managing their hunters. Thus, the governance of hunting ungulates is an important field in Europe. This is one reason why a recent Bern Convention instrument on hunting was drafted as a charter (Brainerd, 2007). A charter is an instrument that confers rights as well as responsibilities on citizens. Therefore, each of the 12 principles in the Charter on Hunting and Biodiversity *make recommendations to government for managing hunters as well as to hunters themselves*.

So how may administrative systems and style of governance affect the ability to deliver effective and sustainable management through influence on hunter prevalence, social trust, regulations and economic incentives? This chapter provides some indications. For a start, it shows that a strong state role in managing ungulates was associated with maintaining high numbers of hunters (Figure 13.3), or at least that hunting ungulates is *not harmed by* management models that emphasise state responsibility.

Determining causation in governance associations, and which mechanisms for managing hunting are most effective, will require experiments. Perhaps this could be part of an adaptive approach to governance. As the trend in EU governance is towards greater devolution, governance experiments, for example through the deliberate taking of different approaches between newly devolved regions, will become more practical and could perhaps be encouraged. Governments also have an important role to play in establishment of social institutions that build trust (Figures 13.3 and 13.4). An expert system designed to support both adaptive management at local level and adaptive governance at higher levels is already being designed (as the Transactional Environment Support System, TESS; Kenward *et al.*, 2009).

Human societies can use ecosystem services, including species hunted for subsistence, in a sustainable way for long periods until technologies or cultures change (Frazier, 2007). However, the 'bushmeat crisis' shows that subsistence hunting can become unsustainable as development occurs

(Brashares *et al.*, 2004). Populations may increase through immigration or introduced resources, technologies may dramatically increase harvest efficiency, and corruption may undermine communal peer pressure (Smith *et al.*, 2003). Under such circumstances, a return to sustainability requires development of new institutions (Beddoe *et al.*, 2009).

Whereas subsistence hunting is primarily a provisioning service from ecosystems, albeit with recreational elements, in developed countries hunting may become important for recreation (Dickson *et al.*, 2009) more than provisioning, and hence as a cultural ecosystem service. Through recreational hunting of ungulates in the absence of other predators, humans also have a regulating role in ecosystems. The hunting of ungulates in Europe, where ungulate populations are increasing generally and ungulate hunter numbers are also stable or increasing in most EU states, shows that recreational hunting does not use ungulate populations unsustainably. However, the current increase of some species, notably wild boar, raises the question of whether current governance is sufficiently adaptive to enable management of culture and habitats that is sustainable in the future.

References

Apollonio, M., Andersen, R. and Putman, R. (2010) Present status and future challenges for European ungulate management. In M. Apollonio, R. Andersen and R. Putman (eds.) *European Ungulates and their Management in the 21st Century*. Cambridge, UK: Cambridge University Press, pp. 578–604.

Beddoe, R., Costanza, R., Farley, J., *et al.* (2009) Overcoming systemic roadblocks to sustainability: the evolutionary redesign of worldviews, institutions, and technologies. *Proceedings of the National Academy of Sciences of the USA* **106**, 2483–2489.

Brainerd, S. (2007) European Charter on Hunting and Biodiversity. Bern Convention, Council of Europe, Strasbourg. Online: www.coe.int/t/dg4/cultureheritage/conventions/Bern/ Recommendations/tpvs07erev_2007.pdf

Brashares, J.S., Arcese, P., Sam, M.K., *et al.* (2004) Bushmeat hunting, wildlife declines, and fish supply in West Africa. *Science* **306**, 1180–1183.

Burnham, K.P. and Anderson, D.R. (2002) *Model Selection and Multimodel Inference*. New York: Springer.

Dickson, B., Hutton, J. and Adams, W.M. (2009) *Recreational Hunting, Conservation and Rural Livelihoods*. Oxford, UK: Wiley-Blackwell.

Frazier, J. (2007) Sustainable use of wildlife: the view from archaeozoology. *Journal for Nature Conservation* **15**, 163–173.

Kenward, R. and Sharp, S. (2008) Use Nationally of Wildlife Resources across Europe (UNWIRE). In B. Manos and J. Papathanasiou (eds.) *GEMCONBIO: Governance and Ecosystem Management for the Conservation of Biodiversity*. Thessaloniki, Greece: Aristotle University of Thessaloniki, pp. 117–123 (EC Contract FP6 - 028827).

Kenward, R.E, Sharp, R., Manos, B., *et al.* (2009) Conservation from use of biodiversity and ecosystem services. In *Proceedings of the XXIX Congress of the International Union of Game Biologists*, held Moscow, 2009, pp. 68–83.

Manos, B. and Papathanasiou, J. (2008) *GEMCONBIO: Governance and Ecosystem Management for the Conservation of Biodiversity*. Thessaloniki, Greece: Aristotle University of Thessaloniki.

Millennium Ecosystem Assessment (2005) *Ecosystems and Human Well-being: Synthesis*. Washington, DC: Island Press.

Morellet, N., Gaillard, J.M., Hewison, A.J.M., *et al.* (2007) Indicators of ecological change: new tools for managing populations of large herbivores. *Journal of Applied Ecology* **44**, 634–643.

Nichols, J.D., Johnson, F.A. and Williams, B.K. (1995) Managing North American waterfowl in the face of uncertainty. *Annual Review of Ecology and Systematics* **26**, 177–199.

Putman, R.J. and Watson, P. (2009) *Developing an impact assessment methodology for use beyond the site scale*. Report to the National Trust, England.

Putman, R.J., Watson, P. and Langbein, J. (in press) Assessing deer densities and impacts at the appropriate level for management: a review of methodologies for use beyond the site scale. *Mammal Review*, in press.

Shea, K. and the NCEAS Working group on population management (1998) Management of populations in conservations, harvesting and control. *Trends in Ecology and Evolution* **13**, 371–375.

Smith, R., Muir, R., Walpole, M., Balmford, A. and Leader-Williams, N. (2003) Governance and the loss of biodiversity. *Nature* **426**, 67–70.

Whittingham, M.J., Swetnam, R.D., Wilson, J.D., Chamberlain, D.E. and Freckleton, R.P. (2005) Habitat selection by yellowhammers *Emberiza citronella* on lowland farmland at two spatial scales: implications for conservation management. *Journal of Applied Ecology* **42**, 270–280.

Williams, B.K., Nichols, J.D. and Conroy, M.J. (2002) *Analysis and Management of Animal Populations*. San Diego, CA: Academic Press.

Index